Springer
Tokyo
Berlin
Heidelberg
New York
Barcelona
Hong Kong
London
Milan
Paris
New Delhi
Singapore

I. Yamamoto
J.E. Casida (Eds.)

Nicotinoid Insecticides and the Nicotinic Acetylcholine Receptor

With 114 Figures, Including 9 in Color

Springer

Izuru Yamamoto
Professor Emeritus, Tokyo University of Agriculture
1-1 Sakuragaoka 1, Setagaya-ku, Tokyo 156-8502, Japan
e-mail: yam-izur@nodai.ac.jp

John E. Casida
Professor of Entomology and
Director of Environmental Chemistry and Toxicology Laboratory
University of California, Berkeley
California 94720-3112, USA
e-mail: ectl@nature.berkeley.edu

ISBN 4-431-70213-x Springer-Verlag Tokyo Berlin Heidelberg New York

Library of Congress Cataloging-in-Publication Data

Nicotinoid insecticides and the nicotinic acetylcholine receptor/
I. Yamamoto, J. E. Casida (eds.)
p. cm.
Includes bibliographical references.
ISBN 4-431-70213-X (hardcover : alk. paper)
1. Nicotinoids. 2. Botanical insecticides. 3. Nicotinic
receptors. I. Yamamoto, Izuru, 1928- . II. Casida, John E.,
1929- .
SB952. N54N535 1999
632'.96--dc21 99-20210
CIP

Printed on acid-free paper

Typesetting: Camera-ready by Authors
Printing & binding: Best-set Typesetter Ltd., Hong Kong
SPIN: 10656057

Preface

Insect pest control has continuously evolved from inorganics to botanicals, to chlorinated hydrocarbons, to organophosphorus compounds and methylcarbamates, then synthetic pyrethroids and most recently synthetic nicotinoids as the major classes. These insecticides allowed high standards of crop protection at minimal cost. A limitation in each new class of compounds is the selection of resistant strains and ultimate control failures and this serves as a driving force to discover and develop replacement compounds to circumvent resistance and overcome problem areas. The nicotinoids now play a critical role in meeting this need.

Three generations of chemicals are involved in the history of nicotinoid insecticides. The first generation was the botanical nicotine used for at least three centuries to control sucking insect pests but largely replaced in the 1940s and 1950s by the more effective organophosphorus compounds and methylcarbamates, some with systemic properties. Synthesis programs based on nicotine as a prototype did not yield compounds that could compete with other synthetic insecticides. The second generation was the nitromethylene type such as nithiazine, discovered by Shell scientists in a screening/optimization program. The nitromethylenes had the potency, selectivity, and systemic properties but lacked the field effectiveness largely because of photolability (so close yet so far from a major commercial product). The third generation required a series of advances made by Bayer researchers starting from nithiazine as the model and enhancing its photostability and potency with a nitroimine and chloropyridyl moiety, respectively, to give imidacloprid, the subject of much of this monograph.

Synthetic nicotinoids are the only major new class of insecticides introduced in the past 40 years. Imidacloprid is currently the top-selling insecticide. Acetamiprid, nitenpyram, and thiamethoxam (CGA 293-343) are increasingly important compounds. The next decade will set the pattern and establish most of the nicotinoids for future use. They will move from controlling only sucking insects to broader scope for chewing insects as well and will undoubtedly grow in number and importance in pest management. This has become the nicotinoid era.

"Nicotinoid insecticides" is the terminology used in the monograph title to include nicotine and the synthetic analogs of discernable structural and conformational similarities and the same mode of action in insects. Nicotinoids are related to nicotine as pyrethroids are to the pyrethrins. Definitions of this type become more diffuse as the knowledge base and diversity of compounds are increased, especially (as with the pyrethroids) if subsequent chemical structures become increasingly dissimilar from the nicotine and imidacloprid prototypes. "Neonicotinoids" as a term emphasized the relationship to nicotine and implied their improved properties; interestingly, neonicotine was the name used in 1931 for the botanical insecticide anabasine. Other names are based on substituents: chloronicotinyls, chloropyridyls, thianicotinyls, nitromethylenes, nitroguanidines, cyanoguanidines, etc.

Nicotinoid insecticide toxicology includes mode of action, metabolism, safety, and resistance. Insecticide safety is a primary consideration and the studies with imidacloprid and others are reassuring for the future. Biodegradability is an intrinsic feature of the effective structural types for now. The safety of the newer synthetic nicotinoid insecticides appears to reside in fundamental differences in the insect nicotinic acetylcholine receptor and that of mammalian nerve and muscle, an intrinsic target-site specificity. There are many closely related natural toxicants and candidate or commercial neuroactive pharmaceuticals, indicating the care necessary to maintain the required safety margin and unique selectivity for insects versus mammals. Insect toxicology and insecticide use are dominated by the specter of resistance, which has already appeared with the nicotinoids. The integration of nicotinoids into diverse pest management programs will help slow this development and prolong the benefits to be gained from this group of highly effective insecticides.

This monograph is based in part on a symposium by the same title held in September 1997 in Las Vegas, Nevada, as a part of the American Chemical Society National Meeting. This symposium was in honor of Professor Izuru Yamamoto on the occasion of his receiving the International Award for Research in Agrochemicals from the Division of Agrochemicals of the American Chemical Society. Each contribution has been revised, expanded, and brought up-to-date. Each author is thanked for a thorough and critical analysis of the relevant topic. The research at Berkeley was supported by Grant R01 ES08424 from the National Institute of Environmental Health Sciences of the National Institutes of Health. We dedicate this book to our students and colleagues who share with us a fascination for the chemistry and action of bioactive compounds and the goal of applying this knowledge for the benefit of all (except the insect pests).

Izuru Yamamoto
Department of Agricultural Chemistry
Tokyo University of Agriculture
1-1 Sakuragaoka, 1-Chome
Setagaya-ku, Tokyo 156-8502, Japan

John E. Casida
Environmental Chemistry and
Toxicology Laboratory
Dept.of Environmental Science
Policy and Management
University of California, Berkeley
CA 94720-3112, USA

Contents

Part 3 Nicotinoid Insecticide Toxicology

Part 1

Nicotine to Nicotinoids

1 Nicotine to Nicotinoids: 1962 to 1997

Izuru Yamamoto
Department of Agricultural Chemistry
Tokyo University of Agriculture
Setagaya-ku, Tokyo 156-0054, Japan

1. Introduction

Nicotine is the active component of tobacco for smoking. It has also a long history as a medicine and an insecticide, although not competitive with modern synthetic insecticides. An excellent review on nicotine as an insecticide was presented by Schmeltz (1971). Due to its remarkable pharmacological properties, the alleged health hazard associated with tobacco smoking and nicotine use as a pesticide, its safety is constantly watched by the public. There have been enormous volumes of literature in chemistry, biochemistry, physiology, pharmacology, toxicology, entomology, botany, medicine, and psychology. However, its mode of insecticidal action was vague until Yamamoto challenged it (Yamamoto et al., 1962; Yamamoto, 1965).

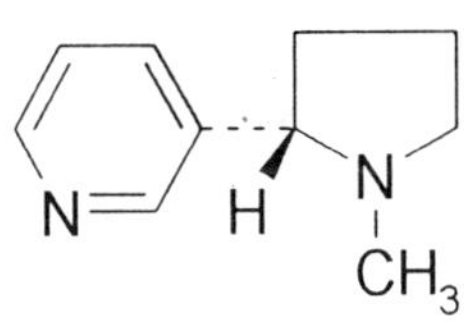

Nicotine

To understand the mode of insecticidal action of nicotine, we need to know that in insects the cholinergic system is located only in the central nervous system, while in mammals it is in both peripheral and central systems. In other words, to exert its insecticidal activity, nicotine must penetrate into the central nervous system and affect the nicotinic acetylcholine receptor (nAChR) directly at the synapse. In mammals, although the central effects of nicotine are associated with tobacco smoking, nicotine is easily accessible to peripheral nAChRs at ganglia and neuromuscular junctions, the latter being the major target for acute poisoning. However, in insects, the neuromuscular junction is not cholinergic and is not affected by nicotine or organophosphorus and carbamate insecticides.

2. Structure – Activity Relationships of Nicotinoids

In 1959, Yamamoto started the study on nicotine first by reading Prof. Metcalf's famous book, *Organic Insecticides* (Metcalf, 1955), which describes that nicotine, nornicotine, and anabasine are insecticidal, while

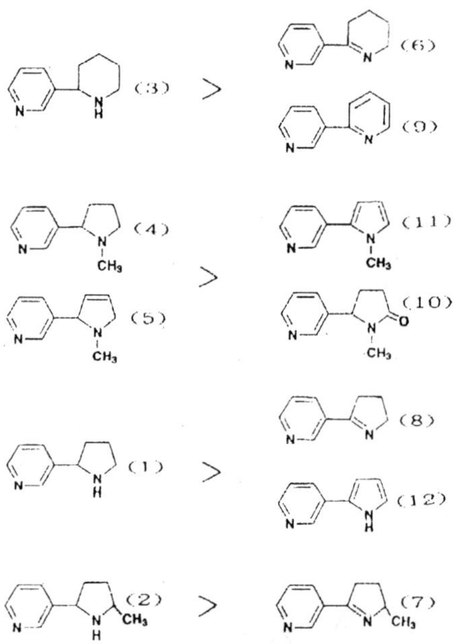

Fig. 1. Comparison of insecticidal activity of similar nicotinoids
(1) Nornicotine; (2) 5`-Methylnornicotine; (3) Anabasine; (4) Nicotine; (5) Dihydronicotyrine; (6) Anabaseine; (7) 5`-Methylmyosmine; (8) Myosmine; (9) 3,2 -Dipyridyl; (10) Cotinine; (11) Nicotyrine; (12) Nornicotyrine

nicotyrine, nornicotyrine, and cotinine, having double bond(s) on the pyrrolidine or piperidine ring, are not. However, dihydronicotyrine, having a double bond, was more insecticidal than nicotine, which was pazzling.

To cover nicotine and the related compounds, the term "nicotinoids" was proposed by Yamamoto et al. (1962). From information that the activity of sulfonamide drugs is affected by ionization, Yamamoto thought that the insecticidal activity of nicotinoids might be affected by ionization or basicity.

2.1. Toxicity of Nicotinoids and the Essential Moiety

As shown in Fig. 1 nicotinoids having a highly basic nitrogen atom seemed more insecticidal than those having a nitrogen atom of low basicity. After surveying hundreds of papers on nicotinoids from the standpoint of the relationship between basicity estimated from the chemical structure and any type of toxicity, Yamamoto found a general rule that all the toxic nicotinoids are provided with 3-pyridylmethylamine moiety with a highly basic nitrogen atom, while those with a nitrogen atom of low basicity are low in toxicity without exception. This rule applied also to the structure-insecticidal activity relationships of pyridylalkylamines (Kamimura et al., 1963; Soeda and Yamamoto, 1968b; Yamamoto et al., 1968). Therefore, the partial structure as shown in Fig. 2 was proposed as the essential moiety of nicotinoids.

Fig. 2. The essential moiety for toxicity of nicotinoids: 3-pyridylmethylamine with a highly basic nitrogen atom that is protonated at physiological pHs

2.2. Ionization and Toxicity

Fig. 3 illustrates the relationships between basicity, ionization, and toxicity. The highly insecticidal group ionizes extensively, while those that ionizes poorly are low in insecticidal activity. The intermediately ionized compound

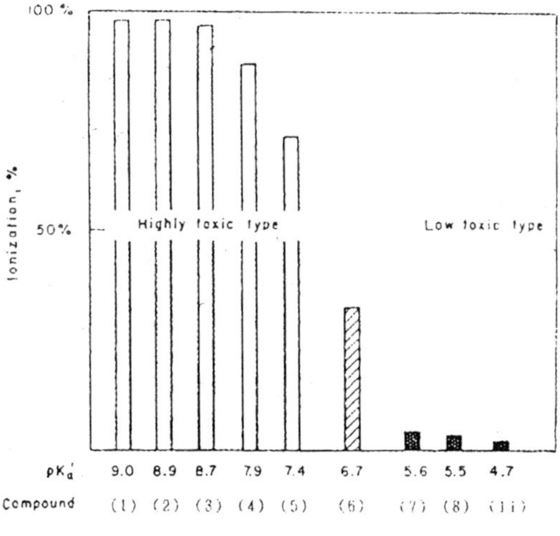

Fig. 3. Correlation of ionization and toxicity of nicotinoids

showed an intermediate activity. It was interpreted that the highly basic nitrogen atom is required for higher ionization. The same relationship was found for a series of 3-pyridylmethylamines (Yamamoto et al., 1968).

2.3. Acetylcholinesterase Inhibition by Nicotinoids

Acetylcholine is the ligand to the ACh receptor and the substrate to acetylcholinesterase (AChE). Although nicotine inhibits AChE competitively only at higher concentrations (Soeda and Yamamoto, 1968a), there was a good correlation among insecticidal activity, AChE inhibition, and ionization (Yamamoto et al., 1968). Comparison of quantitative structure-activity relationships (SAR) for insecticidal activity and AChE inhibition indicated that there is a certain similarity between the receptor and AChE inhibition site, and the latter can be used as a model for the former (Fujita et al., 1970, 1971). The above-mentioned essential moiety is also required for AChE inhibition.

2.4. Common Feature of the nAChR Ligands

By ionization, nicotine mimics acetylcholine (ACh) with respect to conformation and electronic makeup (Fig.4). All the nAChR ligands share the same feature, that is, the presence of an electron-donating group (unshared electron pair), which provides hydrogen bonding to the receptor at 5.9 Å apart from the positively charged center. Nicotine, having a 3-pyridyl ring, satisfies this condition, but the 2- and 4-pyridyl isomers do not. ACh uses the carbonyl group for hydrogen bonding at nAChRs and an ethereal oxygen atom at muscarinic AChRs (Beers and Reich, 1970).

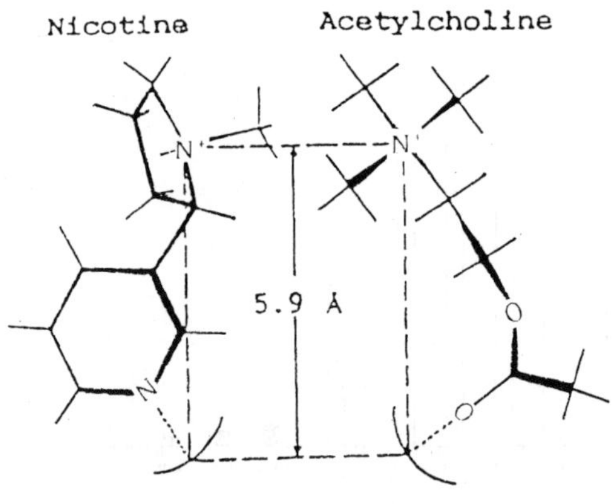

Fig. 4. The feature of nAChR ligands as illustrated with acetylcholine and nicotine

3. Mode of Insecticidal Action of Nicotinoids

Based on SAR, the mode of insecticidal action of nicotine was proposed by Yamamoto et al. (1962) as a working hypothesis. Schmeltz (1971) commented
"Yamamoto and co-workers have painted a pretty picture of nicotine's insecticidal action" (see Fig. 11). The free nicotine, not the sulfate, penetrates into the insect body, where it is mostly ionized. Only a fraction in the free form can reach the target area in the central nervous system because of the presence of the ion-impermeable barrier. Nicotine then ionizes again and the ionic form interacts with nAChR. This scheme has been cited in many textbooks for the past 30 years, because no one has challenged this subject. However, the essential moiety described here was also confirmed later by binding (Tomizawa and Yamamoto, 1992) and neurophysiological studies (Yamamoto et al., 1968).

4. Design of Nicotine Analogs

4.1. pK_a and Insecticidal Activity

Supposing that the ionization percent(%) is the parameter of intrinsic activity and the free base percent(%) is the parameter of target penetrability, their multiplication would represent the overall insecticidal activity as a rough approximation. From the pH of the insect body fluid (about 7) and pK_a, the ionization percent(%) is calculated and thus the overall insecticidal activity can be predicted. The 100% ionized nicotine methiodide, which is as active as nicotine in mammals, was poor in insecticidal activity even though injected, and low basic nicotyrine ionizes poorly and was poorly insecticidal as predicted. The optimization is obtained when pH and pK_a are the same, and dihydronicotyrine (pK_a 7.21 at 37℃) is almost ideal, giving the highest activity as predicted.

4.2. Synthesis of Analogs

Yamamoto's group synthesized many compounds with the aid of support from industry. The compounds shown in Fig. 5 were more or less insecticidal (Yamamoto, 1979). The results indicates the validity of the essential moiety, but none was competitive with conventional synthetic insecticides. A bis-thiocyanate compound gave nereistoxin analogs metabolically.

The cause of the rather low or limited insecticidal activity of nicotine and its analogs was interpreted to be based on their contradictory properties: as the mode of insecticidal action of nicotine indicates, it should be ionized for interaction with nAChRs, but such ionization limits its penetration to the target site through the ion-impermeable barrier. Attempts were made to introduce substituents on the basic nitrogen atom of nornicotine or anabasine, which reduce the basicity and would be removed after penetration into the target site, but we failed to determine suitable substituents. However, it was reported by other workers that the acylated nornicotine mixture found as cuticular components of several species of the *Nicotiana* plant gave far better activity than nicotine to the first instar tobacco hornworms, but not to other insects (Addor, 1995).

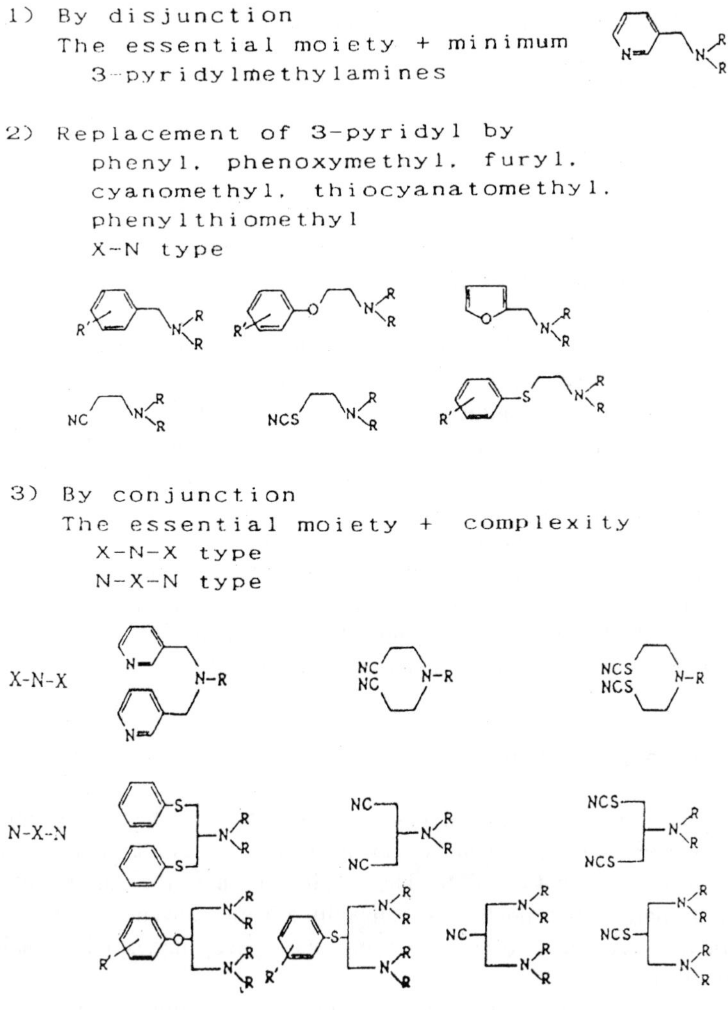

Fig. 5. Designed nicotine-like compounds

5. Development of Neonicotinoids

Nithiazine was invented as the first nitromethylene insecticide by Soloway et al.(1978), but it was not commercially successful because of its photoinstability. Phylogenic depiction of nicotinoid development is shown in Fig. 6. Introduction of a 3-pyridylmethyl group on nitromethylene heterocycles, made by Kagabu and others (Shiokawa et al., 1986; Iwata and Takase, 1993), remarkably improved the insecticidal activity. The new lead would be regarded as the hybrid of 3-pyridylmethylamines and nitromethylene heterocycles, and further optimization resulted in the invention of imidacloprid, which is reasonably persistent. Several analogs of imidacloprid followed, which include nitenpyram (Kashiwada, 1996), acetamiprid (Matsuda and Takahashi, 1996), and others. Based on the essentiality of a chlorine atom present at the 6-position of 3-pyridyl group for

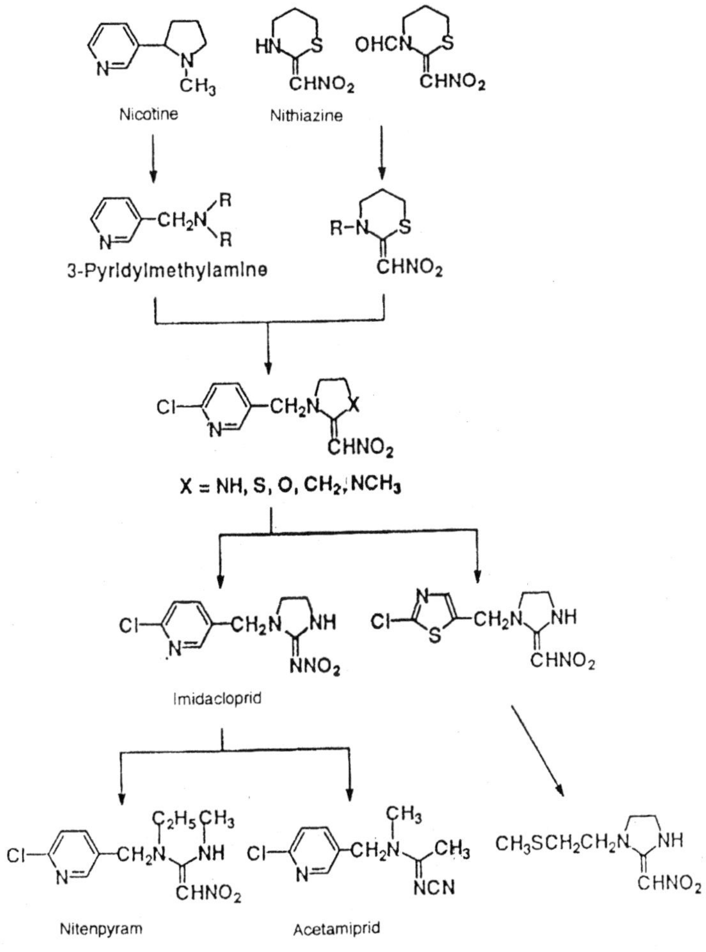

Fig. 6. Phylogeny of nicotinoid development

these commercial products, the term "chloronicotinyl" was proposed (Leicht, 1996). However, because there are and will be compounds having no chloropyridyl group, and based on the mode of action studies discussed later, Yamamoto proposed the term "neonicotinoids" to cover all these compounds (Tomizawa and Yamamoto, 1993). Many terms have appeared in the literature, such as nitromethylenes, chloronicotinyls, nicotinoids, and neonicotinoids. These may be interrelated as shown in Fig. 7. The term "nicotinoids" in a broad sense covers all types, while it indicates nicotine-like compounds in a narrow sense.

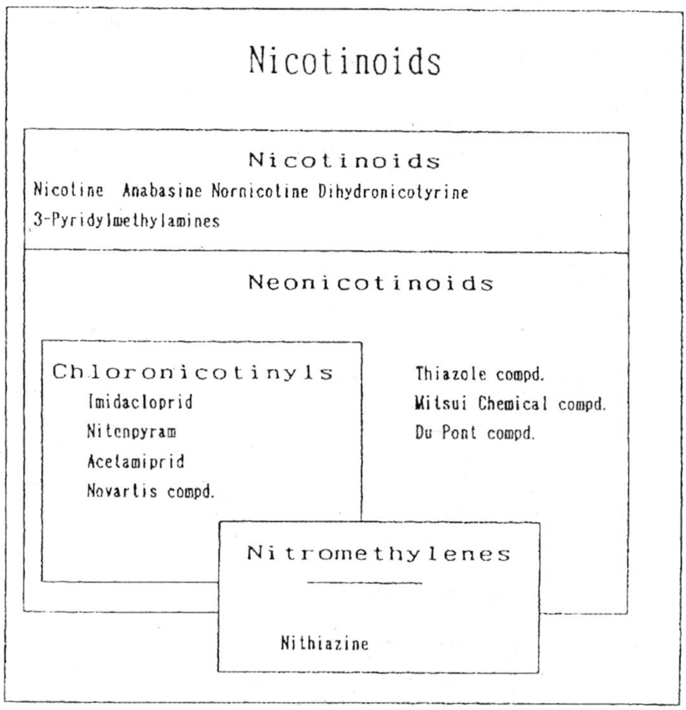

Fig. 7. Naming and relation of various nicotinoid-related insecticides

6. Mode of Action of Neonicotinoids

When the chemical structure of imidacloprid was disclosed, Yamamoto was impressed by the similarity between imidacloprid and nicotine, which seemingly share the same partial structure that he proposed as the essential moiety for nicotinoids.

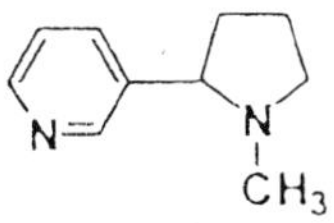

Nicotine

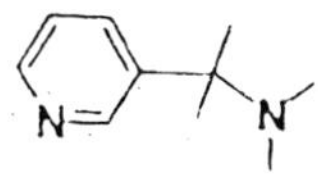

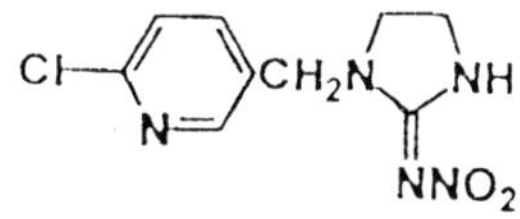

Imidacloprid

6.1. The Site of Action for Nicotine and Imidacloprid

Indeed, it was shown that nicotine and imidacloprid interact with the same site, that is, the ACh recognition site of the nAChR (Tomizawa and Yamamoto, 1992, 1993; Bai et al., 1991). This is the α-bungarotoxin (α-BGT) binding site, and the radioreceptor assay using the radiolabeled α-BGT was adopted for SAR study.

At first the preparation from housefly heads was used for the binding study, but later the preparation from honeybee heads was used throughout the study because of the commercial availability of the bee. Both showed almost the same binding profile for many ligands (Tomizawa and Yamamoto, 1992).

6.2. SAR of Imidacloprid-Related Compounds

The SAR was studied using imidacloprid and 19 related compounds, and the results are shown in Fig. 8. Although the detailed SAR analysis should be referred to Tomizawa and Yamamoto (1993), the essential structural feature of imidacloprid-related compounds could be extracted from this SAR, and the interaction with the nAChR is modeled as shown in Fig. 9.

The 3-position of the pyridine ring is connected with a positively charged nitrogen atom having one carbon in between. The presence of two carbons or no carbon in between, the 2- or 4-position of pyridine ring for connection, or the lack of strong electron withdrawing groups shown by X decrease the binding affinity as well as insecticidal activity. Liu et al. (1993) and Nishimura et al. (1994) reached the same conclusion by different approaches. Several apparent anomalies are explainable as discussed later. This SAR is similar to that for nicotinoids. The difference is that in nicotine the nitrogen atom is ionized, bearing a full positive charge to provide the site of interaction with the receptor. On the other hand, the corresponding nitrogen atom in imidacloprid analogs is not ionized, and seemingly bears a partial positive charge due to the neighboring electron withdrawing group.

	Compound	K_i (μM)[a)]	Insecticidal activity [b)]	Essential moiety [C)]	
(1)	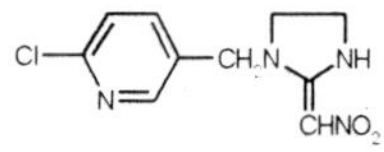	0.021	+++	Yes	6-Cl-PMNI
(2)	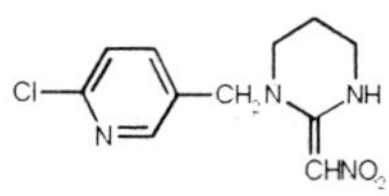	0.028	+++	Yes	
(3)	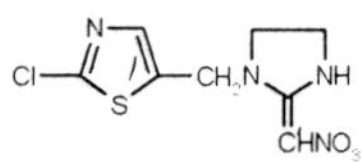	0.081	+++	Yes	
(4)	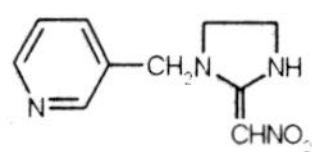	0.083	+++	Yes	PMNI
(5)	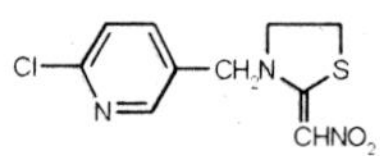	0.116	+++	Yes	S-Analog of (1)
(6)	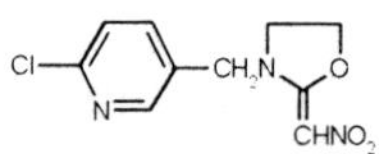	0.308	+++	Yes	
(7)	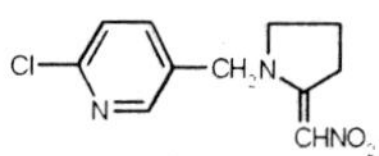	0.359	+++	Yes	
(8)	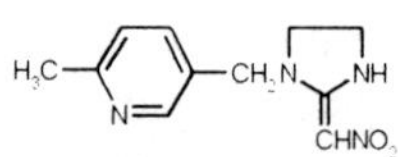	0.473	+++	Yes	6-Me-PMNI
(9)	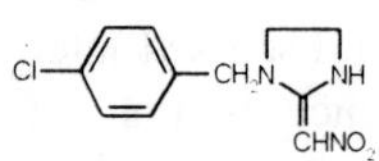	1.1	+	No	
(10)	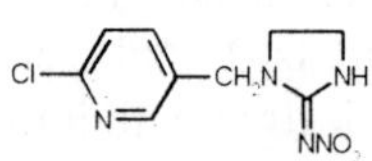	1.5	+++	Yes	Imidacloprid

(Continued)

	Compound	K_i (μM)[a)]	Insecticidal activity [b)]	Essential moiety [C)]	
(11)	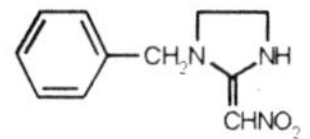	2.46	+	No	
(12)	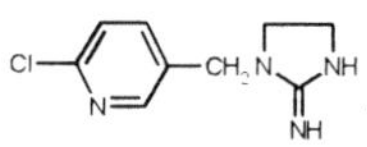	3.08	-	No	
(13)	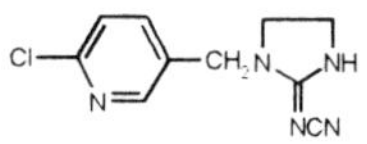	6.88	+++	Yes	
(14)	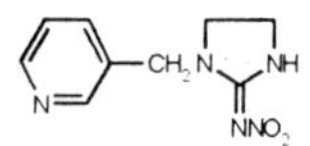	7.29	-	Yes	De-Cl-imidacloprid
(15)	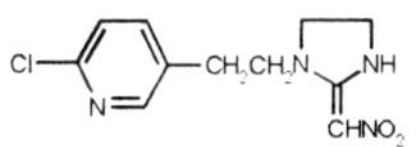	11.1	+	No	
(16)	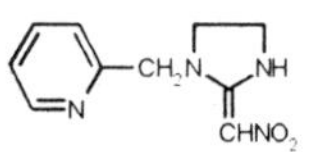	12.1	+	No	
(17)	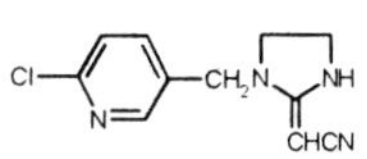	17.6	-	Yes	
(18)	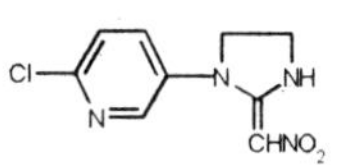	92.1	++	No	
(19)	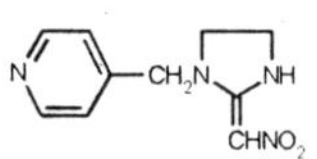	136	++	No	
(20)	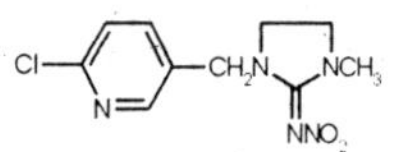	164.	+++	Yes	*N*-Me-imidacloprid

Fig. 8. Structure-activity relationships of imidacloprid related compounds
a) K_i (μM), inhibition constant in binding study using α-BGT as a probe; its reciprocal is binding affinity. b) LC_{90} to the green rice leafhopper (*Nephotettics cincticeps*).
b) +++ <1.6 ppm, ++ 1.6–200 ppm, + 200–1000 ppm, – >1000 ppm c) See Fig. 9

6.3. The ^{15}N-NMR Evidence for the Partial Positive Charge

There was a definite contrast between nicotinoids and neonicotinoids about the shift of the ^{15}N-NMR signals of the concerned nitrogen atom. Among nicotinoids it ranged between -324.1 and -353.3 ppm (average. -338.5) and among neonicotinoids between -288.7 and -293.2 ppm (average. -291.3) (nitromethane, 0 ppm). This implies that the nitrogen atom of neonicotinoids bears a partial positive charge (Yamamoto et al., 1995) as depicted in Fig. 9.

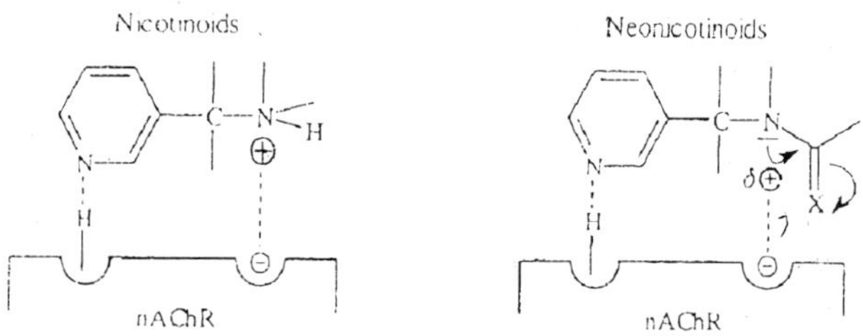

Fig. 9. The essential structural moieties of nicotinoids and neonicotinoids for interacting with nAChR

7. Selective Toxicity of Nicotinoids and Neonicotinoids

From the SAR, imidacloprid can be regarded as the analog of nicotine with an additional feature. In brief, they are relatives. However, the selective toxicity between mammals and insects is quite different (Table 1). Imidacloprid is really safe and excellent as an insecticide, while nicotine is more mammalicidal rather than insecticidal. The puzzle was solved as follows.

Table 1. Toxicity of nicotine and imidacloprid

Insecticide	LD_{50} (mg/kg)	
	Rat (oral)	Housefly (injection)
Nicotine	53	272
Imidacloprid	450	22

7.1. Binding Affinity to Vertebrate and Insect nAChRs

The interaction of nicotinoids and neonicotinoids with vertebrate and insect nAChRs is shown in Table 2 (Yamamoto et al., 1995). The source of nAChRs are rat brain, the *Torpedo* electric organ, and honeybee heads. The preparation from *Torpedo* electric organ and rat brain are regarded as the models of the AChRs for the mammalian peripheral and central nervous systems, respectively. The probes were [^{3}H] α-BGT for *Torpedo* and honeybee and [^{3}H]nicotine for rat brain. The results can be correlated to the structural features in terms of the presence of 3-pyridylmethylamine (3-PMA) moiety, ionization, and positive charge on the nitrogen atom. It is deduced that to interact with the vertebrate nAChR, enough ionization is the key factor, while to interact with the insect nAChR, the presence of the 3-pyridylmethylamine moiety having a full or partial positive charge on the nitrogen atom is essential. For example, TMA (tetramethylammonium salt) and nithiazine lack the 3-pyridylmethylamine moiety and nicotyrine is not ionized, giving a weak interaction with insect nAChR. However, the completely ionized TMA without 3-PMA interacts strongly with vertebrate peripheral nAChR.

7.2. The Cause of Low Mammalian Toxicity of Neonicotinoids

With the nAChR preparation from rat whole brain, imidacloprid interacted poorly as compared with nicotine and anabasine. It was further confirmed that imidacloprid interacts poorly with the individual mammalian brain nAChRs. More than 90% of the high-affinity nicotine binding sites in the brain are $\alpha 4/\beta 2$ receptors, which are α-BGT-insensitive, and the rest are mostly the α-BGT -sensitive neuronal nAChRs, which contain $\alpha 7$ subunits. Imidacloprid did not activate nAChRs of the $\alpha 4/\beta 2$ subtype expressed in *Xenopus* oocytes even at higher doses, while a weak activation of nAChRs of the $\alpha 7$ type was observed (Yamamoto et al., 1997). Thus, low toxicity of neonicotinoids in vertebrates results from the insensitivity of both brain and peripheral nAChRs due to their difference in the receptor types compared to insects. In other words, the partial positive charge in neonicotinoids can distinguish the insect nAChR from the vertebrate nAChR

It can be generalized that (1) there is a difference between vertebrate (both central and peripheral) and insect (central) nAChRs; (2) ionized nicotinoids make no distinction; (3) neonicotinoids with partial positive charge make a distinction; and (4) brain and peripheral nAChRs in mammals are low in sensitivity to neonicotinoids.

Table 2. Relationships between structure of nicotinoids and binding affinity to nAChRs of vertebrate and insect preparations.

	IC_{50} (μM)[a]					Relative binding	
Ligand	Rat whole brain[b]	*Torpedo* electric organ[c]	Honeybee head[c]	Presence of 3-PMA[d]	Ionization	Vertebrate	Insect
Nicotine	0.0096	23.1	1.25	Yes	Yes	Strong	Strong
Anabasine-	0.285	193	1.98	Yes	Yes	Strong	Strong
Nicotyrine	>1000	>1000	>1000	Yes	No	Weak	Weak
TMA	---	197	542	No	Yes	Strong	Weak
Imidacloprid	0.984	1060	1.95	Yes	No*	Weak	Strong
Nitenpyram	---	>300	61.3	Yes	No*	Weak	Strong
Acetamiprid	---	>300	8.83	Yes	No*	Weak	Strong
Nithiazine	109	1594	1364	No	No*	Weak	Weak

a) Reciprocal of IC_{50} = binding affinity.
b) [^{3}H]nicotine as probe.
c) [^{3}H] α-BGT as probe.
d) 3-Pyridylmethylamine moiety.
* Bearing partial positive charge.

8. The Difference Between Vertebrate and Insect nAChRs

For demonstration of the difference of nAChRs in vertebrates and insects, [^{3}H]α-BGT and [^{3}H]Phencyclizine ([^{3}H]PCP) were used as the probes for interaction with the ACh recognition (AR) site and non-competitive binding (NCB) site, respectively, and the interaction of many ligands with *Torpedo* nAChR was first studied as the reference (Tomizawa et al., 1995a) to be compared with the interaction with honeybee nAChR (Tomizawa et al., 1995b).

8.1. The Interaction of Ligands with *Torpedo* nAChR

The ligand interaction profile from *Torpedo* nAChR coincided with the results reported by many workers. Qualitatively, group A, including nicotinoids (nicotine, anabasine), neonicotinoids (imidacloprid, acetamiprid, 6-Cl-PMNI), carbachol, and cytisine interacted with the AR site as the agonists and activated (opened) the receptor's ion channel. Group B, including nithiazine, nitenpyram, nereistoxin, *d*-tubocurarine, coniine, and dimethylphenylpiperazinium (DMPP) interacted not only with the AR site, but also with the NCB site in the ion channel when the channel was opened by an agonist. Group C including PCP, *N*-[1-(2-thienyl)cyclohexyl]piperidine (TCP), lobeline, mecamylamine, ketamine, and chlorpromazine was the noncompetitive blocker, which bound with the NCB site at the resting state of the receptor and bound more when the channel was opened by an agonist. Trimethaphan bound with the NCB site only when the channel was opened.

8.2. The Interaction of Ligands with Honeybee nAchRs

In *Torpedo* nAChR, PCP or TCP specifically bound with the NCB site, and nicotine and anabasine are specific agonists, but in insects they bound with both AR and NCB sites. Therefore, the system to classify the ligands into agonosts, competitive blockers, and noncompetitive blockers using [^{3}H] α-BGT and [^{3}H]PCP as probes cannot be applied in the case of insect nAChR. Accordingly, ligands were grouped into three types: (1) those having higher affinity to the [^{3}H]α-BGT binding site, the neonicotinoids (imidacloprid, acetamiprid, 6-Cl-PMNI), carbachol, cytisine, lobeline, trimethaphan, and DMPP; (2) those having higher affinity to the [^{3}H]PCP binding site, that is, nereistoxin, chlorpromazine, ketamine, mecamylamine, and coniine; and (3) those having high affinity to both binding sites, the nicotinoids (nicotine, anabasine), a neonicotinoid (nitenpyram), PCP, TCP, and *d*-tubocurarine. The profile obtained from insect nAChR was quite different from that from *Torpedo*, which implies the difference between vertebrate and insect nAChRs.

9. Translocation to Target Site

The low mammalian toxicity of neonicotinoids and high mammalian toxicity of nicotinoids are explainable from the binding study, as stated. The question arises why nicotinoids are rather limited and neonicotinoids are strong in insecticidal activity. It was hypothesized formerly that the ionization of nicotinoids limits the target penetrability (Yamamoto et al., 1962), while neonicotinoids without ionization are easily penetrable (Yamamoto et al., 1995).

More information was obtained by the study of translocation of the injected compounds into the housefly head (Yamamoto et al., 1997). The amount of the compound bound to the nAChR in the head was regarded as the penetrability of the compound into the CNS. As shown in Table 3, imidacloprid and acetamiprid had almost six times more penetrability into the CNS than nicotine. However, 6-Cl-PMNI, which is the nitromethylene counterpart of imidacloprid, was far poorer in penetration. It was noticed that there was a correlation between higher hydrophobicity and higher penetrability.

Table 3. Translocation of nicotine and neonicotinoids to the housefly central nervous system

Compound	K_i (μM)	Relative translocation	Log P
Nicotine	1.48	1.00	- 0.29
6-Cl-PMNI	0.02	2.32	- 0.19
Imidacloprid	1.55	5.58	0.59
Acetamiprid	6.96	5.65	0.80

10. Structural Factors Contributing to Insecticidal Activity of Neonicotinoids.

So far we have focussed on the similarity then the dissimilarity between nicotine and imidacloprid as representative of nicotinoids and neonicotinoids, respectively. However, there are differences in insecticidal performance among various neonicotinoids and some apparent anomalies for the essential moiety idea. What are the factors that contribute to these ? In Table 4, as the first approximation, binding affinity can be regarded as the parameter of intrinsic activity, synergistic ratio as the metabolic extent, and log P (hydrophobicity) as the target penetrability.

10.1. Nicotine vs. Imidacloprid

Both nicotine and imidacloprid shared the same level of intrinsic activity, but imidacloprid was 12 times higher in insecticidal activity than nicotine, although far more metabolizable than nicotine. The superiority of imidacloprid resulted from nonionization, higher hydrophobicity, and thus penetrability into the target site.

10.2. Imidacloprid vs. 6-Cl-PMNI

Replacing the nitromethylene group of 6-Cl-PMNI by nitroimino group leads to imidacloprid. Imidacloprid was far inferior to 6-Cl-PMNI in binding affinity, but remarkably higher in hydrophobicity. The extent of metabolism was similar for both compounds. The net result in imidacloprid is an excellent insecticidal activity, although 6-Cl-PMNI is still higher in insecticidal activity because of its strong intrinsic activity.

Table 4. Comparison of nicotine and neonicotinoids in terms of intrinsic activity, insecticidal activity, metabolism in insect and penetrability into the target site

Compound	Binding to nAChR[a] K_i (μM)	Insecticidal activity[b] (LD_{50} μg/g) PPP-	PPP+	Ratio	Log P
Nicotine	1.48	272	126	2.2	- 0.29
PMNI	0.08	>50	35.2	-	- 1.02
6-Me-PMNI	0.47	-	-	-	- 0.59
De-Cl-imidacloprid	7.29	>20	>20	-	- 0.19
6-Cl-PMNI	0.02	2.32	0.005	464	- 0.19
N-Me-imidacloprid	164	23.3	2.41	9.7	0.26
Imidacloprid	1.55	22.3	0.065	343	0.59
S-Analog of 6-Cl-PMNI	0.116	7.64	0.003	2547	0.65
Acetamiprid	6.96	17.0	0.02	850	0.80
Nithiazine	1075	2.84	0.44	6.5	- 0.60

[a] Honeybee. [b] Housefly. PPP: *O*-propyl-*O*-(3-propynyl)phenylphosphonate

10.3. Imidacloprid vs. De-Cl-imidacloprid

All the commercial "chloronicotinyl" insecticides have a chlorine atom at the 6-position on the 3-pyridyl group De-Cl-imidacloprid was far lower in insecticidal activity, although the binding affinity was not as poor. Introduction of a chlorine atom increased binding affinity and hydrophobicity, that is, penetrability to the target site.

10.4. 6-Cl-PMNI vs. PMNI vs. 6-Me-PMNI

Introduction of a chlorine atom in PMNI increased both binding affinity and hydrophobicity, resulting in higher insecticidal activity. However, introduction of a methyl group in PMNI, although decreasing binding affinity, increased hydrophobicity and resulted in increased insecticidal activity. Although not shown here, 6-Me-PMNI was as insecticidal as 6-Cl-PMNI for green rice leafhopper.

10.5. The Importance of Hydrophobicity

Although the chlorine atom was important in chloronicotinyl insecticides, it could be replaced by a methyl group, which also increased hydrophobicity and thus penetrability into the target site. Imidacloprid or 6-Cl-PMNI are not effective against lepidopteran insects, probably because of poor integument penetration. When the NH on the 3-position of the imidazolidine ring of 6-Cl-PMNI was replaced by S to produce the S analog of 6-Cl-PMNI, the hydrophobicity increased to a level higher than that of imidacloprid. The compound showed essentially the same level of insecticidal activity to green rice leafhopper, a sucking insect, as imidacloprid (Tomizawa and Yamamoto, 1993), but with an additional feature, that is, an excellent activity to lepidopteran pests as claimed by Shiokawa et al. (1992). This probably results from the increased hydrophobicity and integument penetrability combined with still higher binding affinity. Also, acetamiprid, although its binding is inferior to that of other neonicotinoids, was far higher in hydrophobicity than imidacloprid and nitenpyram, and effective against lepidopteran and other pests (Matsuda and Takahashi, 1996). Compound 13 in Fig. 8, although relatively low in binding affinity, was high in insecticidal activity due to higher hydrophobicity.

10.6. N-Me-Imidacloprid vs. Imidacloprid

There are two apparent anomalies in Table 4. Introduction of a methyl group on the 3-position of the imidazoline ring of imidacloprid gives N-Me-imidacloprid. Its binding affinity decreased greatly and its log P as well. Nevertheless, the insecticidal activity was the same as imidacloprid. Most probably this form is metabolized to imidacloprid, as implied from the *in vitro* metabolic conversion (Yamamoto et al., 1997).

Nithiazine is known to interact with nAChR (Sattelle et al., 1989), but its low binding affinity could not explain its high excitatory activity to insects.

10.7. Factors Affecting Insecticidal and Selective Action

All in all, it is concluded that the insecticidal and selective actions of neonicotinoids are affected by binding affinity to nAChR, hydrophobicity, and metabolism. For strong insecticidal activity, compounds must have a certain level of binding affinity to nAChR (K_i below 7.0 μM) combined with higher log P, while compounds with lower log P must have very potent binding affinity. Some compounds with higher hydrophobicity become broad in insecticidal spectrum by means of improved integument as well as target penetrability.

11. Comparison of Nicotine and Imidacloprid as the Representatives of Nicotinoids and Neonicotinoids, respectively

The information obtained from these studies is integrated into Figs. 10 and 11 (Yamamoto et al., 1995). In mammals, the effect of imidacloprid on any nAChR is weak, while nicotine affects mostly peripheral nAChRs, resulting in higher toxicity. In insects, nAChR is present only in the CNS. Imidacloprid, once it has entered into the body (by sucking or injection), is easily accessible to the target site while nicotine is not. Imidacloprid itself is poor in integument penetration and less effective on lepidopteran insects, but other neonicotinoids like acetamiprid with higher hydrophobicity are effective to them.

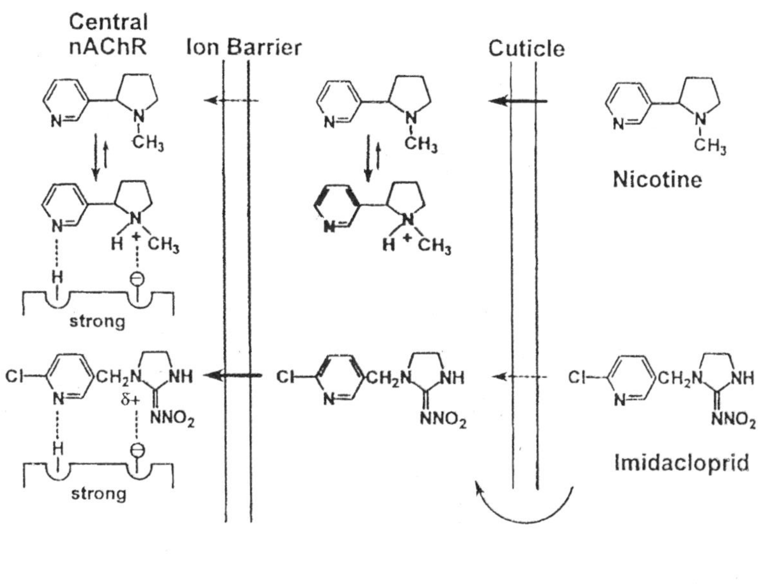

Fig. 10. Action of nicotine and imidacloprid in vertebrates

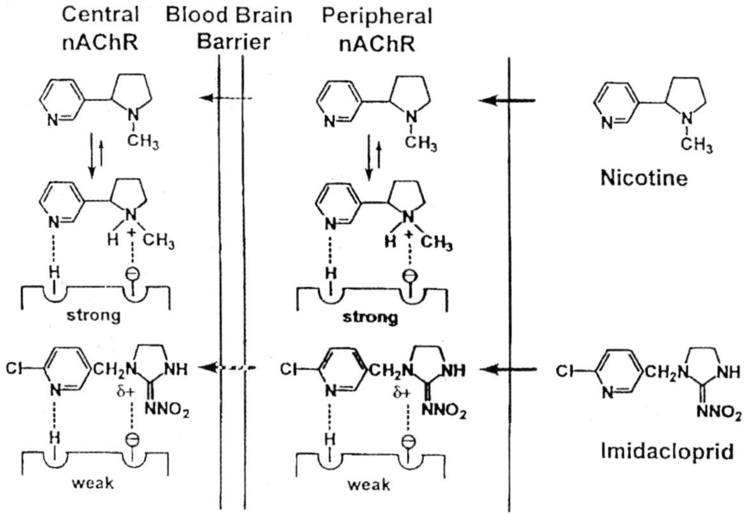

Fig. 11. Action of nicotine and imidacloprid in insects

12. Future Scope

12.1. Search for Channel Blockers

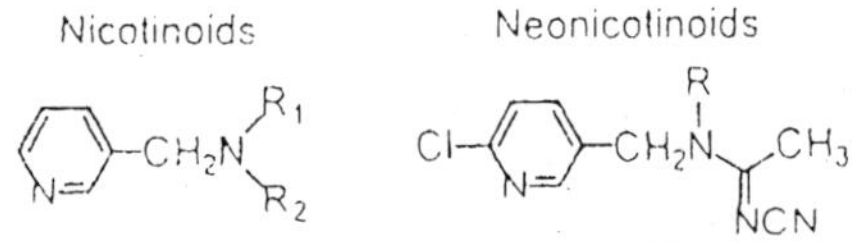

R_1	R_2
H	CH_3
H	C_2H_5
H	n-C_3H_7
H	n-C_4H_9
CH_3	CH_3
C_2H_5	C_2H_5
n-C_3H_7	n-C_3H_7
n-C_4H_9	n-C_4H_9

	R
*	CH_3
	C_2H_5
	n-C_3H_7
	n-C_4H_9

* acetamiprid

Fig. 12. Structures of 3-pyridylmethylamines and acetamiprid homologs

In 3-pyridylmethylamine and acetamiprid type compounds (Fig. 12), the interaction with AR and NCB sites of nAChR changed when alkyl substituents (indicated by R) were changed (Tomizawa et al., 1996; Matsuo et al., 1997). In insects, change from a mono-methyl to mono-butyl group in the 3-pyridylmethylamine type shifted the binding site from an AR to an NCB site. The same trend was obtained in di-alkyl compounds. Particularly, *N,N*-dibutyl-3-pyridylmethylamine showed remarkable selective binding affinity to the NCB site, which gave these IC_{50} (μM): >1000 ([^{3}H]α-BGT as probe) and 0.64 ([^{3}H]PCP as probe), while the mono-methyl compound produced 340 and 430 and the di-methyl compound gave 130 and 55, respectively. In the acetamiprid type, similar shifting of binding site was observed. While acetamiprid (R = CH_3) was selective to AR site [IC_{50} (μM): 8.0 ([^{3}H]α-BGT as probe) and 310 ([^{3}H]PCP as probe)], its *N*-*n*-butyl analog showed excellent NCB site selectivity ($\geqq$300 and 27, respectively) in insects, but bound poorly with the NCB site of *Torpedo* nAChR. Such selectivity was not found in the 3-pyridylmethylamine type. The binding effect was reflected in the symptoms in insects; binding to the NCB site correlated with anesthetic effect. As mentioned in 8.2, nereistoxin was also selective to the insect NCB site, causing paralysis of insects without excitation. Just as in the case of nereistoxin derivatives, exploration in further structural modification of neonicotinoids may lead to newer selective insecticides.

12.2. New Idea for Essential Moiety

Based on the idea of the importance of partial positive charge, we designed compounds not covered by patents or those having carbon, phosphorus, or sulfur instead of the concerned nitrogen atom. So far, they are not competitive in insecticidal activity. It would be possible that the strong electron-withdrawing group is involved in providing not only the partial positive charge on the nitrogen atom, but also the additional binding by itself (Yamamoto, 1996).

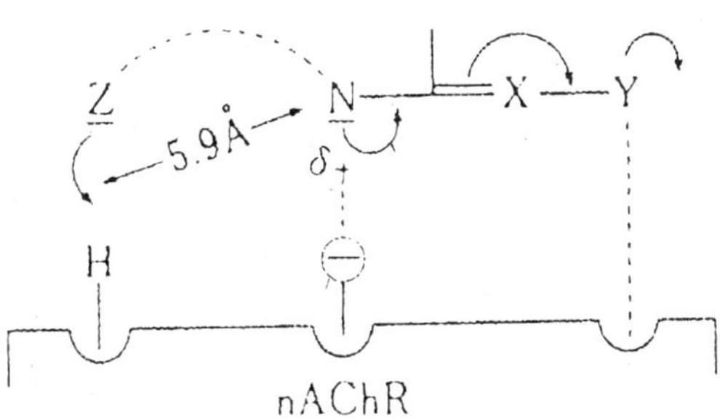

12.3. Nicotinoids As Psychoactive Drugs

It was claimed that nicotine is the addicting agent in tobacco smoking. Nicotine may be also a useful drug in improving cognitive function in schizophrenics, reversing inattentiveness in patients with attention deficit disorder and improving memory in Alzheimer's patients (Ember, 1994). But there are many side effects of nicotine mostly based on the action on peripheral nervous systems. Compounds with the positive effects of nicotine without its harmful side effects are seriously needed, and many analogs of nicotine or 3-pyridylmethylamines have been already examined. However, these compounds are considered more active to the peripheral than to the central nervous system. Information obtained from the insect world on central penetrability of neonicotinoids may bring some insight to design more brain-directed drugs.

12.4. AChR Ligands As Leads

There are many compounds that interact with nAChRs such as trimethaphan, cytisine, β-erythroidine, and strychnine. They are seemingly different from nicotinoids and neonicotinoids in chemical structure, but share the common essential feature as the nAChR ligand, which makes hydrogen bonding about 5.9 Å apart from the center of a positive charge with the receptor. On the other hand, the ligands that interact with muscarinic AChR share the distance of 4.4 Å instead of 5.9 Å (Beers and Reich, 1970). Someday they will be the leads for newer insecticides. Spinosad, a recently developed insecticide of microbial origin (a naturally occurring mixture of spinosin A and D), affects nAChR in a different way and has a different chemical structure (a macrocyclic lactone with sugar moiety) from any nicotinoids and opened a new dimension in AChR interacting agents (Salgado, 1997).

12.5. Nitrogeneous Insecticides

There have been a few insecticides that interact with nAChRs: the nicotine and cartap groups as well as nithiazine. The chlordimeform group is closely related to the nicotine group in SAR, but interacts with the octopamine receptor. The success of neonicotinoids will accelerate the development of more nitrogeneous insecticides after pyrethroids.

12.6. Natural Products As Leads

It is estimated that the total number of plant chemicals may amount to 400,000 or more, but so far only about 10,000 secondary metabolites in plants have been chemically defined, and only a few has been used directly.

As leads, a few compounds have been investigated and the rest await further investigation.

13. Conclusion

Development of nicotinoids from natural nicotine to synthetic nicotinoids is an excellent case history to illustrate the potentiality of natural products as the prototype of newer insecticides and the importance of the study on mode of action.

Acknowledgment

I wish to express my appreciation to Prof. John E. Casida and late Prof. Ryo Yamamoto, who led me to insect toxicolgy and chemistry and gave me constant encouragement, and to all who have worked with me.

References

Addor RW (1995) Insecticides. In: Godfrey CRA (ed) Agrochemicals from natural products. Marcel Dekker, New York, pp 10-11

Bai D, Lummis SCR, Leicht W, Breer H, Sattelle DB (1991) Actions of imidacloprid and a related nitromethylene on cholinergic receptors of an identified insect motor neurone. Pestic Sci 33: 197-204

Beers WH, Reich E (1970) Structure and activity of acetylcholine. Nature (London) 228: 917-922

Ember LR (1994) The nicotine connection. Chem Eng News Nov 28

Fujita T, Nakajima M, Soeda Y, Yamamoto I (1971) Physicochemical properties of biological interest and structure of nicotine and its related compounds. Pestic Biochem Physiol 1: 151-162

Fujita T, Yamamoto I, Nakajima M (1970) Analysis of the structure-activity relationship of nicotine-like insecticides using substituent constants. In: O'Brien RD, Yamamoto I (eds) Biochemical toxicology of insecticides. Academic Press, New York, pp 21-32

Iwata K, Takase I (1993) Admire® - a new systemic insecticide. Agrochem Jpn No. 63: 15-17

Kamimura H, Matsumoto A, Miyazaki Y, Yamamoto I (1963) Relation of structure to toxicity of pyridylmethylamines. Agric Biol Chem 27: 684-688

Kashiwada Y (1996) Bestguard® (nitenpyram, TI-304)- a new systemic insecticide. Agrochem Jpn, No. 68: 18-19

Leicht W (1996) Imidacloprid -a chloronicotinyl insecticide. Biological activity and agricultural significance. Pflanz-Nachr Bayer 49: 71-84

Liu M-Y, Lanford J, Casida JE (1993) Relevance of [^{3}H]imidacloprid binding site in house fly head acetylcholine receptor to insecticidal activity of 2-

nitromethylene- and 2-nitroimino-imidazolidines. Pestic Biochem Physiol 46: 200-206

Matsuda M, Takahashi H (1996) Mospilan® (acetamiprid, NI-25)- a new systemic insecticide. Agrochem Jpn, No. 68: 20-21

Matsuo H, Tomizawa M, Yamamoto I (1998) Structure-activity relationships of acyclic nicotinoids and neonicotinoids for insect nicotinic acetylcholine receptor/ion channel complex. Arch Insect Biochem Physiol 37: 17-23

Metcalf R (1955) Organic insecticides. Interscience, New York

Nishimura K, Kanda Y, Okazawa A, Ueno T (1994) Relationship between insecticidal and neurophysiological activities of imidacloprid and related compounds. Pestic Biochem Physiol 50: 51-59

Salgado VL (1997) The modes of action of spinosad and other insect control products. Down to Earth 52: 35-43

Sattelle DB, Buckingham SD, Wafford KA, Sherby SM, Bakry NM, Eldefrawi AT, Eldefrawi ME, May TE (1989) Actions of the insecticide 2-(nitromethylene)-tetrahydro-1,3-thiazine on insect and vertebrate nicotinic acetylcholine receptors. Proc R Soc Lond B237: 501-514

Schmeltz I (1971) Nicotine and other tobacco alkaloids. In: Jacobson M, Crosby DG (eds) Naturally occurring insecticides. Marcel Dekker, New York, pp 99-136

Shiokawa K, Moriya K, Shibuya K, Hattori Y, Tsuboi S, Kagabu S (1992) 3-(6-Chloronicotinyl)-2-nitromethylene-thiazolidine as a new class of insecticides acting against Lepidoptera species. Biosci Biotechnol Biochem 56:1364-1365

Shiokawa K, Tsuboi S, Kagabu S, Moriya K (1986) European Patent EP 0 192 060 A1

Soeda Y, Yamamoto I (1968a) Inhibition of house fly head cholinesterase by nicotine. Agric Biol Chem 32: 568-573

Soeda Y, Yamamoto I (1968b) Relation of structure to toxicity of pyridylalkylamines. Agric Biol Chem 32: 747-752

Soloway SB, Henry AC, Kollmeyer WD, Padgett WM, Powell JE, Roman SA, Tieman CH, Corey RA, Horne CA (1978) Nitromethylene heterocycles as insecticides. In: Shankland DL, Hollingworth RM, Smyth T (eds) Pesticide and venom neurotoxicity. Plenum, New York, pp 153-158

Tomizawa M, Matsuo H, Miyamoto T, Yamamoto I (1996) *N,N*-Dialkyl substituents of 3-pyridylmethylamines discriminate between α -bungarotoxin and noncompetitive blocker sites of nicotinic acetylcholine receptors of *Torpedo* and insects. J. Pestic Sci 21: 412-418

Tomizawa M, Otsuka H, Miyamoto T, Yamamoto I (1995a) Pharmacological effects of imidacloprid and its related compounds on the nicotinic acetylcholine receptor with its ion channel from the *Torpedo* electric organ. J Pestic Sci 20: 49-56

Tomizawa M, Otsuka H, Miyamoto T, Eldefrawi M E, Yamamoto I (1995b) Pharmacological characteristics of insect nicotinic acetylcholine receptor with its ion channel and the comparison of the effect of nicotinoids and neonicotionoids. J Pestic Sci 20: 57-64

Tomizawa M, Yamamoto I (1992) Binding of nicotinoids and the related compounds to the insect nicotinic acetylcholine receptor. J Pestic Sci 17: 231-236

Tomizawa M, Yamamoto I (1993) Structure-activity relationships of nicotinoids and imidacloprid analogs. J Pestic Sci 18: 91-98

Yamamoto I (1965) Nicotinoids as insecticides. Adv Pest Control Res 6: 231-260

Yamamoto I (1979) Pesticide design- insecticides. In: Yamamoto I, Fukami J (eds) Pesticide Designm: strategy and tactics. Soft Science, Tokyo. pp 891-928 (in Japanese)

Yamamoto I (1996) Neonicotinoids - action and design. Plant Prot 50: 240-245 (in Japanese)

Yamamoto I, Kamimura H, Yamamoto R, Sakai S, Goda M (1962) Studies on nicotinoids as an insecticide: relation of structure to toxicity. Agric Biol Chem 26: 709-716

Yamamoto I, Soeda Y, Kamimura H, Yamamoto R (1968) Cholinesterase inhibition by nicotinoids and pyridylalkylamines - its significance to mode of action. Agric Biol Chem 32: 1341-1348

Yamamoto I, Tomizawa M, Saito T, Miyamoto T, Walcott EC, Sumikawa K (1998) Structural factors contributing to insecticidal and selective actions of neonicotinoids. Arch Insect Biochem Physiol 37: 24-32

Yamamoto I, Yabuta G, Tomizawa M, Saito T, Miyamoto T, Kagabu S (1995) Molecular mechanism for selective toxicity of nicotinoids and neonicotinoids. J. Pestic Sci. 20: 33-40

2 Nicotine and Other Insecticidal Alkaloids

István Ujváry
Plant Protection Institute, Hungarian Academy of Sciences, Herman Ottó út 15, H-1022 Budapest, Hungary

1 Introduction

Alkaloids, with more than 10,000 described, are one of the most diverse and prominent groups of natural products with pharmacological and toxicological importance (Harborne 1993). Plant extracts containing insecticidal alkaloids as bioactive constituents have played an important role in the abatement of insects of agricultural and public health importance for centuries. While the direct use of these substances has recently diminished, they continue to serve as leads for synthetic analogs and are also indispensable biochemical tools in mode of action studies. In recent years, the function of these alkaloids in the host plant has begun to unfold: it is now generally believed that the ecological role of these compounds, often acting in concert with other nonalkaloidal substances, is to provide a chemical defense against predators and pathogens in a sustained manner through multiple biological mechanisms (Wink 1993a; Brown and Trigo 1995).

The history and progress of the research on natural insecticides, botanicals in particular, have been surveyed from time to time (Jacobson and Crosby 1971; Klocke 1989; Addor 1995). This paper focuses on alkaloids with insecticidal properties. This structural restriction obviously excludes some other natural anti-insect agents of practical significance, but it is hoped that the importance of this segment of natural product research will be demonstrated. The compounds are grouped according to their first reported isolation: those from plants, arthropods and other animals, and marine organisms as well as microbial sources.

The following definition for an alkaloid is adopted: "*an alkaloid is a cyclic organic compound containing nitrogen in a negative oxidation state which is of limited distribution among living organisms*" (Pelletier 1983). This permits the inclusion of not only alkali-like compounds (e.g., nicotine, steroid alkaloids) but amides (e.g., lipophilic isobutylamides) as well. Compounds with a nitrogen atom in a positive oxidation state (e.g., nitro compounds) are excluded.

2 Botanical Alkaloids with Insecticidal Properties

2.1 Tobacco Alkaloids

The tobacco plant has a long and fascinating history. After its introduction in 1559 to Spain and Portugal from the Americas where it had been long cultivated by the Indians, tobacco reached France and Italy where it came into use mostly for smoking but since the late seventeenth century has been used as an important insecticide or insect repellent. In America, tobacco was recommended as an insecticide by Yates in New York State in 1814 (Schmeltz 1971).

The main bioactive components of the tobacco plants *Nicotiana tabacum* and *N. rustica* (Solanaceae) are pyridine alkaloids (reviewed by Enzell et al. 1977; Fodor and Colasanti 1985) and the structure of the most important ones is given in Fig. 1.

Nicotine (**1**) is the principal bioactive alkaloid found in tobacco extract. After World War II, nicotine consumption for insecticide use amounted to more than 2,500 tons worldwide with most of it from waste of the tobacco industry. Since then, due to the appearance of more selective and cheaper substitutes, the global production of nicotine for insect control has been declining steadily and was less than 200 tons by the early 1980s.

Nicotine (**1**), probably the most familiar of all alkaloids, was first isolated by Posselt and Reimann in 1828 (Posselt and Reimann 1828) who named it after *Jean Nicot*, who introduced tobacco to the French court around 1560. The structure of **1** was determined by Pinner (1893). In a pure state, **1** is a colorless, nearly odorless liquid of low volatility. There are several reviews on the chemistry (Jackson 1941; Fodor and Colasanti 1985), biosynthesis and metabolism (Leete 1983; Bush and Crowe 1989), and insecticidal activity (Shepard 1951a; Metcalf 1955; Schmeltz 1971) of **1** and related tobacco alkaloids. Structural studies on the conformation and reactivity of **1** and its derivatives have been summarized (Seeman 1984).

Nicotine has a rapid vapor action and is highly toxic to mammals (mouse oral LD_{50} = 24 mg/kg). The levorotatory isomer, i.e., (*S*)-nicotine (**1**), is two to three times more active than the other enantiomer. Illustrative topical LD_{50} values are 4 and 140 μg/g for *Bombyx mori*, and male *Periplaneta americana*, respectively; in a fumigant assay it has an LC_{50} value of 0.0032 mg/l for *Aphis gossypii*. A compilation of acute toxicological data for insects and other animals has been published (Negherbon 1959) while the pharmacology of antipodal nicotine isomers has been summarized by Brossi and Pei (1998).

Nicotine exerts its effect in both vertebrates and invertebrates by binding to a subset of cholinergic receptors, the nicotinic acetylcholine receptors (nAChRs). To explore the structural requirements for insecticidal activity and to develop compounds with improved mammalian toxicity, Yamamoto (1965) investigated a series of nicotine analogs, the nicotinoids, but no practically useful compound emerged. Research at the Bayer laboratory in Japan, however, has led to the dis-

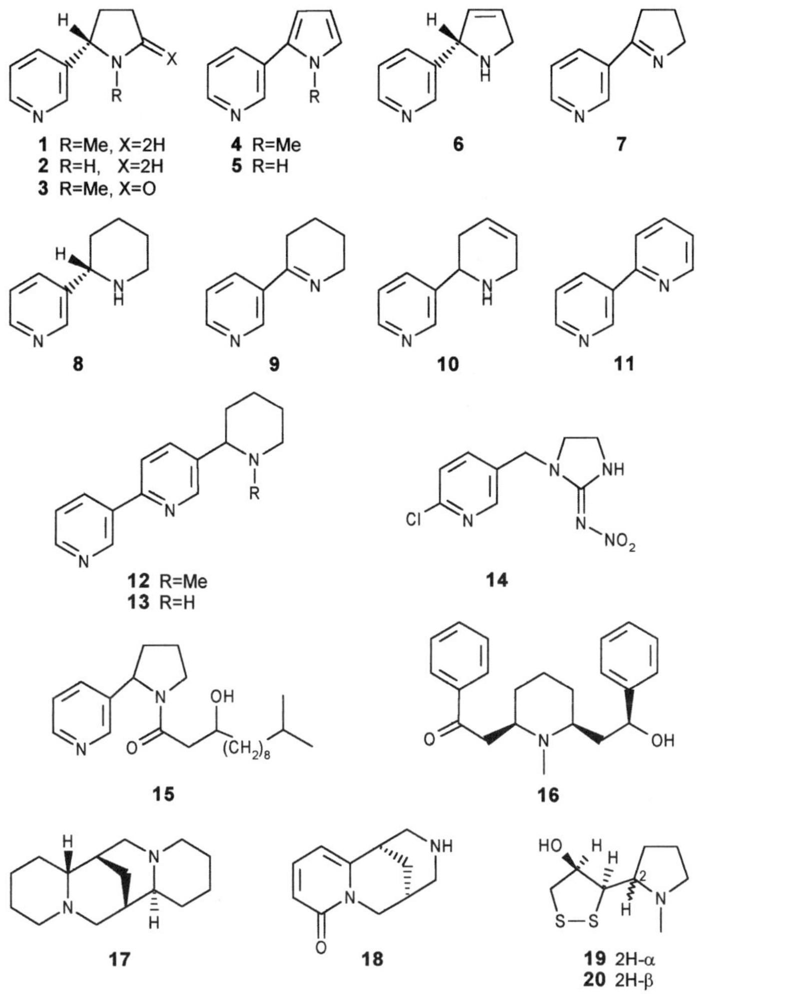

Fig. 1. *Nicotiana* alkaloids and related insecticidal compounds

covery of imidacloprid (**14**), a representative of a new group of insecticides, the neonicotinoids (see Chaps. 1 and 4). Nicotine is usually formulated either as a 40% solution of its sulfate (Black Leaf 40) or in a dust carrier, used especially against aphids. As of this writing (1998), a nicotine-based fumigant and spray or dust products are undergoing reregistration in the United States.

Nornicotine (**2**) usually accompanies nicotine, from which it is derived by demethylation, an essentially irreversible process in leaves. Nicotine and **2** are also found in leaves of the Australian shrub *Duboisia hopwoodii*, used by the Aborigines to alleviate pain, hunger, and fatigue by chewing quids (pituri). A nornicotine-rich variety of this scarce desert plant was used to stupefy fish and game by poisoning water holes.

Cotinine (**3**) is a trace constituent in certain tobacco species but one of the main metabolites of **1** in both insects and mammals. Nicotyrine (**4**), nornicotyrine (**5**), myosmine (**7**), and isonicoteine (2,3'-bipyridyl) (**11**) are also minor components found in fresh or cured tobacco. While dihydronicotyrine (**6**) is insecticidal, compounds **3**, **4**, **5**, and **7** are of moderate or low toxicity. A summary on relative insecticidal activity of these nicotinoids is discussed in detail by Yamamoto (1965).

Smith et al. (1930) identified 3-(2-piperidyl)pyridine as an insecticidal trace constituent of synthetic dipyridyl oil and called it neonicotine. About the same time, Orechoff and Menschikoff (1931) isolated the levorotatory **8** from the toxic Asian plant *Anabasis aphylla* (Chenopodiaceae) and named it anabasine. Anabasine, which is present in small amounts in *Nicotiana* species, as obtained for practical purposes from the American *N. glauca* is mostly racemic. The sulfate of **8** has been a major botanical insecticide in the former Soviet Union. Anabaseine (**9**) and anatabine (**10**) are other minor *Nicotiana* alkaloids, while the bipyridyl derivative anabasamine (**12**) was found in *A. aphylla* (Mukhamedzhanov et al. 1968).

The average nicotine content of the leaves of *N. tabacum* and *N. rustica* is 2% to 6% dry weight. Gas chromatographic analysis of 60 greenhouse-grown *Nicotiana* species has shown large species-dependent differences in the leaves, with nicotine being the predominant alkaloid of 33 species, nornicotine of 24 species, and anabasine of two species (Saitoh et al. 1985).

The ecological role of nicotine in tobacco has not been understood until very recently. Baldwin (1996, 1997) has provided evidence for the generally held view that nicotine constitutes an important chemical defense mechanism of the plant. Wounding of the leaves of *N. sylvestris* elicits a local jasmonic acid level that through a signal-transduction cascade triggers a de novo nicotine synthesis (Baldwin et al. 1997). Airborne, exogenous methyl jasmonate acts similarly on *N. attenuata* (Baldwin 1996). The resulting elevated concentrations of nicotine are sufficient to protect induced tissues from nicotine-adapted herbivores in laboratory trials.

It is known that the tobacco-feeding hornworm *Manduca sexta* can tolerate high levels, up to 1 mM, of nicotine in its blood. Murray et al. (1994) have recently characterized a blood-brain barrier that protects the cholinergic nervous system from dietary nicotine. In the larva, dietary nicotine is pumped by a powerful transport system into the lumen of the malphigian tubules where it is excreted unmetabolized. This nonmetabolic resistance mechanism is similar to the P-glycoprotein multidrug pump responsible for xenobiotic excretion and multidrug resistance observed in other organisms. Interestingly, however, *M. sexta* is susceptible to *N*'-hydroxyacylnornicotines, tridecanoyl amide **15** being the major component found in cuticular extracts of *N. stocktonii* and other wild *Nicotiana* species (Severson et al. 1988). The extract is at least 50 times more toxic to *M.*

sexta larvae than pure nicotine when applied topically to *M. sexta* larvae. The structure of these acyl derivatives resembles that of the insecticidal isobutylamides (see **2.4**).

2.2 Lobeline

The leaves and tops of the Indian tobacco, *Lobelia inflata* (Campanulaceae), native to the eastern and southeastern region of North America, was smoked by the American Indians as a substitute for authentic tobacco and also to treat asthma and bronchitis. It is also a powerful emetic. The structure of the active ingredient, lobeline (**16**) (Fig. 1), was determined in 1929 by Wieland and associates (Wieland and Dragendorff 1929). Lobeline is a flexible molecule with no obvious structural resemblance to nicotine, but it is nevertheless a nicotine-like agonist (Barlow and Johnson 1989) with similar pharmacological features. Its insecticidal properties have not been fully explored.

2.3 Quinolizidine Alkaloids

These compounds, often referred to as lupine alkaloids, occur in plants of the Leguminosae and particularly in the genus *Lupinus* (Kinghorn and Balandrin 1984). Several of the bitter-tasting lupine alkaloids, and thus the seeds where they are stored, are toxic to grazing animals like sheep and cattle. They have been shown to be nAChR agonists, affect K^+ channels and glutamate receptors, and disturb protein biosynthesis. Besides being acutely toxic, the bitter lupine alkaloids have deterrent, antifeedant, and growth regulatory properties for a wide range of insect species (Wink 1993b). The tetracyclic sparteine (**17**) (Fig. 1), first isolated in 1851 and now commonly obtained from *Cytisus scoparius*, is toxic to humans and livestock, and it was once used for the treatment of cardiac problems and as an oxytocic. Cytisine (**18**) occurring, e.g., in the laburnum tree, *Cytisus laburnum*, is a widely used pharmacological tool. Both sparteine and cytisine are insecticidal (Tattersfield et al. 1926). Structural similarities between **1** and **18** have been analyzed (Barlow and Johnson 1989).

Guinesine A (**19**) and B (**20**) (Fig. 1) are novel sulfur-containing pyrrolidine alkaloids obtained from the bark of the Brazilian plant *Cassipourea guianensis* (Rhizophoraceae) with LD_{50} values of 5.12 and 1.10 μg/insect, respectively, against *Chilo suppressalis* (Kato et al. 1989). Their structural similarity to nicotine and nereistoxin (see later) suggests nAChRs as molecular targets. Analogs have also been synthesized and found to be insecticidal against *C. suppressalis* and *Tetranychus urticae* (Mitsudera et al. 1990c).

2.4 Unsaturated Amides

Plants of the families Aristolochiaceae, Compositae, Piperaceae, and Rutaceae contain various olefinic isobutylamides while related acetylenic compounds are prevalently found in Asteraceae (for a chemosystematic overview, see Greger 1988). These substances are notorious for their pungency and strong sialagogue effect; nevertheless, many of them possess useful pharmacological properties (e.g., bronchodilation, local anesthesia) and antimicrobial and insecticide activities. Some of them are typical ingredients of spices (e.g., of black pepper, *Piper nigrum*). The continuous interest in the chemistry and biological activity of insecticidal alkamides has prompted a stream of reviews on the subject (Jacobson 1971; Elliott 1985; Su 1985; Addor 1995). Because these substances are prone to isomerization and polymerization, often synthesis could provide an unequivocal proof for their structure, and special synthetic methodologies furnishing individual stereoisomers of the polyunsaturated system in high purity for spectral characterization and comparative biological studies had to be developed (e.g., Crombie and Fisher 1985a,b; Miyakado et al. 1985a, 1989; Crombie et al. 1987; Strunz and Finlay 1994). Many of these amides have rapid paralyzing ("knockdown") action accompanied with lethal toxicity to many insect pests even to pyrethroid-resistant strains (see, e.g., Elliott et al. 1986), and besides being insecticidal themselves, several members of the group also act as synergists (e.g., Miyakado et al. 1980, 1989). The structures of the most important natural and synthetic amides are shown in Fig. 2.

Pellitorine (**21**), or "pyrethrin" as it was originally called by Parisel who first isolated it in 1834, is the main bioactive ingredient of the roots of a North African plant, *Anacyclus pyrethrum* DC, and is now named as pellitory after its use in medicine. Jacobson (1949) proposed a 2,5-decadieneamide structure for pellitorine but this was soon corrected by Crombie (1952) and Jacobson (1953) for a conjugated 2,4-decadieneamide system as in **21**. The knockdown activity of **21** on *Musca domestica* approaches that of natural pyrethrins, and it is lethal to a wide range of pest insects (Jacobson 1971; Su 1985).

Affinin (**24**) was isolated from the roots of the Mexican medicinal plant *Heliopsis longipes* as well as from *Spilanthes* species, hence the alternative name spilanthol. The vinylog neoherculin (**25**), obtained from the bark of the North American "toothache tree" *Zanthoxylum clava-herculis*, is in fact structurally identical with echinacein and sanshool-I isolated independently as an antibacterial and anti-inflammatory agent from American *Echinacea* spp. and from the Japanese "asakura sansho" plant *Z. piperitum* with anthelmintic properties. The structure of the related mosquito larvicide herculin from *Z. clava-herculis* (Jacobson 1948) has not been fully resolved. Tetrahydroanacyclin (**26**), a potent insecticide, is a partially reduced synthetic analog of the corresponding 8,10-acetylene derivative, anacyclin, a less toxic minor ingredient of the roots of *A. pyrethrum*. Recently, Gbewonyo et al. (1993) examined five insecticidal amides isolated from *Piper guineense* root extract and reported kalecide (**23**) as the most

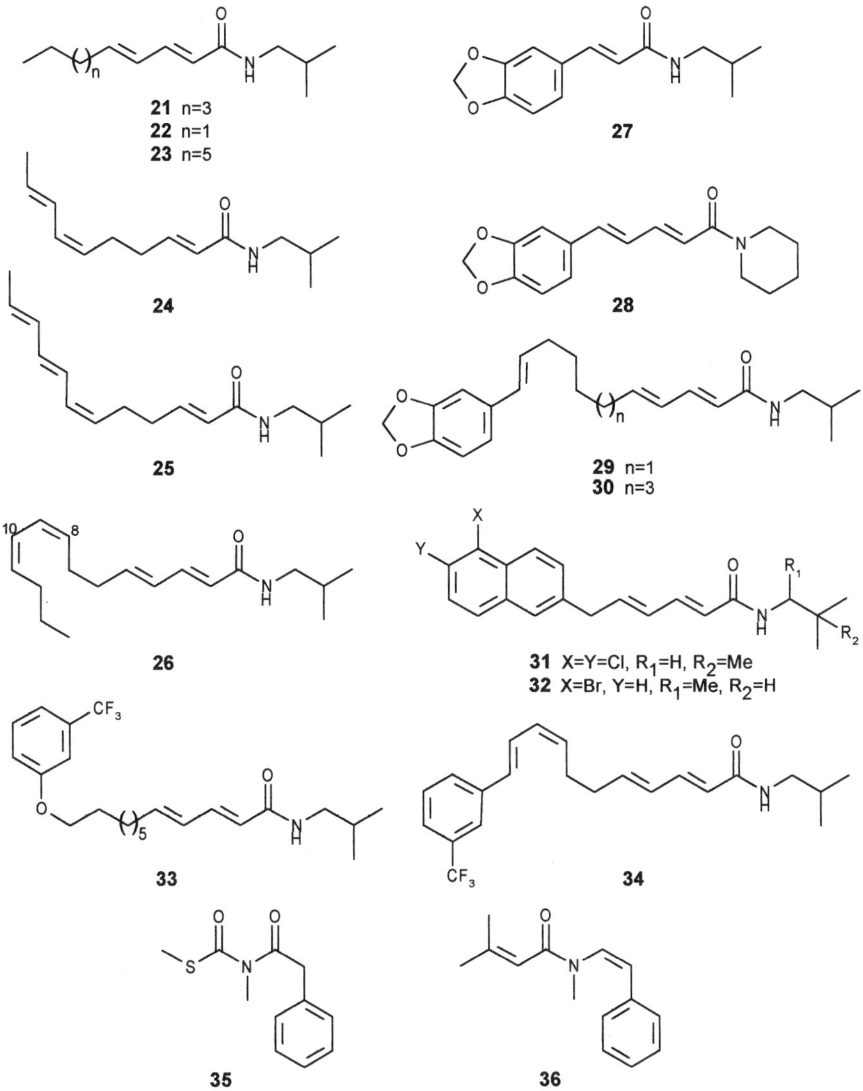

Fig. 2. Representative natural and synthetic insecticidal amides

active knockdown and kill agent with ED_{50} values of 0.06 and 1.01 μg/insect, respectively, against *M. domestica*.

From five isobutylamides isolated from the East African medicinal tree *Fagara macrophylla*, methylenedioxyphenyl-containing fagaramide (**27**) and its vinylog piperlonguamine (structure not shown), pellitorine (**21**), and the octadienamide **22** are strongly insecticidal to *Pectinophora gossypiella, Heliothis virescens, H. zea, Spodoptera frugiperda*, and *Culex pipiens* (Kubo et al. 1984). Piperine (**28**), the major pungent and purported insecticidal principle of pepper fruits, was shown to be inactive against *Callosobruchus chinensis* (Miyakado et al. 1980, 1985a) and *M. domestica* (Miyakado et al. 1989), although it did synergize **21** and pyrethrins. In examining the synergism of *P. nigrum* ingredients on male adult *C. chinensis*, Miyakado et al. (1980, 1989) observed that pipercide (**29**) with two co-occurring amides 10,11-dihydro-**29** and guineensine (**30**) in 1:1:1 ratio resulted in a twofold synergism (LD_{50} = 0.11 μg/insect) with regard to 10,11-dihydro-**29**, the latter being the most active natural amide (LD_{50} = 0.23 μg/insect) in this study.

Early structure-activity relationship (SAR) studies at the Australian CSIRO laboratory dealt with the chain length, as well as the position and configuration of the double bonds in structures related to **24**, **25**, and **26** (Meisters and Wailes 1966a) and led to the conclusion that "the number and distribution of double bonds necessary for appreciable insecticidal activity towards the housefly impart great instability to the amides" (Meisters and Wailes 1966b). A SAR study with more than 220 analogs was disclosed in a series of seven papers (Elliott et al. 1987, 1989). Using **26** as a lead compound, the Rothamsted group established that (a) (2*E*,4*E*)-6-aryl-2,4-hexadienamides, preferably with halogenated aromatic groups (e.g., **31** and **32**), are optimal replacements of the terminal dienoyl chain; and (b) 1,2- or 2,2-dimethylpropylamides, having a reduced propensity to rearrange into the inactive 6-aryl-3,5-hexadienamides, are preferred over the analogous isobutylamides. Amide **31** was the most active in unsynergized topical assays against adult *M. domestica* and *Phaedon cochleariae*, having a relative activity of 4% and 36%, respectively, of that of the standard bioresmethrin (Elliott et al. 1987).

Researchers at Sumitomo, using **29** and 10,11-dihydro-**29** as lead compounds, systematically varied the carbon chain and amine moiety and identified **33** as the most active member of the series with notable knockdown effect and lethal activity against susceptible and pyrethroid resistant insects (Miyakado et al. 1985a,b, 1989). The trifluoromethyl derivative **33** has a topical LD_{50} = 0.043 μg/insect for *C. chinensis*, being at least twice as potent as the natural pyrethrin mixture and equivalent to fenitrothion. Compound **33** is also toxic to *M. domestica* and *C. suppressalis*, but inactive against *Spodoptera litura*.

In studies carried out at the Wellcome Research Laboratories, a similar SAR pattern emerged (Blade 1990). From a wide range of structurally different compounds, including fluorinated analogues as well as proinsecticidal derivatives such as thioamides, the tetraene **34**, benzyl ethers related to **29** and **30**, as well as tetralins related to **32** were the best with insecticidal activity reaching a maximum of 40% of that of permethrin against *M. domestica* and *P. cochleariae*.

The similar symptoms and electrophysiological responses produced by **34** on *M. domestica* showed that the activities of this potent insecticide (LD_{50} = 3.6 μg/insect) resembled those of DDT and allethrin (Blade et al. 1985), indicating the involvement of the Na^+ channels (for a review, see Bloomquist 1996). Studies on *P. americana* synaptosomal preparations showed that representative isobutylamides antagonized the effect of the Na^+ channel activator veratridine while deltamethrin had a potentiating effect (Nicholson et al. 1985). Recent binding studies by Ottea et al. (1990) have showed that BTG-502 (**32**) and related synthetic isobutylamides inhibit the specific binding of [^{3}H]batrachotoxinin A 20α-benzoate to sodium channels, further corroborating that isobutylamides affect the voltage-sensitive Na^+ channel domain at a site distinct from that of the pyrethroids and DDT. These findings not only explain the unique biological properties of this class of compounds but also make them challenging leads for insecticide design to combat pyrethroid resistance. In spite of extensive efforts, however, no commercial compounds have been developed.

Recently, a range of biogenetically related thiocarbamates and amides, isolated from shrubs of the *Glycosmis* genera (Rutaceae) distributed from Southeast Asia to North Australia, has been shown to possess strong antifungal and insecticidal activities (Greger et al. 1996). For example, against *Spodoptera littoralis* larvae ritigalin (**35**) (see Fig. 2) had an LC_{50} of 0.05 μmol/dm^2 in the contact toxicity assay and an EC_{50} of 0.43 μmol/g in the growth inhibition test. In the amide series, dehydrothalebanin B (**36**) had the highest insecticidal activity.

2.5 Veratrum Alkaloids

Liliaceous plants are well known both for their poisonous properties and also for their medicinal value. Powdered rhizomes of the white hellebore (*Veratrum album*), indigenous in Europe and Asia, and the "green hellebore" or Indian poke (*Veratrum viride*), growing in the eastern part of North America, were used to cure herpes, toothache, rheumatism, and catarrh. Drugs from these plants have also been important hypotensive agents (Kupchan and Flacke 1967). Preparations from hellebore roots were once commercial insecticides, although often of variable quality, against hemipteran and homopteran pests of fruits and vegetables (Fisher 1940; Shepard 1951b). The crushed seeds of *Schoenocaulon officinale*, indigenous in the northern region of South America and Central America, had been known as sabadilla powder and were used by the American Indians in pre-Columbian times as an insecticide (Crosby 1971). With the advent of cheap synthetics in the 1950s, these botanicals lost importance. Nevertheless, thanks to its low persistence and compatibility with beneficial insects (Bellows and Morse 1993), sabadilla reappeared in the late 1970s. It is now used against thrips in citrus and avocado (Hare and Morse 1997), and has been recommended as an alternative to conventional synthetic insecticides for *H. virescens* control (Yoshida and Toscano 1994).

Although chemical studies were initiated by Pelletier and Caventou in 1819 (Pelletier and Caventou 1820), it was only in 1954 when the structure of the readily available alkamine cevine, i.e., the 3α-epimer of **37**, was elucidated (Barton et al. 1954), and the structure of its veratrate **39** (i.e., veratridine) was confirmed by X-ray crystallography (Codding 1983). The main insecticidal components in sabadilla are the steroidal alkaloids cevadine (**38**), veratridine

(**39**), and cevacine (**40**), which are esters of the virtually nontoxic veracevine (**37**) (Fig. 3). 3-*O*-Vanilloylveracevine (**41**), a minor sabadilla component, is 25fold less toxic than **39** to *M. domestica* but comparable to **38** and **39** to *P. americana* nymphs (Ujváry and Casida 1997). To date, more than 350 related natural and synthetic alkaloids have been characterized (Greenhill and Grayshan 1992).

Various *Veratrum* and *Solanum* alkaloids were examined for toxicity towards *M. domestica* by Bergmann et al. (1958). From a recent SAR study with 5 natural and 68 semisynthetic *Veratrum* alkaloids (Ujváry et al. 1991), the 3,5-dimethoxy-

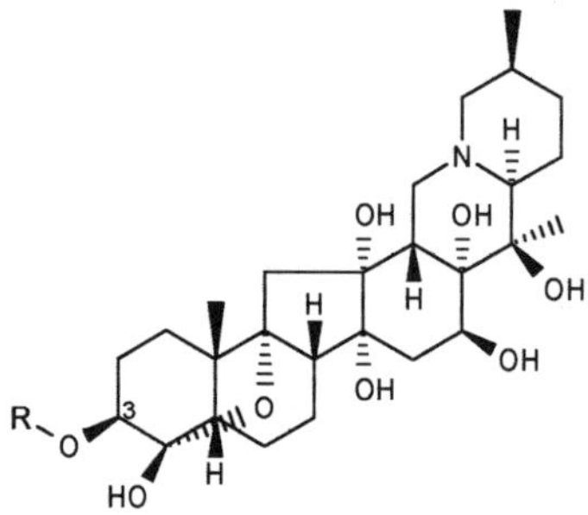

37 R=H
38 R=angeloyl
39 R =3,4-dimethoxybenzoyl
40 R=Ac
41 R=vanilloyl
42 R=3,5-dimethoxybenzoyl

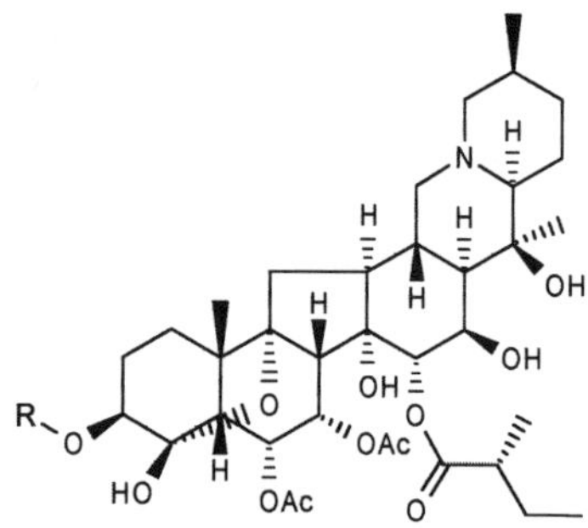

43 R=(-)-2-hydroxy-2-methylbutyryl
44 R=(+)-*threo*-2,3-dihydroxy-2-methylbutyryl

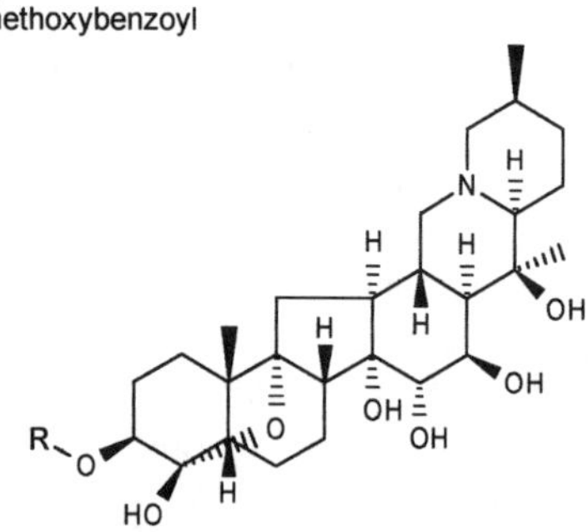

45 R=angeloyl
46 R=H
47 R=Ac

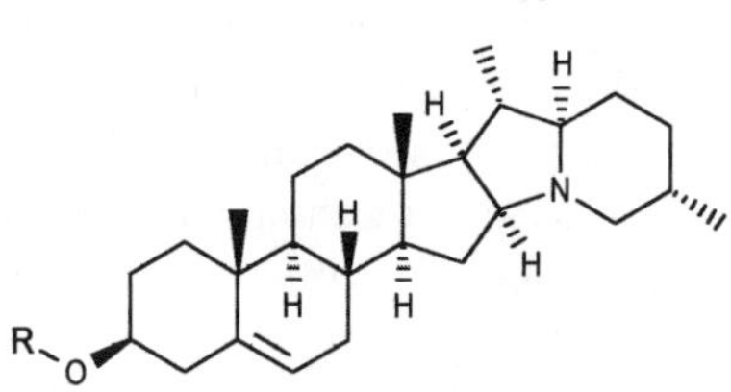

48 R=Glcβ3(Rhaβ2)Galβ1-
49 R=Rhaβ4(Rhaβ2)Glcβ1-

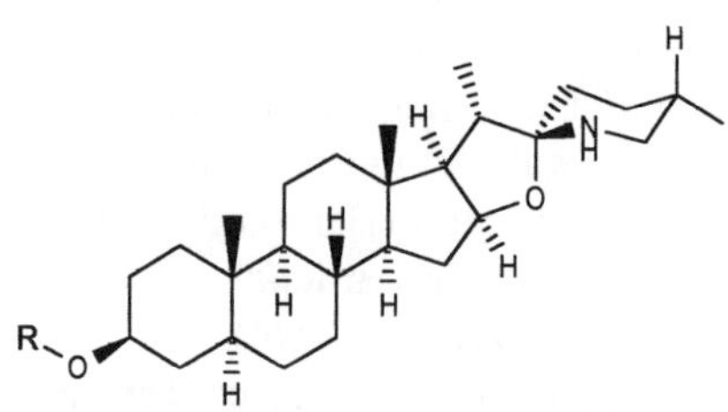

50 R=Glcβ2(Xylβ3)Glcβ4Galβ1-

Fig. 3. Representative insecticidal steroid alkaloids of the *Veratrum* and *Solanum* group

benzoyl ester **42** emerged as the most potent compound with a topical LD_{50} of 6.6 μg/g on adult *M. domestica* pretreated with piperonylbutoxide (PBO), a 4.7-fold improvement over **39** and with a simultaneous increase in insect vs. mouse selectivity.

The neurotoxicity of **39** and related *Veratrum* alkaloids is attributable to their activation of the Na^+ channels of excitable nerve membranes with a binding site distinct from those of pyrethroids and the lipophilic isobutylamides (Bloomquist 1996). Structure-based design based on a putative pharmacophore model of Na^+ channel toxins (Codding 1983; Kosower 1983) led to nonsteroid mimics of **39** with negligible insecticidal activity (Dell et al. 1990; Ujváry et al. 1995).

An investigation on the ecological role of *Veratrum* alkaloids during the life cycle of *Rhadinoceraea nodicornis*, a specialist sawfly feeding on *V. album* leaves, has recently shown that the larvae degrade or secrete most plant alkaloids including protoveratrine A (**43**) and B (**44**) but sequester 3-*O*-angeloylzygadenine (**45**) via alkamine **46** to 3-*O*-acetylzygadenine (**47**), which is then stored to provide a highly potent chemical defense against predatory ants (Schaffner et al. 1994; Gfeller et al. 1995).

2.6 Solanum Alkaloids

Several genera of the Solanaceae and Liliaceae contain *Solanum* alkaloids typically occurring in plants as glycosides conjugated to an oligosaccharide residue at the 3-OH group (Schreiber 1968). Because of their poisonous effect, there has been considerable interest especially in α-solanine (**48**), α-chaconine (**49**), and α-tomatine (**50**) (Fig. 3) (Sharma and Salunkhe 1989), but it was exactly the fear of these toxic properties that precluded their application as insecticides.

The phagorepellency of *Solanum* alkaloids towards certain herbivores, believed to be one of the ecological functions of steroid alkaloids of Solanaceae (Levinson 1976; Wink 1993a), is not necessarily a sole or dominant defensive factor responsible for insect resistance of these plants. The lack of feeding and growth inhibitory effects on *Leptinotarsa decemlineata* of varying foliar content of **50** in tomato plant challenges the conclusions of previous experiments that used alkaloids *ex planta*; nevertheless, **48**, **49**, and, to some extent, **50** did show insecticidal activity in various bioassays at natural concentration levels (Barbour and Kennedy 1991). Their mode of action could involve the inhibition of acetylcholinesterase (AChE) (Roddick 1989) or the disruption of cell membranes (Keukens et al. 1995).

2.7 Physostigmine (Eserine)

The pharmacohistory of the Calabar bean of *Physostigma venenosum* (Fabaceae), a woody vine growing in the rain forest of West Africa, is a brilliant example of how a deadly poison became an essential tool for the elucidation of chemical neurotransmission, and how it could be transformed into healing medicines (Holmstedt 1972) and also, although indirectly, into a major class of insecticides, the *N*-methylcarbamates (Kuhr and Dorough 1976). The Efik people of Old Calabar (now in Nigeria) have used the bean, or a potion of it, for their trial by ordeal and for divination of witches (the Efik word *esere* stands for both the bean and the poison ordeal). Structure determination of the deadly poison by Stedman and Barger (1925) revealed a hydrogenated pyrroloindole ring system and a carbamate moiety, the latter being an exceptionally rare functional group in nature.

Physostigmine (**51**) (Fig. 4), by virtue of inhibition of AChE, is toxic to adult *Locusta migratoria* when injected with a medium lethal dose of 20-25 μg per locust (Hopf 1952). Although the primary mode of action of **51** is the inhibition of AChE, recent studies have consistently shown that binding to nAChR may contribute to its insecticidal properties (for references and discussion, see Liu et al. 1995).

2.8 Ryanodine

Based on the reported insecticidal properties of the stem wood dusts of *Ryania speciosa* (Flacourtiaceae), a tropical tree growing in Central America and the Amazon Basin, Rogers et al. (1948) isolated an alkaloid, ryanodine (for a historical review, see Crosby 1971). Degradative studies by Wiesner (1972) and subsequent X-ray crystallography (Srivastava and Przybylska 1970) established the structure of the complex diterpene ryanodine (**53**), the pyrrolecarboxylate of ryanodol (**52**) (Fig. 4). Reexamination of commercial ryania by Waterhouse et al. (1984) revealed 9,21-didehydroryanodine (**55**) as the most abundant insecticidal component. Nine other natural ryanoids have also been characterized in Casida's laboratory (Jefferies et al. 1992). Compound **52** has recently been found in *Persea indica* (Lauraceae) from the Canarian Archipelago (González-Coloma et al. 1993). The total synthesis of **52** has been accomplished (Deslongchamps et al. 1990, and subsequent papers). Ryania is a safe botanical insecticide and was once used in orchards and ornamentals, but as of this writing it is not currently registered in the United States.

Extensive SAR studies pursued by the Berkeley group with more than 50 natural and synthetic ryanoids defined the structural requirements for insecticidal activity and selective toxicity (Waterhouse et al. 1987; Jefferies et al. 1993; Jefferies and Casida 1994) and recent work uncovered several new derivatives with improved insecticidal activity (Jefferies et al. 1997). Compound **56**, the most potent in this series, was about 2 to 4 fold more active (LD_{50} = 0.11 μg/g) than **53** or **55** on injection and 29fold more (LD_{50} = 0.41 μg/g) when applied topically to *M. domestica* pretreated with PBO. A 2-fold improvement in insect vs. mammalian receptor selectivity was also observed for **56**.

The pharmacological and toxicological properties of ryanodine result from the activation of Ca^{2+} release of sarcoplasmic reticulum of skeletal and cardiac muscle of both vertebrates and invertebrates (Bloomquist 1996; Schmitt et al. 1996; for a review, see Sutko et al. 1997). However, the observed differences in selective toxicity and muscle receptor binding of esters (**53** and **55**) and those of non-esters (**52** and **54**) suggest that these two types of compounds may act on entirely different targets, i.e., the former on Ca^{2+} and the latter on K^{+} channel receptors (Ushwerwood and Vais 1995).

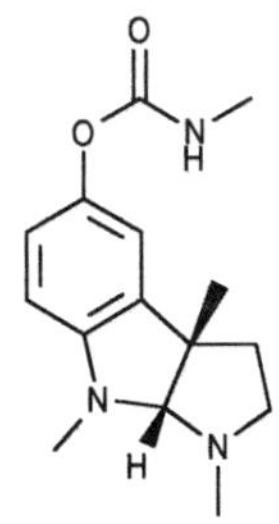

51

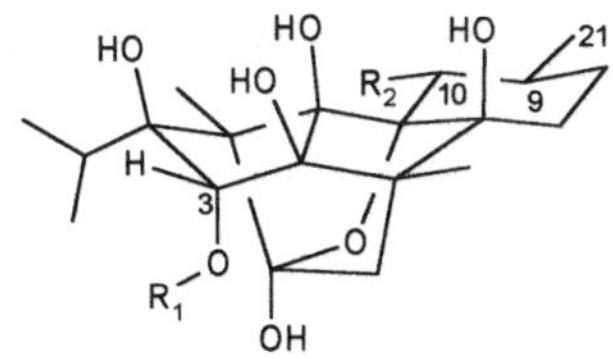

	R_1	R_2	C_9-C_{21}
52	H	OH	CH-CH_3
53	2-pyrroleCO	OH	CH-CH_3
54	H	OH	$C{=}CH_2$
55	2-pyrroleCO	OH	$C{=}CH_2$
56	2-pyrroleCO	H	$C{=}CH_2$

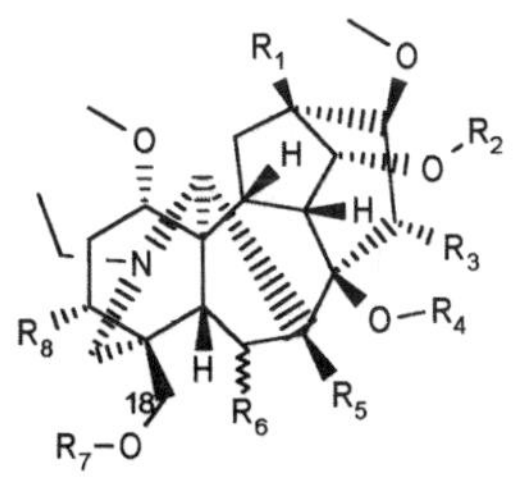

57 R_1=R_3=R_8=OH, R_2=Bz, R_4=Ac, R_5=H, R_6=α-OMe, R_7=Me
58 R_1=R_3=R_4=R_8=H, R_2=Me, R_5=OH, R_6=β-OMe, R_7=2-((*S*)-Methylsuccinimido)PhCO
59 R_1=R_3=R_4=R_8=H, R_2=Me, R_5=OH, R_6=β-OH, R_7=2-((*S*)-Methylsuccinimido)PhCO
60 R_1=R_3=OH, R_2=Bz, R_4=Ac, R_5=R_8=H, R_6=α-OMe, R_7=2-((*S*)-Methylsuccinimido)PhCO

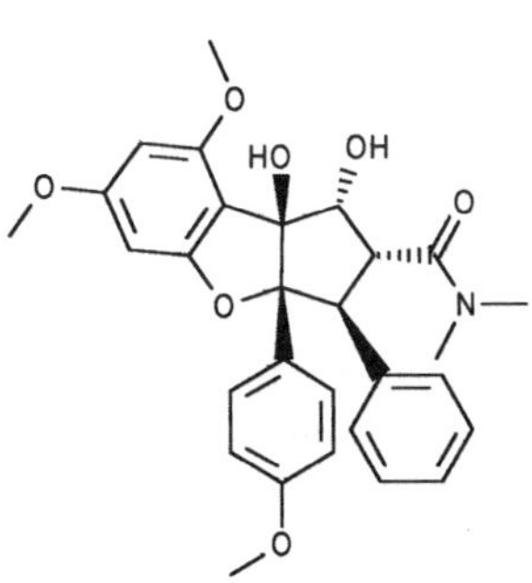

61

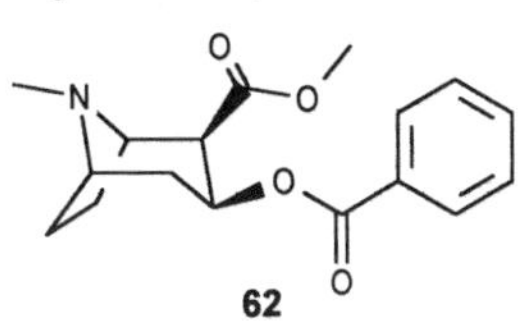

62

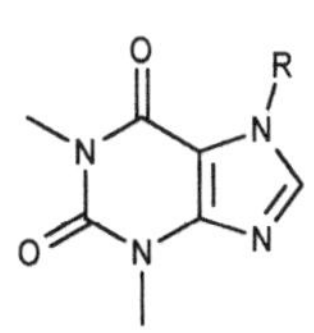

63 R=Me
64 R=H

Fig. 4. Various botanical insecticides and analogs

2.9 Aconitum and Delphinium Alkaloids

The medicinal and poisonous properties of the popular garden plants *Aconitum* and *Delphinium* (Ranunculaceae) have fascinated various civilizations. Preparations from the roots and leaves of the *Aconitum* spp. (aconites or monkshoods) were recommended for the treatment of gout, neuralgia, hypertension, and rheumatism. The use of the plant to poison arrows and baits for killing wolves gave the name *lycoctonum*, from the ancient Greek λυκο–κτονο, meaning wolf-slayer. *Aconitum* decoctions were also used for execution-by-suicide in ancient times. Although serious poisons of grazing livestock, *Delphinium* plants (or larkspurs) have a more peaceful history. Tinctures from the crushed seeds of *D. staphisagria*, sometimes mixed with sabadilla extract, have been used to kill head lice.

The diterpenoid alkaloid-rich fraction of these plants is responsible for the various pharmacological activities (Pelletier and Mody 1979; Benn and Jacyno 1983; Jacyno 1996). The complete structure of aconitine (**57**) (Fig. 4), the major and one of the most toxic C_{19}-norditerpene alkaloid (intraperitoneal LD_{50} to mouse is ~0.35 mg/kg) isolated from *A. napellus*, was elucidated after decades of extensive chemical studies (Birnbaum et al. 1971). The structure of methyllycaconitine (**58**), the principal toxic alkaloid of many *Delphinium* spp. but not found in *Aconitum* species, was fully assigned only in 1981 (Pelletier et al. 1981); its intraperitoneal LD_{50} value is ~18 mg/kg for mouse.

Despite the structural similarities of **57** and **58**, their biological activities are different: aconitine affects the cardiovascular system, while methyllycaconitine shows a more pronounced action on the skeletal neuromuscular junction. In studies with insects, **57** paralyzed *M. domestica* larvae only on injection (ED_{50} = 0.6 μg/larva) with symptoms similar to that of Na^+ channel neurotoxins (Bloomquist and Miller 1986). Methyllycaconitine (**58**), however, was insecticidal against a wide range of insects including *M. domestica* as well as lepidopteran species not affected by nicotine such as *Spodoptera eradiana* and *H. virescens* (Jennings et al. 1986). Recent studies with a series of *Delphinium* alkaloids have defined the structural requirements for potency and selectivity of binding to rat vs. housefly nAChRs (Kukel and Jennings 1994, 1995). Although **58** strongly inhibited α-bungarotoxin binding to rat and housefly neural membranes (the respective IC_{50} values were 2.0 and 1.0 nM), glaudelsine (**59**), the most potent and selective inhibitor, had IC_{50} values of 16 and 0.042 nM, respectively. The importance of the side chain at the C-18 carbon atom in determining receptor selectivity was further demonstrated by Hardick et al. (1995, 1996), who converted the Na^+ channel activator **57** into the 2-(methylsuccinimido)benzoate analog **60**, which is a potent nAChR ligand.

2.10 Rocaglamide

Rocaglamide (**61**) (see Fig. 4), a novel antileukemic natural product with a tetrahydrocyclopentabenzofuran core, was first isolated from the roots and stems of *Aglaia elliptifolia* (Meliaceae) (King et al. 1982). The recent total synthesis of (-)-**61** by Trost et al. (1990) established the absolute configuration of the natural compound. A related tropical ornamental tree, the slow-growing *Aglaia odorata*, is a favorite source of fragrance in China and a traditional anti-inflammatory agent, heart stimulant, and expectorant throughout Southeast Asia. Screening of extracts from the twigs of this plant yielded a powerful insecticidal and antifeedant ingredient identified also as **61** (Satasook et al. 1992; Janprasert et al. 1993). Physiological and behavioral experiments with **61** on *Peridroma saucia* larvae revealed a powerful growth inhibitory activity with an EC_{50} = 1.37 ppm in artificial diet. In insecticidal assays, LD_{50} values of 0.32 and 0.34 μg/larva for topical and oral administration, respectively, were obtained, although lethal effects developed slowly. Compound **61** inhibited feeding and caused death of *S. litura* larvae at an LC_{50} = 4.80 ppm in a no-choice leaf-disk test. Rocaglamide remarkably inhibited the growth of *Ostrinia nubilalis* larva even below 0.1 ppm, thus being more effective than the limonoid azadirachtin from *Azadirachta indica* (Meliaceae) (Ewete et al. 1996). Recently, strong insecticidal activity against *S. littoralis* larvae was observed for new rocaglamide derivatives isolated from *A. odorata* flowers (Güssregen et al. 1997) and *Aglaia duppereana* twigs (Nugroho et al. 1997). Several synthetic studies (Trost et al. 1990; Davey et al. 1992) should provide important technical background for future SAR studies that so far have been limited to naturally occurring congeners.

2.11 Cocaine

Two South American woody shrubs, *Erythroxylon coca* and *E. truxillense* (Erythroxylaceae), have been considered as divine plants by the Indians of Bolivia and Peru for two millennia. Thanks to its exceptional psychotropic/stimulant and local anesthetic properties, the pharmacology of cocaine (**62**) (Fig. 4), the plants' main alkaloid, is now well understood. There is, however, not much known about the ecological role of this tropane derivative. Based on observations that *Erythroxylum* trees are relatively pestfree, Nathanson et al. (1993) have recently demonstrated that **62** protects tomato leaves from feeding *Manduca sexta* larvae at concentrations that occur naturally in coca leaves (3-10 mM). Light emission tests with firefly lanterns suggested that the toxic effect of **62** was caused by its inhibition of the (re)uptake of octopamine, a unique insect neurotransmitter and neurohormone. In mammals, by contrast, **62** affects reuptake of dopamine.

2.12 Methylxanthines

Caffeine (**63**), in coffee beans of *Coffea arabica* (Rubiaceae), and theophylline (**64**) (Fig. 4), in leaves of *Camelia sinensis* (Theaceae), are the most regularly used stimulants but their ecological roles in the plants are basically unknown. Nathanson (1984) has recently showed that these two methylxanthines inhibit the feeding of larvae of *M. sexta*, *Tenebrio* spp., and *Culex* spp., thus could play a natural protective role. Caffeine (**63**) killed *M. sexta* larvae at concentrations found in tea leaves (0.68% to 2.1%) and coffee beans (0.8% to 1.8%). The mode of action is associated with the inhibition of phosphodiesterases, which is also responsible for their physiological action in vertebrates. In addition, **63** is a phagorepellent (Levinson 1976) and an inhibitor of eel AChE (I_{50} = 87 μm) although about 270fold weaker than physostigmine (**51**) (Karadsheh et al. 1991).

2.13 Isoquinoline-Type Alkaloids

Many species of the custard apple family (Annonaceae), growing mainly in tropical and subtropical regions, have large delicious fruits, and extracts from various parts of the tree are known to have medicinal value as well as being insecticides and parasiticides. Anonaine (**65**) (Fig. 5) was first isolated from the bark of *Annona reticulata* as an insecticide with activity comparable to rotenone against aphids (Harper et al. 1947; Crosby 1971). It has also been obtained from other *Annona* species along with other close structural relatives (Leboeuf et al. 1982). Recently, an alkaloidal fraction from *A. squamosa* have shown larvicidal and development regulatory activity for the malaria vector *Anopheles stephensi* (Saxena et al. 1993). Information on the insecticidal mode of action of these compounds is lacking, but **65** is a smooth muscle relaxant by modulating adrenoreceptors and voltage-gated Ca^{2+} channels in the rat (Chuliá et al. 1995). It should be mentioned that another group of pharmacologically important compounds from *Annona* spp., the bistetrahydrofuran-containing acetogenins, have elicited considerable interest recently as insecticides (McLaughlin et al. 1997). The plant *Cocculus tribolus* DC. contains the aporphine-type isoboldine (**66**), which inhibits the feeding of *Trimeresia miranda* and *Prodenia* (*Spodoptera*) *litura* larvae at 200 ppm (Wada and Munakata 1967, 1968). For berberine (**67**), a widely distributed benzophenanthridine alkaloid from Papaveraceae with insecticidal and insect growth regulatory effect (Devitt et al. 1980), Schmeller et al. (1997) have recently discussed the structural aspects involved in its biological activity.

2.14 Dioncophyllines

Lianas of the rain forests of Africa, India, and Southeast Asia, such as the carnivorous *Triphyophyllum peltatum* (Dioncophyllaceae) and the taxonomically related *Ancistrocladus* species (Ancistrocladaceae), are used in folk medicine and have been shown to contain various naphthylisoquinolines (Bringmann and Pokorny 1995). Dioncophylline A (**68**) (see Fig. 5), an axially chiral biaryl alkaloid isolated from *T. peltatum* (Bringmann et al. 1990a,b), is one of the most common bioactive component exhibiting fungicidal, antimalarial, molluscicidal activity as well as being insect growth retardant or feeding deterrent (Bringmann et al. 1992) and having larvicidal (François et al. 1996) activities. According to a recent SAR study, benzylation of the phenolic OH group of **68** resulted in a 10-fold improvement in growth inhibitory activity against *S. littoralis* larva (Bringmann et al. 1997).

2.15 *Erythrina* Alkaloids

Studies prompted by the observation that the plant *Cocculus tribolus* is rarely attacked by insects led to the isolation, from its leaves, of an insecticidal component cocculolidine (**69**) (Fig. 5) (Wada et al. 1966, 1967, 1968; Wada and Munakata 1967). The insecticidal activity of **69** is between that of pyrethrin and anabasine (**8**) (Fig. 1) against adults of the leafhopper *Nephotettix bipunctatus cincticeps*, but it is less active against *C. chinensis* and inactive against *Myzus persicae* (Wada and Munakata 1967). Structurally, cocculolidine belongs to the curare-like *Erythrina* alkaloids (Dyke and Quessy 1981), among which erysodine (**70**) has been shown to be a competitive and highly selective antagonist of nicotine at the rat neuronal nAChR (Decker et al. 1995).

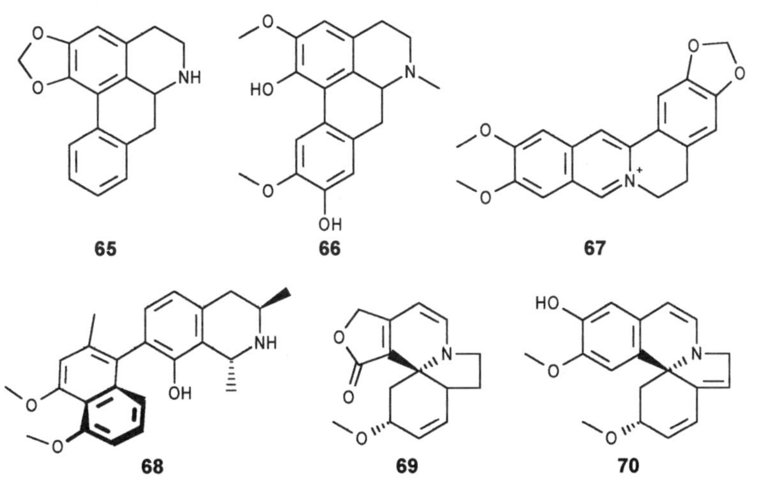

Fig. 5. Isoquinoline derivative insecticides

2.16 *Stemona* Alkaloids

The roots of *Stemona tuberosa* Lour., *S. sessilifolia*, and *S. japonica* (Stemonaceae) have long been used in Chinese traditional medicine as antitussive, antitubercular, and anthelmintic agents (Götz and Strunz 1973; on chemical aspects, see Jacobi and Lee 1997). In addition, extracts and dried plant materials are used throughout Eastern Asia as insecticides for zooparasites of man and cattle. Over 20 *Stemona* alkaloids have been characterized and nearly all possess the unusual perhydroazaazulene skeleton such as in tuberstemonine (**71**) (Fig. 6). Stemofoline (**72**), isolated from the leaves and stem of *S. japonica*, when mixed with an artificial diet is an effective insect growth inhibitor and kills *Bombyx mori* larva at 50 ppm but is inactive against *Mamestra brassicae* larvae at 100 ppm (Sakata et al. 1978). Synthetic efforts towards stemofoline-based insecticides have recently been reported (Thomas 1994). Neurophysiological studies with crayfish indicate that **71** acts by inhibiting glutamate-induced responses at the neuromuscular junction (Shinozaki and Ishida 1985).

2.17 Trypterygium Alkaloids

The powdered root of the thunder god vine, *Trypterygium wilfordii* Hook (Celastraceae), found in the mountainous areas of southern China has been used to control chewing insects for hundreds of years. Preparations from *T. forrestii* and the Chinese bittersweet, *Celastrus angulatus*, have also been used to protect plants from insect damage. Extracts from these species have traditionally been used for the treatment of rheumatism, skin disorders, cancer, and inflammatory disease as well as male contraceptives. The remarkable insecticidal properties of *T. wilfordii* prompted intense studies (Swingle et al. 1941). Beroza (1951) isolated wilfordine as the insecticidal principle with low mammalian toxicity (Beroza and Bottger 1954), the structure of which was established as the nicotinoyl dihydroagarofuran **73** (Fig. 6) (Yamada et al. 1978). A recent report describes a synthetic approach towards **73** (Thomas 1994). Beroza (1951) isolated another component wilforine, the structure of which was deduced to be a similar polyester **74** (Yamada et al. 1978). Compound **74** was recently also obtained from the tropical tree *Maytenus rigida* and showed potent antifeedant activity against *Locusta migratoria* and *Pieris rapae* (Delle Monache et al. 1984). The chemistry, chemotaxonomy, and pharmacology of the Celastraceae have been surveyed (Smith 1977; Brüning and Wagner 1978; Li Ya et al. 1990).

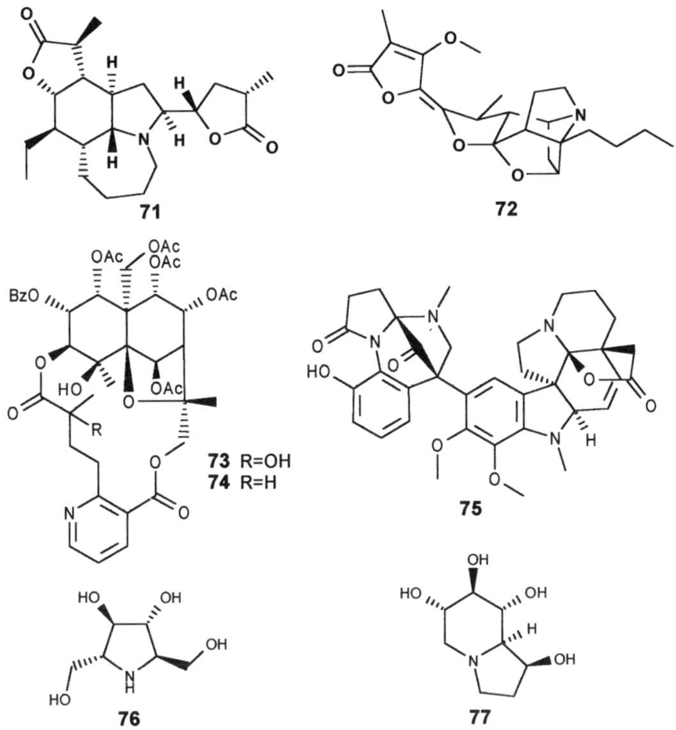

Figure 6. Miscellaneous botanical insecticides

2.18 Haplophyton Alkaloids

In their systematic work aimed at isolating novel insecticides, researchers at Merck analyzed the Mexican cockroach plant, *Haplophyton cimicidum* (Apocynaceae), and reported that one of the principal components, haplophytine, was highly toxic to a wide range of pest insects such as *Blatella germanica* (contact LD_{50} = 18 μg/g) (Rogers et al. 1952). Yates et al. (1973), revising their earlier proposal (Rae et al. 1967), established the structure of haplophytine as a bisindole alkaloid with an aspidospermane skeleton (**75**) (Fig. 6). The observed AChE inhibitory activity found for crooksiine (structure not shown), a closely related insecticidal alkaloid isolated from *H. crooksii*, suggests a possible similar mode of action for these structural analogs (Mroue and Alam 1991).

2.19 Polyhydroxy Alkaloids

An ecologically interesting group of alkaloids from *Astragalus*, *Oxytropis*, and *Swansonia* plants (Leguminosae) cause neurological and developmental abnormalities of grazing livestock (Winchester 1992; Molyneux 1993). These hydroxylated pyrrolidine- and piperidine-related compounds, as "sugar-shaped weapons," are transition-state analogs inhibiting carbohydrate- or glycoprotein-metabolizing enzymes. Certain polyhydroxy alkaloids are also insect antifeedants, indicating a general defensive role for these compounds (Fellows et al. 1986). In a recent comparative study with 16 representative compounds, significant differences have been found in the inhibition of glycosidases not only between various insect species but between insects and mammals (Scofield et al. 1995; see also Campbell et al. 1987). For insects, the pyrrolidine fructose analog **76** (Fig. 6), isolated from the leaves of *Derris elliptica* (Fabaceae), inhibited both α-glucosidase and β-glucosidase, while a glucose analog, castanospermine (**77**), obtained from seeds of *Castanospermum australe* (Fabaceae), was a strong inhibitor only of β-glucosidase.

3 Insecticidal Alkaloids from Terrestrial Animals

3.1 Arthropodal Alkaloids

Certain insects when threatened or injured secrete a powerful mixture of defensive chemicals, many of them alkaloidal in nature. These compounds, synthesized in specialized exocrine glands, have relatively simple structures and are fairly volatile. Since there are numerous excellent reviews on this extensively researched area (Jones and Blum 1983; Numata and Ibuka 1987; Blum 1996; King and Meinwald 1996), only selected examples will be presented to demonstrate the structural diversity of these compounds (Fig. 7).

The mono- and bicyclic nitrogenous venoms of *Solenopsis*, *Monomorium*, and *Megalomyrmex* ants are the most studied of the arthropodal alkaloids. The typical venom components are isomeric (*cis* or *trans*) 2,5-dialkylpyrrolidines, 2,5-dialkylpyrrolines, 2,6-dialkylpiperidines, 3,5-dialkylpyrrolizidines and, for a few *Solenopsis* and *Monomorium* species, 3,5-dialkylindolizidines (Numata and Ibuka 1987; Blum 1992). One of the alkyl chains is often methyl, while the other is a C_4-C_{14} alkyl or alkenyl. For example, compound **78** is the major piperidine alkaloid of the fire ant *Solenopsis xyloni*. The individual components, besides being insecticidal on contact, possess diverse biological activities (Jones and Blum 1983; Blum 1992, 1996), so the ant venom, usually a cocktail of several of these alkaloids, induces a multitude of physiological responses. The practical potential of the various ant secretions as models for pesticides has been discussed (Toia 1990).

Recently, novel tricyclic pyrroloindolizine derivatives, exemplified by myrmicarin 215A (**79**), have been identified from the poisonous secretion of the African ant *Myrmicaria opaciventris* (Schröder et al. 1996). Interestingly, anabaseine (**9**) (Fig. 1), a tobacco alkaloid, was discovered in the venomous glands of

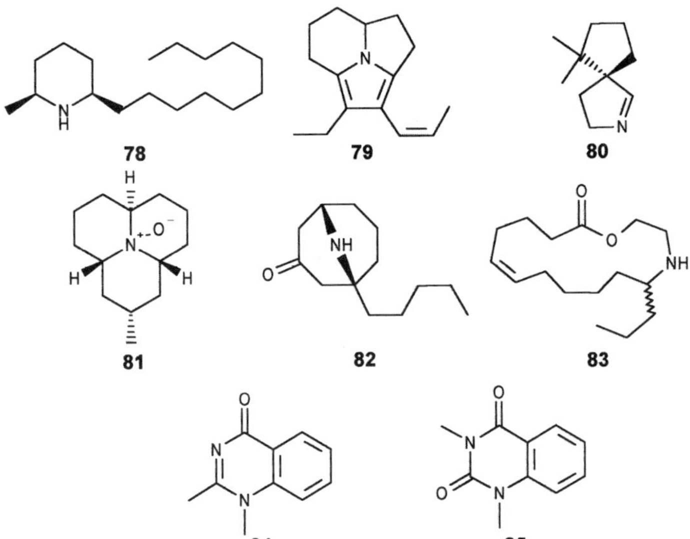

78 79 80 81 82 83 84 85

Fig. 7. Selected arthropod alkaloids with insecticidal or related activity

Aphaenogaster fulva and *A. tennesseensis* (Wheeler et al. 1981), while nicotine (**1**) was detected in the defensive secretion of the harvestman, *Sclerobunus robustus* (Ekpa et al. 1984). A unique spirocyclic pyrrolizidine, polyzonamine (**80**) is a powerful repellent identified from the defensive secretion of the millipede *Polyzonium rosalbum* (Smolanoff et al. 1975). When molested by predators, the European ladybugs, *Coccinella septempunctata*, secrete bitter-tasting droplets ("reflex bleeding") that contain a perhydroazaphenalene *N*-oxide, coccinelline (**81**), as the main ant deterrent active at 0.1% (Tursch et al. 1971). Two species of ladybugs of the genus *Adalia* contain a homotropane derivative, adaline (**82**), that is repellent to various insects (Tursch et al. 1973; Brown and Moore 1982). Both enantiomers of **82** deter ants from feeding at millimolar concentrations (Hill and Renbaum 1982). During chemical studies of the defensive secretions from the glandular hairs of pupae of the Mexican bean beetle, *Epilachna varivestis*, a novel structural group, the azamacrolides, were discovered (Attygalle et al. 1993). The most abundant ant deterrent, comprising >90% of the defensive secretion, is epilachnene (**83**).

When disturbed, the pill millipede, *Glomeris marginata*, discharges a viscous noxious secretion fortified with homologous quinazolinones termed the glomerins (Meinwald et al. 1966; Schildknecht et al. 1967). The 1,2-dimethyl derivative, glomerin (**84**), is moderately toxic to spiders and ants but highly toxic to mice (intraperitoneal LD_{50} = 9-18 mg/kg) (Schildknecht et al. 1967). Another insect-derived alkaloid is the recently identified quinazolinedione derivative **85**, which is the sex pheromone of the pale-brown chafer, *Phyllopertha diversa*, a turf pest in Japan (Leal et al. 1997).

3.2 Amphibian Alkaloids

Amphibians are a rich source of alkaloids with remarkable pharmacological activity and structural diversity and with a presumed ecological role as protectants from pathogens and predators, including insects (Daly and Spande 1986). While some of the compounds are well established biochemical tools used in drug and pesticides research, their insecticidal activity has not been studied and they have rarely been considered as new insecticide leads.

Batrachotoxin (**86**) (Fig. 8) was isolated from the skin of the Colombian poison arrow frog, *Phyllobates aurotaenia* (Albuquerque et al. 1971), and is one of the most widely used pharmacological tools for the study of voltage-sensitive Na^+ transport in nerve and muscle of both vertebrates and invertebrates. Activating Na^+ channels like the botanicals **39** and **57**, the steroid alkaloid **86** is a potent insecticide against both sensitive and super-*kdr* resistant flies with an injected ED_{50} of ~2 ng/fly in a paralysis assay (Pauron et al. 1989).

Recently, Bargar et al. (1995) have carried out a structure-insecticidal activity relationship study by varying the side chain of pumiliotoxin 251D (**87**) (Fig. 8), one of the indolizidine derivatives isolated from the Panamanian frog *Dendrobates pumilio*. The toxicity of 13 synthetic analogs, determined as LD_{50} by injection to *H. virescens* larvae, ranged from 0.32 to 10 μg/larva, which compared well to that of the frog toxin **87** (LD_{50} = 0.15 μg/larva). The fluorinated **88** was the best analog in the series. The mode of action of pumiliotoxins has been shown to involve ACh receptors and Ca^{2+}-dependent ATPase (Witkop and Gössinger 1983). They also activate Na^+ channels by binding to an alkaloid-binding domain common with batrachotoxin (**86**), veratridine (**39**), and aconitine (**57**) (Gusovsky et al. 1992).

Noranabasamine (**13**) (Fig. 1), a desmethyl analog of the plant alkaloid **12**, was recently isolated in trace amounts from the skin of three *Phyllobates* species (Tokuyama and Daly 1983).

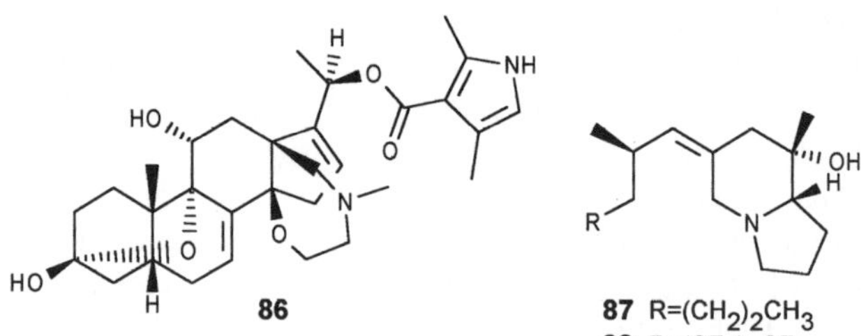

86

87 R=$(CH_2)_2CH_3$
88 R=$(CF_2)_3CF_3$

Fig. 8. Amphibian alkaloids and a fluorinated analog with insecticidal activity

4 Insecticidal Alkaloids from Marine and Other Aquatic Organisms

Until recently the immense potential of marine natural products as prototype insecticidal agents has not been fully explored (El Sayed et al. 1997). There are, however, some notable exceptions worth mentioning here. The marine annelid *Lumbriconereis heteropoda* ("isome" in Japanese), used as a fish bait along the southern and western seashores of Japan, contains an insect-paralyzing factor, nereistoxin (**89**) (Fig. 9) (Okaichi and Hashimoto 1962). This relatively simple compound has rapid knockdown activity and was the lead for the development of novel systemic insecticides such as cartap (**90**) (Konishi 1972), used against a wide range of sucking and chewing insects. These synthetic compounds are in fact proinsecticides that are converted metabolically (Sakai and Sato 1972) or (photo)chemically (Tsao and Eto 1989) into the natural product **89** which, in the target organism, is transformed again into a reducing agent capable of reacting with disulfide bonds of nAChR (Xie et al. 1996). The structurally related charatoxin **91**, isolated from the freshwater skunkweed alga (Anthoni et al. 1980), is a less effective insecticide; nevertheless, it has also served as a lead compound for new 1,2-dithiolane (Jacobsen and Pedersen 1983; Mitsudera et al. 1990a) and 1,3-dithiane (Mitsudera et al. 1990b) insecticides. Although the toxicity symptoms of **91** resemble those seen for **89**, studies with honey bee brain preparations have shown that the mechanism of action of these compounds on nAChR receptors differs (Sherby et al. 1986).

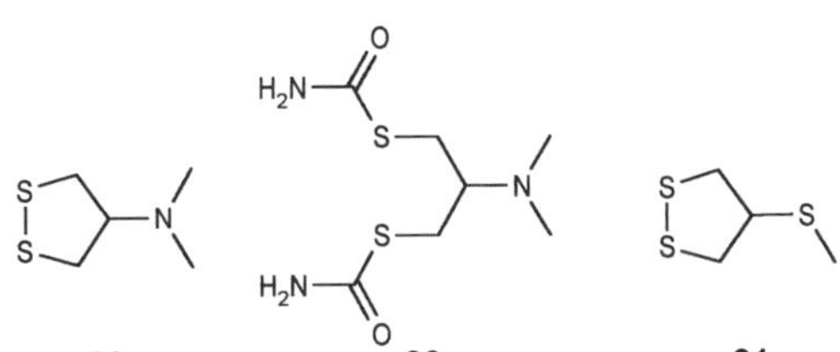

89 **90** **91**

Fig. 9. 1,2-Dithiolane and derived insecticides

5 Insecticidal Alkaloids from Microorganisms

Microorganisms are probably the most abundant reservoirs for metabolic diversity and thus hold a "low-risk" approach for the discovery of novel insecticidal substances represented by some important examples in Fig. 10.

In the search for new antimicrobial agents, researchers at Meiji Seika Kaisha reported the first new chlorinated arylnitropyrroles, the pyrrolomycins, isolated from Actinomycetes (Ezaki et al. 1981). Subsequently, pyrrolomycin C (**92**) and two of its analogs were characterized from a *Streptomyces* species by the same group (Koyama et al. 1983). In 1987, three companies reported the isolation of the tricyclic analog, dioxapyrrolomycin (**93**) (Carter et al. 1987; Nakamura et al. 1987; Yano et al. 1987). Compound **93** is a moderate broad-spectrum insecticide and miticide (Addor et al. 1992) having unfavorable mammalian toxicity with an oral LD_{50} of ~13 mg/kg to mice (Carter et al. 1987; Yano et al. 1987). An extensive SAR study at American Cyanamid ultimately led to the development of AC-303,630 or pyrrol (**94**), active against a wide range of insects and mites, and effective against pyrethroid-resistant *Heliothis* species with acceptable mammalian toxicity and low phytotoxicity (Addor et al. 1992; Kuhn 1997). Compound **94** is actually a proinsecticide and exerts its toxicity via the metabolically *N*-dealkylated pyrrole, a potent uncoupler of oxidative phosphorylation in mitochondria (Black et al. 1994).

The sclerotia of *Aspergillus* species have proven to be rich sources of structurally novel alkaloids, many of them possessing notable insecticidal and antifeedant activity (Gloer 1995). Among the 70 new anti-insectan mycotoxins known, the indole diterpenoid nominine (**95**), isolated as the major organic-soluble component of the sclerotia of *Aspergillus nomius*, is remarkable for its high activity (Gloer et al. 1989). In feeding experiments, **95** caused 40% mortality and 96% weight reduction of *Heliothis zea* larvae at 100 ppm, a potency substantially higher than that of rotenone and comparable to that of permethrin and malathion in this assay. As the co-occurring and structurally related compounds aspernomine (Staub et al. 1992) and 14-hydroxypaspalinine (Staub et al. 1993) (structures not shown) are also highly active orally, the incorporation of the fungus into an "attracticidal bait" for practical purposes has been proposed (Wicklow et al. 1994).

Due to their potent anthelmintic and antinematodal properties, the chemistry and biological activity of the structurally complex dioxepineindole alkaloid paraherquamide A (**96**), isolated first from *Penicillium paraherquei* (Yamazaki et al. 1981) and later from *P. charlesii* (Ondeyka et al. 1990), and related natural (Ondeyka et al. 1990; Liesch and Wichmann 1990; Blanchflower et al. 1991) and synthetic (Blizzard et al. 1989, 1991; Lee and Clothier 1997) derivatives have been examined in detail. In tests on *Lucilia seracata* and *Aedes aegypti* larvae, the LC_{50} values were 50 ppm for both insects (Ondeyka et al. 1997). A close structural relative, sclerotiamide (**97**), isolated recently from sclerotia of *A. sclerotium*, also showed high mortality and reduced growth rate of *H. zea* at 200 ppm in diet (Whyte et al. 1996).

Recently, novel dithiolopyrrolone (or pyrrothine) derivatives, the xenorhabdins, were isolated from *Xenorhabdus* spp. symbiont bacteria of insect pathogenic nematodes belonging to the families Heterorhabditidae and Steinernematidae (McInerney et al. 1991). In a larval feeding assay against *Heliothis*

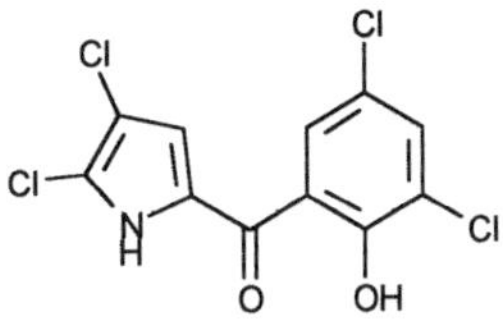

92

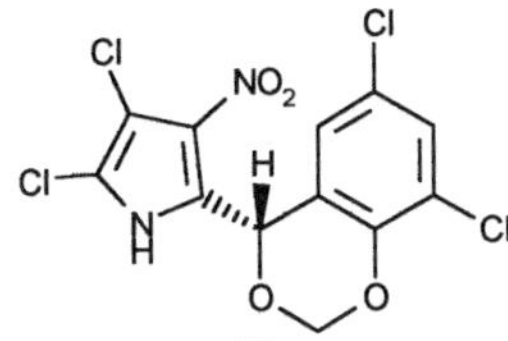

93

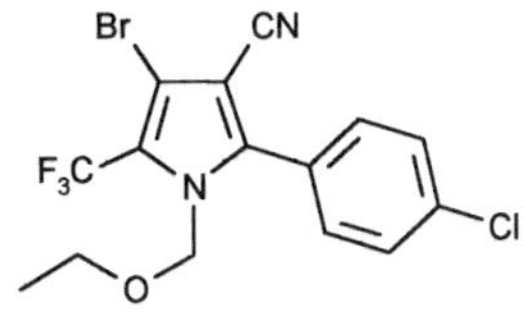

94

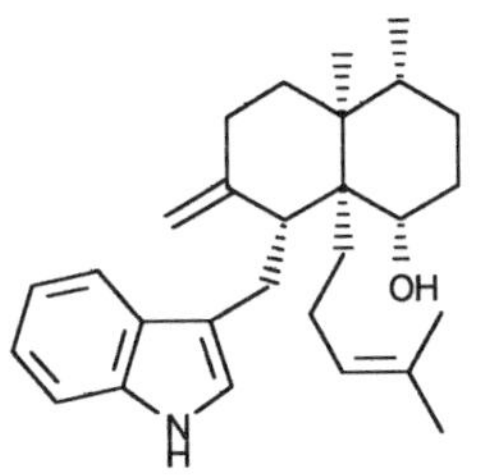

95

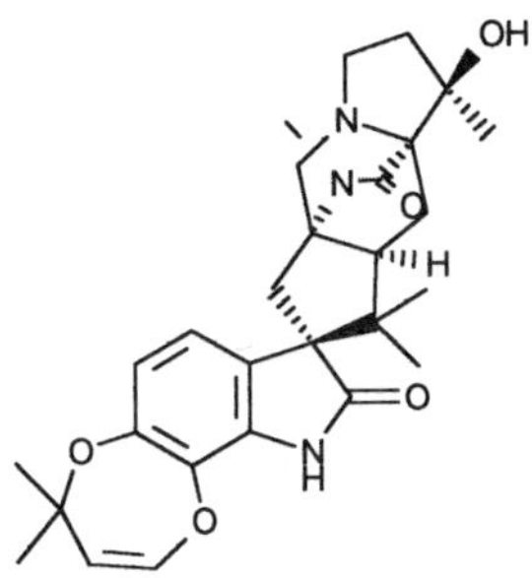

96

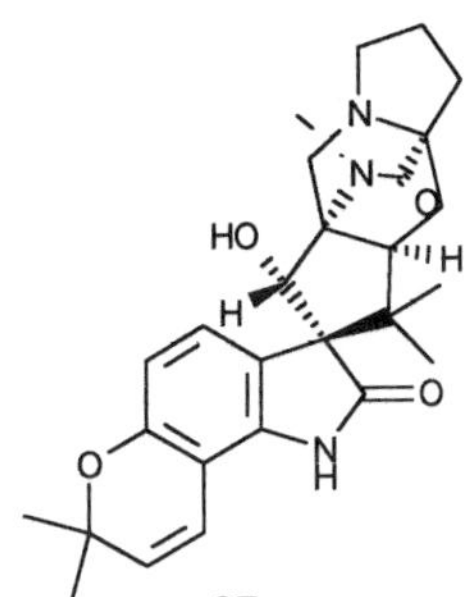

97

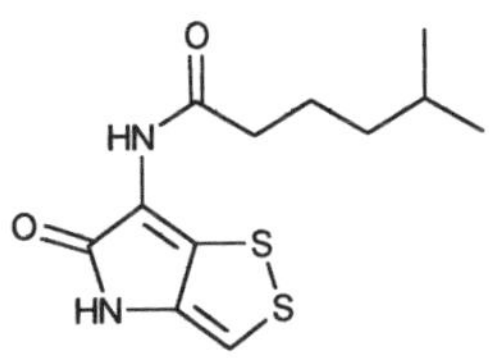

98

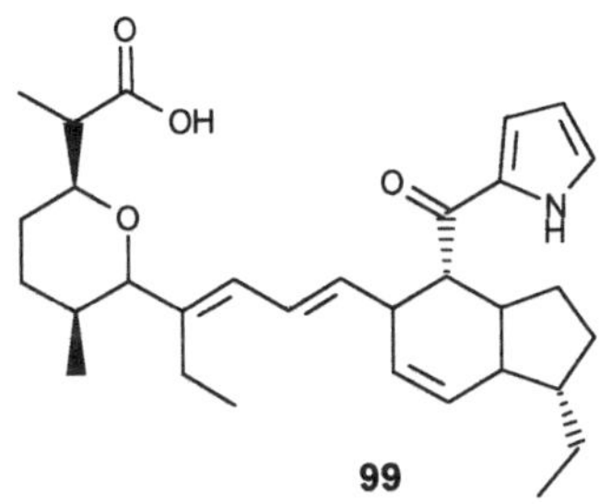

99

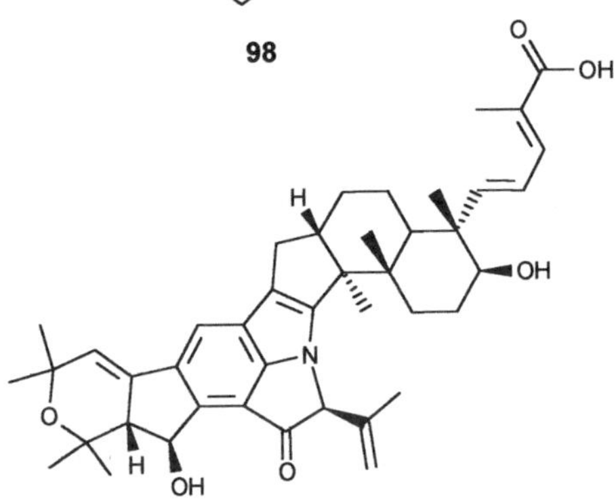

100

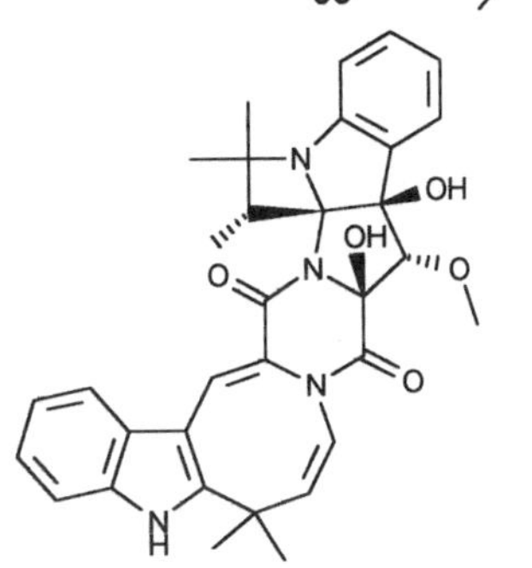

101

Fig. 10. Insecticidal alkaloids of microbial origin

punctigera larvae, xenorhabdin 2 (**98**) showed modest insecticidal (LC_{50} = 59.5 mg/cm^2) and growth inhibitory activity.

The insecticidal activity of the antibiotic indanomycin (**99**) was recently discovered during screening of *Streptomyces griseofuscus* extracts (Zhang et al. 1997). In addition to completely killing fourth-instar larvae of *A. aegypti* at 20 ppm in artificial diet, **99** caused significant weight reduction of *M. sexta* and

Lymantria dispar larvae, but was less active on *H. zea* at 100 ppm. Screening for insecticidal activity of fermentation products of *Nodulisporium* sp., an endophytic fungus isolated from an unidentified woody plant, afforded nodulisporic acid A (**100**) (Ondeyka et al. 1997). This novel indole terpene mycotoxin had LC_{50} values of 0.3 ppm and 0.5 ppm against larvae of *L. seracata* and *A. aegypti*, respectively, being more active than paraherquamide but about 10,000 fold less active than the macrocyclic lactone ivermectin in this assay.

Penicillium simplicissimum isolates, when grown on the insoluble residue of whole soybean (okara), produce a series of novel azocinoindole alkaloid insecticides, the okaramines (Hayashi et al. 1989, 1991), from which the octacyclic okaramine B (**101**) was more active than either physostigmine or rotenone in feeding experiments against *B. mori* (LD_{50} = 0.2 μg/g) and *Spodoptera exigua* larvae.

Microorganisms have also been found to produce certain insecticides that are identical or closely related to botanicals discussed earlier. Thus, nigragillin, i.e., 1-[(2*E*,4*E*)-2,4-hexadienoyl]-2,4,5-trimethylpiperazine, isolated from *Aspergillus niger* strains, was highly toxic with strong knockdown activity to *B. mori* larvae when either mixed with the diet (>20 ppm) or applied topically (≥5 μg/g) (Isogai et al. 1975). Interestingly, physostigmine (**51**) (see Fig. 4) has also been found as a metabolite of *Streptomyces* strains (Murao and Hayashi 1986).

6 Conclusions

Research on alkaloids contributes to our understanding of their ecological role and provides essential information on the structural requirements for their insecticidal activity. However, the development of novel insecticides of commercial importance based on these prototypes is not readily predictable. Alkaloids are typically produced as a cocktail of metabolically related compounds and occasionally co-occur with the other nonalkaloidal substances, all modulating the toxicological (biological) properties of an individual component. Consequently, it is not unreasonable to assume that a single natural compound is not optimized for a particular biological activity. There is much room - and hope - for chemical modifications and structure optimizations to develop synthetic compounds with improved insecticidal activity and environmental safety, as has been demonstrated by some of the examples discussed in this chapter.

Acknowledgment

Part of this chapter was prepared while the author enjoyed the hospitality of the Food Animal Protection Research Laboratory, USDA-ARS, College Station, Texas, while on leave in 1997.

References

Addor RW (1995) Insecticides. In: Godfrey CRA (Ed) Agrochemicals from natural products. Marcel Dekker, New York, pp 1–62

Addor RW, Babcock TJ, Black BC, Brown DG, Diehl RE, Furch JA, Kameswaran V, Kamhi VM, Kremer KA, Kuhn DG, Lovell JB, Lowen GT, Miller TP, Peevey RM, Siddens JK, Treacy MF, Trotto SH, Wright DP Jr (1992) Insecticidal pyrroles: discovery and overview. In: Baker DR, Fenyes JG, Steffens JJ (Eds) Synthesis and chemistry of agrochemicals. III. American Chemical Society, Washington, DC, pp 283–297

Albuquerque EX, Daly JW, Witkop B (1971) Batrachotoxin: chemistry and pharmacology. Science 172:995–1002

Anthoni U, Christophersen C, Madsen JØ, Wium-Andersen S, Jacobsen N (1980) Biologically active sulphur compounds from the green alga *Chara globularis*. Phytochemistry 19:1228–1229

Attygalle AB, McCormick KD, Blankespoor CL, Eisner T, Meinwald J (1993) Azamacrolides: a family of alkaloids form the pupal defensive secretion of a ladybird beetle (*Epilachna varivestis*). Proc Natl Acad Sci USA 90:5204–5208

Baldwin IT (1996) Methyl jasmonate-induced nicotine production in *Nicotiana attenuata*: inducing defenses in the field without wounding. Entomol Exp Appl 80:213–220

Baldwin IT, Zhang Z-P, Diad N, Ohnmeiss TE, McCloud ES, Gladys GY, Schmelz EA (1997) Quantification, correlations, and manipulations of wound-induced changes in jasmonic acid and nicotine in *Nicotiana sylvestris*. Planta (Heidelb) 201:397–404

Barbour JD, Kennedy GG (1991) Role of steroidal glycoalkaloid α-tomatine in host-plant resistance of tomato to Colorado potato beetle. J Chem Ecol 17:989–1005

Bargar TM, Lett RM, Johnson PL, Hunter JE, Chang CP, Pernich DJ, Sabol MR, Dick MR (1995) Toxicity of pumiliotoxin 251D and synthetic analogs to the cotton pest *Heliothis virescens*. J Agric Food Chem 43:1044–1051

Barlow RB, Johnson O (1989) Relations between structure and nicotine-like activity: X-ray crystal structure analysis of (-)-cytisine and (-)-lobeline hydrochloride and a comparison with (-)-nicotine and other nicotine-like compounds. Br J Pharmacol 98:799–808

Barton DHR, Jeger O, Prelog V, Woodward RB (1954) The constitutions of cevine and some related alkaloids. Experientia (Basel) 10:81–90

Bellows TS Jr, Morse JG (1993) Toxicity of insecticides used in citrus to *Aphytis melinus* DeBach (Hymenoptera: Aphelinidae) and *Rhizobius lophanthae* (Blaisd.) (Coleoptera: Coccinellidae). Can Entomol 125:987–994

Benn MH, Jacyno JM (1983) The toxicogy and pharmacology of diterpenoid alkaloids. In: Pelletier SW (Ed) Alkaloids. Chemical and biological perspectives. Vol 1. John Wiley & Sons, New York, pp 153–210

Bergmann ED, Levinson ZH, Mechoulam R (1958) The toxicity of *Veratrum* and *Solanum* alkaloids to housefly larvae. J Insect Physiol 2:162–177

Beroza M (1951) Alkaloids from *Tripterygium wilfordii* Hook - wilforine and wilfordine. J Am Chem Soc 73:3656–3659

Beroza M, Bottger GT (1954) The insecticidal value of *Tripterygium wilfordii*. J Econ Entomol 47:188–189

Birnbaum KB, Wiesner K, Jay EWK, Jay L (1971) Configuration of the ring *A* methoxyl in delphinine and aconitine. Tetrahedron Lett 867–870

Black BC, Hollingworth RM, Ahammadsahib KI, Kukel CD, Donovan S (1994) Insecticidal action and mitochondrial uncoupling activity of AC-303,630 and related halogenated pyrroles. Pestic Biochem Physiol 50:115–128

Blade RJ (1990) Some aspects of synthesis and structure–activity in insecticidal lipid amides. In: Crombie L (Ed) Recent advances in the chemistry of insect control. II. The Royal Society of Chemistry, London, pp 151–169

Blade RJ, Burt PE, Hart RJ, Moss MDV (1985) The action of insecticidal isobutylamide compounds on the insect nervous system. Pestic Sci 16:554

Blanchflower SE, Banks RM, Everett JR, Reading CJ (1991) New paraherquamide antibiotics with anthelmintic activity. J Antibiot (Tokyo) 44:492–497

Blizzard TA, Marino G, Mrozik H, Fisher MH, Hoogsteen K, Springer JP (1989) Chemical modifications of paraherquamide. 1. Unusual reactions and absolute stereochemistry. J Org Chem 54:2657–2663

Blizzard TA, Margiatto G, Mrozik H, Schaeffer JM, Fisher MH (1991) Chemical modification of paraherquamide. 4. 1-*N*-Substituted analogs. Tetrahedron Lett 32:2441–2444

Bloomquist JR (1996) Ion channels as targets for insecticides. Annu Rev Entomol 41:163–190

Bloomquist JR, Miller TA (1986) Sodium channel neurotoxins as probes of the knockdown resistance mechanism. NeuroToxicology (Little Rock) 7:217–224

Blum MS (1992) Ant venoms: chemical and pharmacological properties. J Toxicol Toxin Rev 11:115–164

Blum MS (1996) Chemistry and toxicology of arthropod alkaloids. In: Blum MS (Ed) Chemistry and toxicology of diverse classes of alkaloids. Alaken, Fort Collins, CO, pp 145–184

Bringmann G, Rübenacker M, Jansen JR, Scheutzow D, Aké Assi L (1990a) Acetogenic isoquinoline alkaloids. 16. On the structure of the dioncophyllaceae alkaloids dioncophylline A ("triphyophylline") and "*O*-methyl-triphyophylline". Tetrahedron Lett 31:639–642

Bringmann G, Jansen JR, Reuscher H, Rübenacker M, Peters K, von Schnering HG (1990b) Acetogenic isoquinoline alkaloids. 17. First total synthesis of (-)-dioncophylline A ("triphyophylline") and of selected stereoisomers: complete (revised) stereostructure. Tetrahedron Lett 31:643–646

Bringmann G, Pokorny F (1995) The naphthylisoquinoline alkaloids. In: Cordell GA (Ed) The alkaloids: chemistry and pharmacology. Vol 46. Academic Press, New York, pp 127–271

Bringmann G, Gramatzki S, Grimm C, Proksch P (1992) Feeding deterrency and growth retarding activity of the naphthylisoquinoline alkaloid dioncophylline A against *Spodoptera littoralis*. Phytochemistry 31:3821–3825

Bringmann G, Holenz J, Wiesen B, Nugroho BW, Proksch P (1997) Dioncophylline A as a growth-retardant agent against the herbivorous insect *Spodoptera littoralis*: structure–activity relationships. J Nat Prod 60:342–437

Brossi A, Pei X-F (1998) Biological activity of unnatural alkaloid enantiomers. In: Cordell GA (Ed) The alkaloids: chemistry and pharmacology. Vol 50. Academic Press, San Diego, pp 109–139

Brown KS Jr, Trigo JR (1995) The ecological activity of alkaloids. In: Cordell GA (Ed) The alkaloids: chemistry and pharmacology. Vol 47. Academic Press, San Diego, pp 227–354

Brown WV, Moore BP (1982) The defensive alkaloids of *Cryptolaemus montrouzieri* (Coleoptera: Coccinellidae). Aust J Chem 35:1255–1261

Brüning R, Wagner H (1978) Übersicht über die Celastraceen-Inhaltsstoffe: Chemie, Chemotaxonomie, Biosynthese, Pharmakologie. Phytochemistry 17:1821–1858

Bush LP, Crowe MW (1989) *Nicotiana* alkaloids. In: Cheeke PR (Ed) Toxicants of plant origin. Vol I. Alkaloids. CRC Press, Boca Raton, FL, pp 87–107

Campbell BC, Molyneux RJ, Jones KC (1987) Differential inhibition by castanospermine of various insect disaccharidases. J Chem Ecol 13:1759–1770

Carter GT, Nietsche JA, Goodman JJ, Torrey MJ, Dunne TS, Borders DB, Testa RT (1987) LL-F42248α, a novel chlorinated pyrrole antibiotic. J Antibiot (Tokyo) 40:233–236

Chuliá S, Ivorra MD, Cavé A, Cortés D, Noguera MA, D'Ocón MP (1995) Relaxant activity of three aporphine alkaloids from *Annona cherimolia* on isolated aorta of rat. J Pharm Pharmacol 47:647–650

Codding PW (1983) Structural studies of sodium channel neurotoxins. 2. Crystal structure and absolute configuration of veratridine perchlorate. J Am Chem Soc 105:3172–3176

Crombie L (1952) The structure of an insecticidal isobutylamide from pellitory root. Chem Ind (Lond) 1034–1035

Crombie L, Fisher D (1985a) Synthons for general routes to natural insecticidal lipid isobutylamides. Tetrahedron Lett 26:2477–2480

Crombie L, Fisher D (1985b) Synthesis of natural polyene isobutylamides. Stereochemistry of the Wittig reactions. Tetrahedron Lett 26:2481–2484

Crombie L, Horsham MA, Blade RJ (1987) Synthetic approaches to isobutylamides of insecticidal interest. Tetrahedron Lett 28:4879–4882

Crosby DG (1971) Minor insecticides of plant origin. In: Jacobson M, Crosby DG (Eds) Naturally occurring insecticides. Marcel Dekker, New York, pp 177–239

Daly JW, Spande TF (1986) Amphibian alkaloids: chemistry, pharmacology, and biology. In: Pelletier SW (Ed) Alkaloids: chemical and biological perspectives. Vol 4. John Wiley & Sons, New York, pp 1–273

Davey AE, Schaeffer MJ, Taylor RJK (1992) Synthesis of the novel anti-leukaemic tetrahydrocyclopenta[b]benzofuran, rocaglamide and related synthetic studies. J Chem Soc Perkin Trans 1 2657–2666

Decker MW, Anderson DJ, Brioni JD, Donnelly-Roberts DL, Kang CH, O'Neill AB, Piattoni-Kaplan M, Swanson S, Sullivan JP (1995) Erysodine, a competitive antagonist at neuronal nicotinic acetylcholine receptors. Eur J Pharmacol 280:79–89

Dell S, Holding MJ, Perrior TR (1990) The design and synthesis of simple putative analogues of insecticidal alkaloids. Abstracts, Seventh International Congress Pesticide Chemistry (IUPAC), Hamburg, Vol I, Abstr 01B-05, p 97

Delle Monache FD, Marini Bettolo GB, Bernays EA (1984) Isolation of insect antifeedant alkaloids from *Maytenus rigida* (Celastraceae). Z Angew Entomol 97:406–414

Deslongchamps P, Bélanger A, Berney DJF, Borschberg H-J, Brousseau R, Doutheau A, Durand R, Katayama H, Lapalme R, Leturc DM, Liao C-C, MacLachlan FN, Maffrand J-P, Marazza F, Martino R, Moreau C, Ruest L, Saint-Laurent L, Saintonge R, Soucy P (1990) The total synthesis of (+)-ryanodol. Part I. General strategy and search for a convenient diene for the construction of a key tricyclic intermediate. Can J Chem 68:115–126

Devitt BD, Philogène BJR, Hinks CF (1980) Effects of veratrine, berberine, nicotine and atropine on developmental characteristics and survival of the dark-sided cutworm, *Euxoa messoria* (Lepidoptera: Noctuidae). Phytoprotection 61:88–102

Dyke SF, Quessy SN (1981) Erythrina and related alkaloids. In: Cordell GA (Ed) The alkaloids: chemistry and pharmacology. Vol XVIII. Academic Press, San Diego, pp 1–98

Ekpa O, Wheeler JW, Cokendolpher JC, Duffield RM (1984) *N,N*-Dimethyl-β-phenylethylamine and bornyl esters from the harvestman *Sclerobunus robustus* (Arachnidae: Opiliones). Tetrahedron Lett 25:1315–1318

Elliott M (1985) Lipophilic insect control agents. In: Janes NF (Ed) Recent advances in the chemistry of insect control agents. The Royal Society of Chemistry, London, pp 73–102

Elliott M, Farnham AW, Janes NF, Johnson DM, Pulman DA, Sawicki RM (1986) Insecticidal amides with selective potency against a resistant (*super-kdr*) strain of houseflies (*Musca domestica* L.) Agric Biol Chem 50:1347–1349

Elliott M, Farnham AW, Janes NF, Johnson DM, Pulman DA (1987) Synthesis and insecticidal activity of lipophilic amides. Part 1. Introductory survey, and discovery of an active synthetic compound. Pestic Sci 18:191–201

Elliott M, Farnham AW, Janes NF, Johnson DM, Pulman DA (1989) Synthesis and insecticidal activity of lipophilic amides. Part 7. Alternative aromatic groups for phenyl in 6-phenylhexa-2,4-dienamides. Pestic Sci 26:199–208

El Sayed KA, Dunbar DC, Perry TL, Wilkins SP, Hamann MT, Greenplate JT, Wideman MA (1997) Marine natural products as prototype insecticidal agents. J Agric Food Chem 45:2735–2739

Enzell CR, Wahlberg I, Aasen AJ (1977) Isoprenoids and alkaloids of tobacco. In: Herz W, Grisebach H, Kirby GW (Eds) Fortschritte der Chemie organischer Naturstoffe. Vol 34. Springer-Verlag, Wien, pp 1–79

Ewete F, Nicol RW, Hengsawad V, Sukumalanand P, Satasook C, Wiriyachitra P, Isman MB, Kahn Y, Duval F, Philogène BJR, Arnason JT (1996) Insecticidal activity of *Aglaia odorata* extract and the active principle, rocaglamide, to the European corn borer, *Ostrinia nubilalis* Hübn. (Lep., Pyralidae). J Appl Entomol 120:483–488

Ezaki N, Shomura T, Koyama M, Niwa T, Kojima M, Inouye S, Itô T, Nida T (1981) New chlorinated nitro-pyrrole antibiotics, pyrrolomycin A and B (SF-2080 A and B). J Antibiot (Tokyo) 34:1363–1365

Fellows LE, Evans SV, Nash RJ, Bell EA (1986) Polyhydroxy-alkaloids as glycosidase inhibitors and their possible ecological role. In: Green MB, Hedin PA (Eds) Natural resistance of plants to pests. American Chemical Society, Washington, DC, pp 72–78

Fisher RA (1940) Insecticidal action of extracts of *Veratrum viride*. J Econ Entomol 33:728–734

Fodor GB, Colasanti B (1985) The pyridine and piperidine alkaloids: chemistry and pharmacology. In: Pelletier SW (Ed) Alkaloids: chemical and biological perspectives. Vol 3. John Wiley & Sons, New York, pp 1–273

François G, Van Looveren M, Timperman G, Chimanuka B, Aké Assi L, Holenz J, Bringmann G (1996) Larvicidal activity of the naphthylisoquinoline alkaloid dioncophylline A against the malaria vector *Anopheles stephensi*. J Ethnopharmacol 54:125–130

Gbewonyo WSK, Candy DJ, Abderson M (1993) Structure–activity relationships of insecticidal amides from *Piper guineense* root. Pestic Sci 37:57–66

Gfeller H, Schlunegger UP, Schaffner U, Boevé J-L, Ujváry I (1995) Analysis of the chemical defense system in an insect larva by tandem mass spectrometry. J Mass Spectrom 30:1291–1295

Gloer JB (1995) Antiinsectan natural products from fungal sclerotia. Acc Chem Res 28:343–350

Gloer JB, Rinderknecht BL, Wicklow DT, Dowd PF (1989) Nominine: a new insecticidal indole diterpene from the sclerotia of *Aspergillus nomius*. J Org Chem 54:2530–2532

Götz M, Strunz GM (1973) Tuberostemonine and related compounds: the chemistry of the *Stemona* alkaloids. In: Wiesner K (Ed) MTP International review of science: organic chemistry, series one. Butterworths, London, pp 143–160

González-Coloma A, Cabrera R, Socorro Monzón AR, Fraga BM (1993) *Persea indica* as a natural source of the insecticide ryanodol. Phytochemistry 34:397–400

Greenhill JV, Grayshan P (1992) The cevane group of *Veratrum* alkaloids. In: Brossi A, Cordell GA (Eds) The alkaloids: chemistry and pharmacology. Vol 41. Academic Press, San Diego, pp 177–237

Greger H (1988) Comparative phytochemistry of the alkamides. In: Lam J, Breteler H, Arnason T, Hansen L (Eds) Chemistry and biology of naturally-occurring acetylenes and related compounds (NOARC). Elsevier, Amsterdam, pp 159–178

Greger H., Zechner G, Hofer O, Vajrodaya S (1996) Bioactive amides from *Glycosmis* species. J Nat Prod 59:1163–1168

Güssregen B, Fuhr M, Nugroho BW, Wray V, Witte L, Proksch P (1997) New insecticidal rocaglamide derivatives from flowers of *Aglaia odorata*. Z Naturforsch 52C:339–344

Gusovsky F, Padgett WL, Creveling CR, Daly JW (1992) Interaction of pumiliotoxin B with an "alkaloid-binding domain" on the voltage-dependent sodium channel. Mol Pharmacol 42:1104–1108

Harborne JB (1993) Introduction to ecological biochemistry. 4th edn. Academic Press, London, p 318

Hardick DJ, Cooper G, Scott-Ward T, Blagbrough IS, Potter BVL, Wonnacott S (1995) Conversion of the sodium channel activator aconitine into a potent α7-selective nicotinic ligand. FEBS Lett 365:79–82

Hardick DJ, Blagbrough IS, Cooper G, Potter BVL, Critchley T, Wonnacott S (1996) Nudicauline and elatine as potent norditerpenoid ligands at rat neural α-bungarotoxin binding sites: importance of the 2-(methylsuccinimido)benzoyl moiety for neuronal nicotinic acetylcholine receptor binding. J Med Chem 39:4860–4866

Hare JD, Morse JG (1997) Toxicity, persistence, and potency of sabadilla alkaloid formulations to citrus thrips (Thysanoptera: Thripidae). J Econ Entomol 90:326–332

Harper SH, Potter C, Gillham EM (1947) *Annona* species as insecticides. Ann Appl Biol 34:104–112

Hayashi H, Takiuchi K, Murao S, Arai M (1989) Structure and insecticidal activity of new indole alkaloids, okaramines A and B, from *Penicillium simplicissimum* AK-40. Agric Biol Chem 53:461–469

Hayashi H, Fujiwara T, Murao S, Arai M (1991) Okaramine C, a new insecticidal indole alkaloid from *Penicillium simplicissimum*. Agric Biol Chem 55:3143–3145

Hill RK, Renbaum LA (1982) Asymmetric syntheses of the ladybug alkaloid adaline and 1-methyl-9-azabicyclo[3.3.1]nonan-3-one. Tetrahedron 38:1959–1963

Holmstedt B (1972) The ordeal bean of Old Calabar: the pageant of *Physostigma venenosum* in medicine. In: Swain T (Ed) Plants in the development of modern medicine. Harvard University Press, Cambridge, MA, pp 303–360

Hopf HS (1952) Studies in the mode of action of insecticides. I. Injection experiments on the role of cholinesterase inhibition. Ann Appl Biol 39:193–202

Isogai A, Horii T, Suzuki A, Murakoshi S, Ikeda K, Sato S, Tamura S (1975) Isolation and identification of nigragillin as a insecticidal metabolite produced by a *Aspergillus niger*. Agric Biol Chem 39:739–740

Jackson KE (1941) Alkaloids of tobacco. Chem Rev 29:123–197

Jacobi PA, Lee K (1997) Total synthesis of (±)-stemoamide. J Am Chem Soc 119:3409–3410

Jacobsen N, Pedersen LEK (1983) Synthesis and insecticidal properties of derivatives of propane-1,3-dithiol (Analogues of the insecticidal derivatives of dithiolane and trithiane from the alga *Chara globularis* Thuillier) Pestic Sci 14:90–97

Jacobson M (1948) Herculin, a pungent insecticidal constituent of southern prickly ash bark. J Am Chem Soc 70:4234–4237

Jacobson M (1949) The structure of pellitorine. J Am Chem Soc 71:366–367

Jacobson M (1953) Pellitorine isomers. II. The synthesis of *N*-isobutyl-*trans*-2,*trans*-4-decadienamide. J Am Chem Soc 75:2584–2586

Jacobson M (1971) The unsaturated isobutylamides. In: Jacobson M, Crosby DG (eds) Naturally occurring insecticides. Marcel Dekker, New York, pp 137–176

Jacobson M, Crosby DG (Eds) (1971) Naturally occurring insecticides. Marcel Dekker, New York, p 585

Jacyno JM (1996) Chemistry and toxicology of the diterpenoid alkaloids. In: Blum MS (Ed) Chemistry and toxicology of diverse classes of alkaloids. Alaken, Fort Collins, CO, pp 301–336

Janprasert J, Satasook C, Sukumalanand P, Champagne DE, Isman MB, Wiriyachitra P, Towers GHN (1993) Rocaglamide, a natural benzofuran insecticide from *Aglaia odorata*. Phytochemistry 32:67–69

Jefferies PR, Toia RF, Brannigan B, Pessah I, Casida JE (1992) Ryania insecticide: analysis and biological activity of 10 natural ryanoids. J Agric Food Chem 40:142–146

Jefferies PR, Casida JE (1994) Ryanoid chemistry and action. In: Hedin PA, Menn JJ, Hollingworth RM (Eds) Natural and engineered pest management agents. American Chemical Society, Washington, DC, pp 130–144

Jefferies PR, Yu P, Casida JE (1997) Structural modifications increase the insecticidal activity of ryanodine. Pestic Sci 51:33–38

Jennings KR, Brown DG, Wright DP Jr (1986) Methyllycaconitine, a naturally occurring insecticide with a high affinity for the insect cholinergic receptor. Experientia (Basel) 42:611–613

Jones TH, Blum MS (1983) Arthropod alkaloids: distribution, functions, and chemistry. In: Pelletier SW (Ed) Alkaloids: chemical and biological perspectives. Vol 1. John Wiley & Sons, New York, pp 33–84

Karadsheh N, Kussie P, Linthicum DS (1991) Inhibition of acetylcholinesterase by caffeine, anabasine, methyl pyrrolidine and their derivatives. Toxicol Lett 55:335–342

Kato A, Ichimaru M, Hashimoto Y, Mitsudera H (1989) Guinesine-A, -B and -C: new sulfur containing insecticidal alkaloids from *Cassipourea guianensis*. Tetrahedron Lett 30:3671–3674

Keukens EAJ, de Vrije T, van den Boom C, de Waard P, Plasman HH, Thiel F, Chupin V, Jongen WMF, de Kruijff B (1995) Molecular basis of glycoalkalod induced membrane disruption. Biochim Biophys Acta Biomembranes 1240:216–228

King AG, Meinwald J (1996) Review of the defensive chemistry of coccinellids. Chem Rev 96:1105–1122

King ML, Chiang C-C, Ling H-C, Fujita E, Ochiai M, McPhail AT (1982) X-Ray crystal structure of rocaglamide, a novel antileukemic 1*H*-cyclopenta[b]benzofuran from *Aglaia elliptifolia*. J Chem Soc Chem Commun 1150–1151

Kinghorn AD, Balandrin MF (1984) Quinolizidine alkaloids of the Leguminosae: structural types, analysis, chemotaxonomy, and toxicological activities. In: Pelletier SW (Ed) Alkaloids. chemical and biological perspectives. Vol 2. John Wiley & Sons, New York, pp 105–148

Klocke JA (1989) Plant compounds as sources and models of insect-control agents. In: Wagner H, Hikino H, Farnsworth NR (Eds) Economic and medicinal plant research. Vol 3. Academic Press, London, pp 103–144

Konishi K (1972) Nereistoxin and its relatives. In: Tahori AS (Ed) Pesticide chemistry. Vol 1. Insecticides. Gordon and Breach, New York, pp 179–189

Kosower EM (1983) A hypothesis for the mechanism of sodium channel opening by batrachotoxin and related toxins. FEBS Lett 163:161–164

Koyama M, Ezaki N, Tsuruoka T, Inouye S (1983) Structural studies on pyrrolomycins C, D and E. J Antibiot (Tokyo) 36:1483–1489

Kubo I, Matsumoto T, Klocke JA, Kamikawa T (1984) Molluscicidal and insecticidal activities of isobutylamides isolated from *Fagara macrophylla*. Experientia (Basel) 40:340–341

Kuhn DG (1997) Structure–activity relationships for insecticidal pyrroles. In: Hedin PA, Hollingworth RM, Masler EP, Miyamoto J, Thompson DG (Eds) Phytochemicals for pest control. American Chemical Society, Washington, DC, pp 195–205

Kuhr RJ, Dorough HW (1976) Carbamate insecticides: chemistry, biochemistry, and toxicology. CRC Press, Cleveland, OH, p 301

Kukel CF, Jennings KR (1994) Delphinium alkaloids as inhibitors of α-bungarotoxin binding to rat and insect neural membranes. Can J Physiol Pharmacol 72:104–107

Kukel CF, Jennings KR (1995) Erratum. Can J Physiol Pharmacol 73:145

Kupchan SM, Flacke WE (1967) Hypotensive *Veratrum* alkaloids. In: Schlittler E (Ed) Antihypertensive agents. Academic Press, New York, pp 429–458

Leal WS, Zarbin PHG, Wojtasek H, Kuwahara S, Hasegawa M, Ueda Y (1997) Medicinal alkaloid as a sex pheromone. Nature (Lond) 385:213

Leboeuf M, Cavé A, Bhaumik PK, Mukherjee B, Mukherjee R (1982) The phytochemistry of the Annonaceae. Phytochemistry 21:2783–2813

Lee BH, Clothier MF (1997) Conversion of marcfortine A to paraherquamide A via paraherquamide B. The first formal synthesis of paraherquamide A. J Org Chem 62:1795–1798

Leete E (1983) Biosynthesis and metabolism of the tobacco alkaloids. In: Pelletier SW (Ed) Alkaloids. chemical and biological perspectives. Vol 1. John Wiley & Sons, New York, pp 85–152

Levinson HZ (1976) The defensive role of alkaloids in insects and plants. Experientia (Basel) 32:408–411

Li Ya, Strunz GM, Calhoun LA (1990) Sesquiterpene alkaloids from *Tripterygium wilfordii* (Hook): a nuclear magnetic resonance study of 1-desacetylwilfordine, 1-desacetylwilfortrine, and 2-debenzoyl-2-nicotinoylwilforine. Can J Chem 68:371–374

Liesch JM, Wichmann CF (1990) Novel antinematodal and antiparasitic agents from *Penicillium charlesii*. II. Structure determination of paraherquamides B, C, D, E, F, and G. J Antibiot (Tokyo) 43:1380–1386

Liu M-Y, Latli B, Casida JE (1995) Imidacloprid binding site in *Musca* nicotinic acetylcholine receptor: interactions with physostigmine and a variety of nicotinic agonists with chloropyridyl and chlorothiazolyl substituents. Pestic Biochem Physiol 52:170–181

McInerney BV, Gregson RP, Lacey MJ, Akhurst RJ, Lyons GR, Rhodes SH, Smith DRJ, Engelhardt LM, White AH (1991) Biologically active metabolites from *Xenorhabdus* spp. Part 1. Dithiolopyrrolone derivatives with antibiotic activity. J Nat Prod 54:774–784

McLaughlin JL, Zeng L, Oberlies NH, Alfonso D, Johnson HA, Cummings BA (1997) Annonaceous acetogenins as new natural pesticides: recent progress. In: Hedin PA,

Hollingworth RM, Masler EP, Miyamoto J, Thompson DG (Eds) Phytochemicals for pest control. American Chemical Society, Washington, DC, pp 117–133

Meinwald YC, Meinwald J, Eisner T (1966) 1,2-Dialkyl-4-(3H)-quinazolinones in the defensive secretion of a millipede (*Glomeris marginata*). Science 154:390–391

Meisters A, Wailes PC (1966a) Synthesis of four geometrical isomers of hexadeca-2,6,8,12-tetraenoic acid. The insecticidal properties of their isobutylamides. Aust J Chem 19:1207–1213

Meisters A, Wailes PC (1966b) The isobutylamides of 7-phenylhepta-2,4-dienoic acid, 7-phenylhepta-2,4,6-trienoic acid, and *p*-(2-phenylethyl)benzoic acid. Aust J Chem 19:1215–1220

Metcalf RL (1955) Nicotine, nornicotine, and anabasine. In: Metcalf R (Ed) Organic insecticides: their chemistry and mode of action. Interscience, New York, pp 1–21

Mitsudera H, Kamikado T, Uneme H, Kono Y (1990a) Synthesis and biological activity of 4-alkylthio-1,2-dithiolanes and related compounds. Agric Biol Chem 54:1719–1722

Mitsudera H, Kamikado T, Uneme H, Manabe Y (1990b) Synthesis and biological activity of 5-alkylthio-1,3-dithianes. Agric Biol Chem 54:1723–1730

Mitsudera H, Uneme H, Okada Y, Numata M, Kato A (1990c) Synthesis of *dl*-guinesines and related compounds. J Heterocycl Chem 27:1361–1367

Miyakado M, Nakayama I, Yoshioka H (1980) Insecticidal joint action of pipercide and co-occurring compounds isolated from *Piper nigrum* L. Agric Biol Chem 44:1701–1703

Miyakado M, Nakayama I, Inoue A, Hatakoshi M, Ohno N (1985a) Chemistry and insecticidal activities of *Piperaceae* amides and their synthetic analogues. J Pestic Sci 10:11–17

Miyakado M, Nakayama I, Inoue A, Hatakoshi M, Ohno N (1985b) Insecticidal activities of phenoxy analogues of dihydropipercide. J Pestic Sci 10:25–30

Miyakado M, Nakayama I, Ohno N (1989) Insecticidal unsaturated isobutylamides. From natural products to agrochemical leads. In: Arnason JT, Philogène BJR, Morand P (Eds) Insecticides of plant origin. American Chemical Society, Washington, DC, pp 173–187

Molyneux RJ (1993) Isolation, characterization and analysis of polyhydroxy alkaloids. Phytochem Anal 4:193–204

Mroue M, Alam M (1991) Crooksiine, a bisindole alkaloid from *Haplophyton crooksii*. Phytochemistry 30:1741–1744

Mukhamedzhanov SZ, Aslanov KA, Sadykov AS, Leontev VB, Kiryukhin VK (1968) The structure of anabasamine. Khim Prir Soedin 4:158–161 (in Russian)

Murao S, Hayashi H (1986) Physostigmine and N^8-norphysostigmine, insecticidal compounds, from *Streptomyces* sp. Agric Biol Chem 50:523–524

Murray CL, Quaglia M, Arnason JT, Morris CE (1994) A putative nicotine pump at the metabolic blood-brain barrier of the tobacco hornworm. J Neurobiol 25:23–34

Nakamura H, Shiomi K, Iinuma H, Naganawa H, Obata T, Takeuchi T, Umezawa H, Takeuchi Y, Iitaka Y (1987) Isolation and characterization of a new antibiotic, dioxapyrrolomycin, related to pyrrolomycins. J Antibiot (Tokyo) 40:899–903

Nathanson JA (1984) Caffeine and related methylxanthines: possible naturally occurring pesticides. Science 226:184–187

Nathanson JA, Hunnicutt EJ, Kantham L, Scavone C (1993) Cocaine as a naturally occurring insecticide. Proc Natl Acad Sci USA 90:9645–9648

Negherbon WO (1959) Nicotine. In: Handbook of toxicology. Vol III: Insecticides. A compendium. WB Saunders, Philadelphia, pp 508–519

Nicholson RA, Botham RP, Blade RJ (1985) The interaction of sodium channel directed neurotoxicants and a novel insecticidal isobutylamide with central nerve terminals prepared from the cockroach (*Periplaneta americana*). Pestic Sci 16:554–555

Nugroho BW, Edrada RA, Güssregen B, Wray V, Witte L, Proksch P (1997) Insecticidal rocaglamide derivatives from *Aglaia duppereana*. Phytochemistry 44:1455–1461

Numata A, Ibuka T (1987) Alkaloids from ants and other insects. In: Brossi A (Ed) The alkaloids: chemistry and pharmacology. Vol 31. Academic Press, San Diego, pp 193–315

Okaichi T, Hashimoto Y (1962) The structure of nereistoxin. Agric Biol Chem 26:224–227

Ondeyka JG, Goegelman RT, Schaeffer JM, Kelemen L, Zitano L (1990) Novel antinematodal and antiparasitic agents from *Penicillium charlesii*. I. Fermentation, isolation and biological activity. J Antibiot (Tokyo) 43:1375–1379

Ondeyka JG, Helms GL, Hensens OD, Goetz MA, Zink DL, Tsipouras A, Shoop WL, Slayton L, Dombrowski AW, Polishook JD, Ostlind DA, Tsou NN, Ball RG, Singh SB (1997) Nodulisporic acid A, a novel and potent insecticide from a *Nodulisporium* sp. Isolation, structure determination, and chemical transformations. J Am Chem Soc 119:8809–8816

Orechoff A, Menschikoff G (1931) Über die Alkaloide von *Anabasis aphylla* L. Ber Dtsch Chem Ges 64:266–274

Ottea JA, Payne GT, Soderlund DM (1990) Action of insecticidal *N*-alkylamides at site 2 of the voltage-sensitive sodium channel. J Agric Food Chem 38:1724–1728

Pauron D, Barhanin J, Amichot M, Pralavorio M, Bergé J-B, Lazdunski M (1989) Pyrethroid receptor in the insect Na^+ channel: alteration of its properties in pyrethroid-resistant flies. Biochemistry 28:1673–1677

Pelletier PJ, Caventou JB (1820) Examen chimique de plusieurs végétaux de la famille des colchicées, et du principe actif qu'ils renferment. [Cévadille (*veratrum sabadilla*); hellébore blanc (*veratrum album*); colchique commun (*colchicum autumnale*)]. Ann Chim Phys 14:69-83

Pelletier SW (1983) The nature and definition of an alkaloid. In: Pelletier SW (Ed) Alkaloids: chemical and biological perspectives. Vol 1. John Wiley & Sons, New York, pp 1–31

Pelletier SW, Mody NV, Varughese KI, Maddry JA, Desai HK (1981) Structure revision of 37 lycoctonine-related diterpenoid alkaloids. J Am Chem Soc 103:6536–6538

Pelletier SW, Mody NV (1979) The structure and synthesis of C_{19}-diterpenoid alkaloids. In: Manske RHF, Rodrigo RGA (Eds) The alkaloids: chemistry and physiology. Vol XVII. Academic Press, New York, pp 1–103

Pinner A (1893) Ueber Nicotin. Die Constitution des Alkaloids. Ber Dtsch Chem Ges 26:292–305

Posselt W , Reimann L (1828) Chemische Untersuchung des Tabaks und Darstellung des eigenthümlichen wirksamen Princips dieser Pflanze. Geiger's Mag Parmac 24:138—161

Rae ID, Rosenberger M, Szabo AG, Willis CR, Yates P, Zacharias DE, Jeffrey GA, Douglas B, Kirkpatrick JL, Weisbach JA (1967) Haplophytine. J Am Chem Soc 89:3061–3062

Roddick JG (1989) The acetylcholinesterase-inhibitory activity of steroidal glycoalkaloids and their aglycones. Phytochemistry 28:2631–2634

Rogers EF, Koniuszy FR, Shavel J Jr, Folkers K (1948) Plant insecticides. I. Ryanodine, a new alkaloid from *Ryania speciosa* Vahl. J Am Chem Soc 70:3086–3088

Rogers EF, Snyder HR, Fischer RF (1952) Plant insecticides. II. The alkaloids of *Haplophyton cimicidum*. J Am Chem Soc 74:1987–1989

Saitoh F, Noma M, Kawashima N (1985) The alkaloid contents of sixty *Nicotiana* species. Phytochemistry 24:477–480

Sakai M, Sato Y (1972) Metabolic conversion of the nereistoxin-related compounds into nereistoxin as a factor of their insecticidal action. In: Tahori AS (Ed) Pesticide chemistry. Vol 1. Insecticides. Gordon and Breach, New York, pp 455–467

Sakata K, Aoki K, Chang C-F, Sakurai A, Tamura S, Murakoshi S (1978) Stemospironine, a new insecticidal alkaloid of *Stemona japonica* Miq. Isolation, structural determination and activity. Agric Biol Chem 42:457–463

Satasook C, Isman MB, Wiriyachitra P (1992) Activity of rocaglamide, an insecticidal natural product, against the variegated cutworm, *Peridroma saucia* (Lepidoptera: Noctuidae). Pestic Sci 36:53–58

Saxena RC, Harshan V, Saxena A, Sukumaran P, Sharma MC, Lakshamana Kumar M (1993) Larvicidal and chemosterilant activity of *Annona squamosa* alkaloids against *Anopheles stephensi*. J Am Mosq Control Assoc 9:84–87

Schaffner U, Boevé J-L, Gfeller H, Schlunegger UP (1994) Sequestration of *Veratrum* alkaloids by specialist *Rhadinoceraea nodicornis* Konow (Hymenoptera, Tenthredinidae) and its ecoethological implications. J Chem Ecol 20:3233–3249

Schildknecht H, Maschwitz U, Wenneis WF (1967) Neue Stoffe aus dem Wehrsekret der Diplopodengattung Glomeris. Naturwissenschaften 54:196–197

Schmeller T, Latz-Brüning B, Wink M (1997) Biochemical activities of berberine, palmatine and sanguinarine mediating chemical defence against microorganisms and herbivores. Phytochemistry 44:257–266

Schmeltz I (1971) Nicotine and other tobacco alkaloids. In: Jacobson M, Crosby DG (Eds) Naturally occurring insecticides. Marcel Dekker, New York, pp 99–136

Schmitt M, Turberg A, Londershausen M, Dorn A (1996) Binding sites for Ca^{2+}-channel effectors and ryanodine in *Periplaneta americana* - possible targets for new insecticides. Pestic Sci 48:375–385

Schreiber K (1968) Steroid alkaloids: the *Solanum* group. In: Manske RHF (Ed) The alkaloids: chemistry and physiology. Vol X. Academic Press, New York, pp 1–192

Schröder F, Franke S, Francke W, Baumann H, Kaib M, Pasteels JM, Daloze D (1996) A new family of tricyclic alkaloids from *Myrmicaria* ants. Tetrahedron 52:13539–13546

Scofield AM, Witham P, Nash RJ, Kite GC, Fellows LE (1995) Castanospermine and other polyhydroxy alkaloids as inhibitors of insect glycosidases. Comp Biochem Physiol 112A:187–196

Seeman JI (1984) Recent studies in nicotine chemistry. Conformational analysis, chemical reactivity studies, and theoretical modeling. Heterocycles (Tokyo) 22:165–193

Severson RF, Huesing JE, Jones D, Arrendale RF, Sisson VA (1988) Identification of tobacco hornworm antibiosis factor from cuticulae of Repandae section of *Nicotiana* species. J Chem Ecol 14:1485–1494

Sharma RP, Salunkhe DK (1989) Solanum glycoalkaloids. In: Cheeke PR (Ed) Toxicants of plant origin. Vol 1. Alkaloids. CRC Press, Boca Raton, FL, pp 179–236

Shepard HH (1951a) Plant products: nicotine and other alkaloids. In: Shepard HH (Ed) The chemistry and action of insecticides. McGraw-Hill, New York, pp 115–143

Shepard HH (1951b) Pyrethrins, rotenone, and miscellaneous plant extractives. In: Shepard HH (Ed) The chemistry and action of insecticides. McGraw-Hill, New York, pp 144–190

Sherby SM, Eldefrawi AT, David JA, Sattelle DB, Eldefrawi ME (1986) Interactions of charatoxins and nereistoxin with the nicotinic acetylcholine receptors of insect CNS and *Torpedo* electric organ. Arch Insect Biochem Physiol 3:431–445

Shinozaki H, Ishida M (1985) Inhibitory actions of tuberstemonine on the excitatory transmission at the crayfish neuromuscular junction. Brain Res 334:33–40

Smith CR, Richardson CH, Shepard HH (1930) Neonicotine and certain other derivatives of the dipyridyls as insecticides. J Econ Entomol 23:863–867

Smith RM (1977) The Celastraceae alkaloids. In: Manske RHF (Ed) The alkaloids: chemistry and physiology. Vol 16. Academic Press, New York, pp 215–248

Smolanoff J, Kluge AF, Meinwald J, McPhail A, Miller RW, Hicks K, Eisner T (1975) Polyzonimine: a novel terpenoid insect repellent produced by a milliped. Science 188:734–736

Srivastava SN, Przybylska M (1970) The crystal structure of ryanodol *p*-bromobenzyl ether. Acta Crystallogr B26:707–715

Staub GM, Gloer JB, Wicklow DT, Dowd PF (1992) Aspernomine: a cytotoxic anti-insectan metabolite with a novel ring system from the sclerotia of *Aspergillus nomius*. J Am Chem Soc 114:1015–1017

Staub GM, Gloer KB, Gloer JB, Wicklow DT, Dowd PF (1993) New paspalinine derivatives with antiinsectan activity from the sclerotia of *Aspergillus nomius*. Tetrahedron Lett 34:2569–2572

Stedman E, Barger G (1925) Physostigmine (eserine). Part III. J Chem Soc 127:247–258

Strunz GM, Finlay H (1994) Concise, efficient new synthesis of pipercide, an insecticidal unsaturated amide from *Piper nigrum*, and related compounds. Tetrahedron 50:11113–11122

Su HCF (1985) *N*-Isobutylamides. In: Kerkut GA, Gilbert LI (Eds) Comprehensive insect physiology, biochemistry and pharmacology. Vol 12. Insect control. Pergamon Press, Oxford, pp 273–289

Sutko JL, Airey JA, Welch W, Ruest L (1997) The pharmacology of ryanodine and related compounds. Pharmacol Rev 49:53–98

Swingle WT, Haller HL, Siegler EH, Swingle MC (1941) A Chinese insecticidal plant, *Tripterygium wilfordii*, introduced into the United States. Science 93:60–61

Tattersfield F, Gimingham CT, Morris HM (1926) Studies on contact insecticides. Part IV. A quantitative examination of the toxicity of certain plants and plant products to *Aphis rumicis* L. (the bean aphis). Ann Appl Biol 13:424–445

Thomas EJ (1994) Approaches to the synthesis of insecticidal compounds. In: Briggs GG (Ed) Advances in the chemistry of insect control. III. Royal Society of Chemistry, Cambridge, UK, pp 223–237

Toia RF (1990) Ant secretions as a source of natural product models for possible pest control agents. In: Casida JE (Ed) Pesticides and alternatives: innovative chemical and biological approaches to pest control. Elsevier Science, Amsterdam, pp 301–309

Tokuyama T, Daly JW (1983) Sterodial alkaloids (batrachotoxins and 4β-hydroxy-batrachotoxins), "indole alkaloids" (calycanthine and chimonanthine) and a piperidinyl-dipyridine alkaloid (noranabasamine) in skin extracts from the Colombian poison-dart frog *Phyllobates terribilis* (Dendrobatidae). Tetrahedron 39:41–47

Trost BM, Greenspan PD, Yang BV, Saulnier MG (1990) An unusual oxidative cyclization. A synthesis and absolute stereochemical assignment of (-)-rocaglamide. J Am Chem Soc 112:9022–9024

Tsao R, Eto M (1989) Chemical and photochemical transformation of the insecticide cartap hydrochloride into nereistoxin. J Pestic Sci 14:47–51

Tursch B, Daloze D, Dupont M, Pasteels JM, Tricot M-C (1971) A defense alkaloid from a carnivorous beetle. Experientia (Basel) 27:1380–1381

Tursch B, Braekman JC, Daloze D, Hootele C, Losman D, Karlsson R, Pasteels JM (1973) Chemical ecology of arthropods, VI, adaline a novel alkaloid from *Adalia bipunctata* L. (Coleoptera, Coccinellidae). Tetrahedron Lett 201–202

Ujváry I, Eya BK, Grendell RL, Toia RF, Casida JE (1991) Insecticidal activity of various 3-acyl and other derivatives of veracevine relative to the *Veratrum* alkaloids veratridine and cevadine. J Agric Food Chem 39:1875–1881

Ujváry I, Polgár L, Darvas L, Casida JE (1995) Non-steroidal analogues of veratridine: model-based design, synthesis and insecticidal activity. Pestic Sci 44:95–102

Ujváry I, Casida JE (1997) Partial synthesis of 3-*O*-vanilloylveracevine, an insecticidal alkaloid from *Schoenocaulon officinale*. Phytochemistry 44:1257–1260

Usherwood PNR, Vais H (1995) Towards the development of ryanoid insecticides with low mammalian toxicity. Toxicol Lett 82/83:247–254

Wada K, Munakata K (1967) An insecticidal alkaloid, cocculolidine from *Cocculus trilobus* DC. Part I. The isolation and the insecticidal activity of cocculolidine. Agric Biol Chem 31:336–339

Wada K, Munakata K (1968) Naturally occurring insect control chemicals. Isoboldine, a feeding inhibitor, and cocculolidine, an insecticide in the leaves of *Cocculus trilobus* DC. J Agric Food Chem 16:471–474

Wada K, Marumo S, Munakata K (1966) An insecticidal alkaloid, cocculolidine from *Cocculus trilobus*. Tetrahedron Lett 5179–5184

Wada K, Marumo S, Munakata K (1967) An insecticidal alkaloid, cocculolidine from *Cocculus trilobus* DC. Part II. The structure of cocculolidine. Agric Biol Chem 31:452–460

Wada K, Marumo S, Munakata K (1968) An insecticidal alkaloid, cocculolidine from *Cocculus trilobus* DC. Part III. The stereochemistry of cocculolidine. Agric Biol Chem 32:1187–1189

Waterhouse AL, Holden I, Casida JE (1984) 9,21-Didehydroryanodine: a new principal toxic constituent of the botanical insecticide *Ryania*. J Chem Soc Chem Commun 1265–1266

Waterhouse AL, Pessah IN, Francini AO, Casida JE (1987) Structural aspects of ryanodine action and selectivity. J Med Chem 30:710–716

Wheeler JW, Olubajo O, Storm CB, Duffield RM (1981) Anabaseine: venom alkaloid of *Aphaenogaster* ants. Science 211:1051–1052

Whyte AC, Gloer JB, Wicklow DT, Dowd PF (1996) Sclerotiamide: a new member of the paraherquamide class with potent antiinsectan activity from the sclerotia of *Aspergillus sclerotiorum*. J Nat Prod 59:1093–1095

Wicklow DT, Dowd PF, Gloer JB (1994) Antiinsectan effects of *Aspergillus* metabolites. In: Powell KA, Renwick A, Peberdy JF (Eds) The genus *Aspergillus*: from taxonomy and genetics to industrial application. Plenum Press, New York, pp 93–114

Wieland H, Dragendorff O (1929) Die Konstitution der Lobelia-Alkaloide. Ann Chem 473:83-102

Wiesner K (1972) The structure of ryanodine. Adv Org Chem 8:295–316.

Wink M (1993a) Allelochemical properties or the *raison d'etre* of alkaloids. In: Cordell GA (Ed) The alkaloids: chemistry and pharmacology. Vol 43. Academic Press, San Diego, pp 1–118

Wink M (1993b) Production and application of phytochemicals from an agricultural perspective. In: van Beek TA, Breteler H (Eds) Phytochemistry and agriculture. Clarendon Press, Oxford, pp 171–213

Winchester B (1992) Natural and synthetic inhibitors of glycosidases. Biochem Soc Trans 20:699–705

Witkop B, Gössinger E (1983) Amphibian alkaloids. In: Brossi A (Ed) The alkaloids: chemistry and pharmacology. Vol XXI. Academic Press, New York, pp 139–253

Xie Y, McHugh T, McKay J, Jones GS Jr, Loring RH (1996) Evidence that a nereistoxin metabolite, and not nereistoxin itself, reduces neuronal nicotinic receptors: studies in the whole chick ciliary ganglion, on isolated neurons and immunoprecipitated receptors. J Pharmacol Exp Ther 276:169–177

Yamada K, Shizuri Y, Hirata Y (1978) Isolation and structures of a new alkaloid alatamine and an insecticidal alkaloid wilfordine from *Euonymus alatus* forma *striatus* (Thunb.) Makino. Tetrahedron 34:1915–1920

Yamamoto I (1965) Nicotinoids as insecticides. In: Metcalf RL (Ed) Advances in pest control research. Vol VI. John Wiley & Sons, New York, pp 231–260

Yamazaki M, Okuyama E, Kobayashi M, Inoue H (1981) The structure of paraherquamide, a toxic metabolite from *Penicillium paraherquei*. Tetrahedron Lett 22:135–136

Yano K, Oono J, Mogi K, Asaoka T, Nakashima T (1987) Pyrroxamycin, a new antibiotic. Taxonomy, fermentation, isolation, structure determination and biological properties. J Antibiot (Tokyo) 40:961–969

Yates P, MacLachlan FN, Rae ID, Rosenberg M, Szabo AG, Willis CR, Cava MP, Behforouz M, Lakshmikantham MV, Zeiger W (1973) Haplophytine. A novel type of indole alkaloid. J Am Chem Soc 95:7842–7850

Yoshida HA, Toscano NC (1994) Comparative effects of selected natural insecticides on *Heliothis virescens* (Lepidoptera: Noctuidae) larvae. J Econ Entomol 87:305–310

Zhang D, Nair MG, Murry M, Zhang Z (1997) Insecticidal activity of indanomycin. J Antibiot (Tokyo) 50:617–620

3 Discovery of the Nitromethylene Heterocycle Insecticides

Willy D. Kollmeyer[1, 2], Roger F. Flattum[3], James P. Foster[3], James E. Powell[1], Mark E. Schroeder[1], and S. Barney Soloway[4]
Biological Sciences Research Center, Shell Agricultural Chemical Company, P.O. Box 4248, Modesto, CA[5], USA
[1]Current address: DuPont Agricultural Products, Stine-Haskell Research Center, P.O. Box 30, Newark, DE 19714, USA.
[2]Corresponding author; phone: (302) 366-5922; fax: (302) 451-4840: e-mail: willy.d.kollmeyer@usa.dupont.com.
[3]Current address: DuPont Agricultural Products, Experimental Station, P. O. Box 402, Wilmington, DE 19880, USA.
[4]Current address: 3401 Mansfield Lane, Modesto, CA 95350, USA.
[5]Purchased by E. I. du Pont de Nemours and Company, 1007 Market Street, Wilmington, DE 19898, USA, on October 1, 1986.

1 Introduction

Structure-activity relationships of the seminal nitromethylene heterocycle (NMH) insecticides have been described previously (Soloway et al. 1979). Rather than recount those observations, this paper describes key events involved in the discovery and exploration of this new class of insecticides along with glimpses of their remarkable chemical and biological properties.

2 Discovery

As part of a program to discover new crop protection chemicals, Shell Development Company's Biological Research Center in Modesto, California, tested research samples from university sources. According to company records, the lead compound 2-(dibromonitromethyl)-3-methylpyridine (**1**) was received from Professor Henry Feuer of Purdue University on May 11, 1970. This compound, although not explicitly described in the literature at that time, originated from his studies on the synthesis of α-nitroalkyl heterocycles (Feuer and Lawrence 1969). In Shell's primary insecticide screen, compound **1** revealed modest but detectable activity against housefly and pea aphid as indicated in Table 1.

By way of explanation, the toxicity indices (TIs) listed in Table 1 are a measure of a test compound's activity relative to a commercial standard, in this case parathion. For example, based upon a relative comparison of the doses required to kill 50% of the test insects (LC_{50}s), a compound with TI of 1 is only 1% as toxic as the standard whereas another compound with TI of 200 is twice as toxic as parathion (TI=[LC_{50} parathion/LC_{50} test compound]100).

Besides testing for a combination of topical and ingestion activity in the primary insecticide screen, selected compounds were also evaluated by direct injection into houseflies, some of which had been pretreated with sesamex, a well-known inhibitor of insect mixed-function oxidase. The purpose of the injection assay was to give a test compound a better chance to reveal intrinsic activity, which might be suppressed because of poor compound uptake or oxidative metabolism. In this case the assay served to further highlight compound **1**. Pretreatment with synergist increased its injection toxicity by more than tenfold, suggesting a more encouraging level of intrinsic activity than might have otherwise been inferred.

At this point compound **1** was not a highly compelling research lead and, in fact, was viewed by some as an analog of the biocide chloropicrin. It did however possess the desirable attributes of (1) structural novelty versus existing insecticide classes and (2) modest biological activity with apparent potential for improvement. More importantly, the Shell research organization had recently abandoned all efforts to find newer organophosphate insecticides, especially vinyl phosphates (unpublished communication, R.W. Whetstone, 1969), and was, therefore, particularly motivated to discover a new and safer class of insecticides.

Table 1. Insecticidal activity of nitromethylene heterocycles

Compound number			Primary screen toxicity index[a]				Housefly injection	
							Unsynergized	Synergized[g]
Text	Shell	Structure	HF[b]	HF +S[c]	PA[d]	CEW[e]	LD_{50}[f] or TI[a]	LD_{50}[f] or TI[a]
1	SD-031588	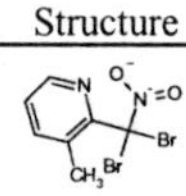	1.7	~1.5	~5	0	1.1[f]	0.09[f]
2	SD-032931	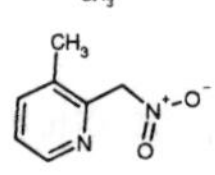	3.2[k]	20[k]	4.5	12	0.4[f]	0.03[f]
3	SD-033420	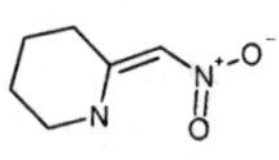	5.4[k]	27[k]	~3[k]	155[k]	0.35[f]	0.018[f, k]
4	SD-033452	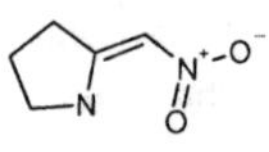	1.3[k]	2.7[k]	~1	89	0.9[f]	~0.12[f]

Table 1. (Continued)

Compound number			Primary screen toxicity index[a]				Housefly injection	
							Unsynergized	Synergized[g]
Text	Shell	Structure	HF[b]	HF +S[c]	PA[d]	CEW[e]	LD_{50}[f] or TI[a]	LD_{50}[f] or TI[a]
5	SD-033488	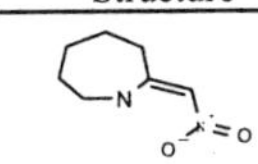	<1	<1	~1[k]	10[k]		0.1[f]
6	SD-033606	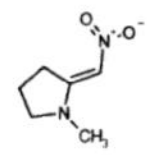	+	5	~1	106		~0.1[f,k]
7	SD-033647	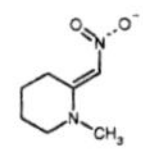	1.6[k]	13	2.7	25	0.3[f, k]	0.04[f, k]
8	SD-033690	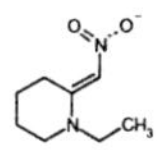	10	205	15[k]	13		0.0026[f,k]
9	SD-034064	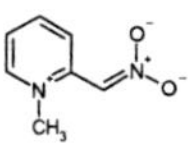	0	+	+	38[k]		0.3[f]
10	SD-035381	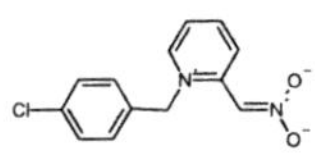	0		91	15		
11	SD-033900	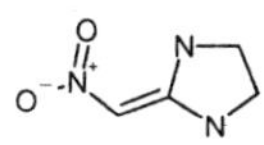	0		+	12		0.55[f]
12	SD-035351	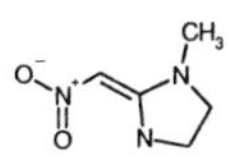	6.7		11	305	10[a, k]	50[a, k]
13	SD-035349	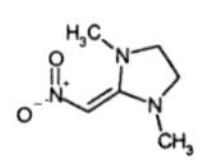	0	0	+	3	0[a]	+
14	SD-035347	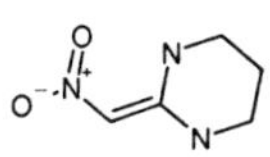	0		+	5.2	1.5[a, k]	+
15	SD-035472	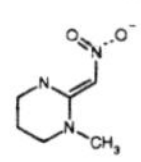	8.2		67	137	22[a, k]	67[a, k]
16	SD-037637	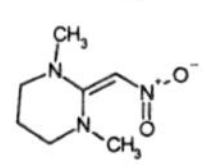	0		0	+	0[a]	0[a]
17	SD-035590	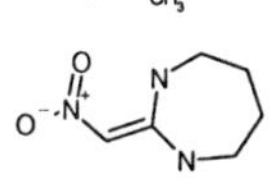	0		+	9.8	1[a, k]	11[a, k]
18	SD-035767	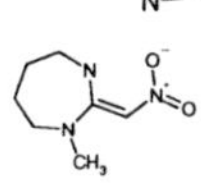	2.8[k]		3.6[k]	30	1.5[a]	>3[a]
19	SD-035651	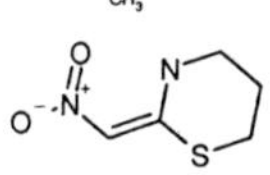	161		24	1662	138[a]	400[a]
20	SD-035790	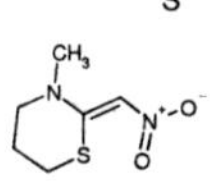	0.8		36	138	1.5[a, k]	33[a, k]

Table 1. (Continued)

Compound number			Primary screen toxicity index[a]				Housefly injection	
							Unsynergized	Synergized[g]
Text	Shell	Structure	HF[b]	HF +S[c]	PA[d]	CEW[e]	LD_{50}[f] or TI[a]	LD_{50}[f] or TI[a]
21a	SD-031603	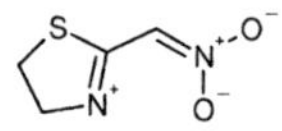	0	+	+	+		
21b	SD-035733	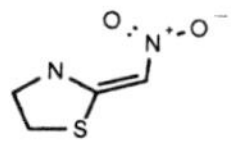	0		1	7		
22	SD-035704	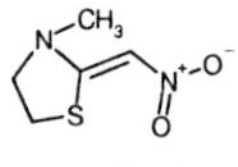	0		2	49		
23	SD-035747	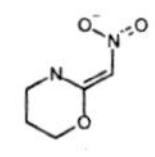	3		3	64	6[a]	14[a]
24	SD-037547	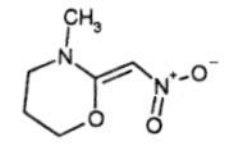	0		+	0	0[a]	0[a]
25	SD-035789	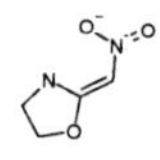	0		+	0	+	+
26	WL-043297	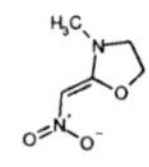				1		
27	SD-035957	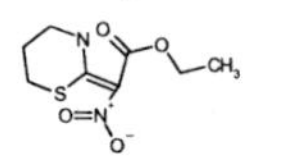	5		+	292[k]	1.5[a, k]	50[a, k]
28	SD-036006	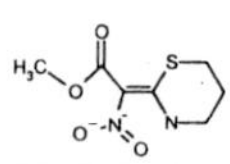	4		+	207	0[a]	105[a]

Data for two-spotted spider mite (*Tetranychus urticae*) and mosquito larvae (*Anopheles albimanus*) were uninteresting and are omitted here.

[a]TI, toxicity index = (LC_{50} parathion/LC_{50} test compound)100; a "+" indicates detectable activity with a toxicity index of <1 but >0.

[b]HF, housefly (*Musca domestica*) sprayed with test compound only.

[c]HF + S, housefly (*Musca domestica*) sprayed with test compound plus 1% sesamex as synergist.

[d]PA, pea aphid (*Acyrthosiphon pisum*).

[e]CEW, corn earworn (*Heliothis zea*).

[f]LD_{50}, lethal dose in *ug*/fly to kill 50% of the houseflies.

[g]Houseflies were pretreated topically with 10 *ug* sesamex 4 hours prior to injection.

[k]Rapid knockdown of treated insects was observed.

3 Chemical Exploration and Structural Evolution of the Lead

Chemical exploration of low-level leads is often influenced by perceptions of the relative ease and scope for further synthesis. In this case, the initial approach and the molecule's reactivity were not congruous, but a highly useful outcome resulted nevertheless. When compound **1** was treated with an iodide salt in an attempt to carry out a halogen exchange reaction, the desired diiodo analog was not obtained. Instead iodide acted as a reducing agent, providing isolable amounts of 3-methyl-2-(nitromethyl)pyridine (**2**) as shown in Eq. 1 (unpublished work, C.H. Tieman, 1970). Fortunately, this structural change was associated with a higher and broader spectrum of insecticidal activity, as shown in Table 1 (Johnson, Reed, Tieman, and Soloway 1974). The corn earworm activity was particularly encouraging because Shell's specific objective was to discover a new cotton insecticide, and being a lepidopteran species, corn earworm (*Heliothis zea*) was a good indicator for this outlet. In addition, synergist treatment with compound **2** continued to enhance housefly toxicity, but now by both topical and injection routes of application.

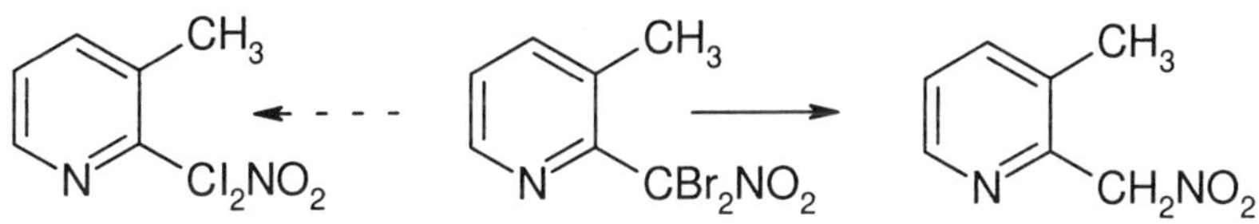

(1)

1 **2**

With these results in hand, the synthesis of a variety of ring-substituted 2-(nitromethyl)pyridines was desired. Unfortunately, the preparation of analogs via the Feuer alkaline nitration of substituted 2-picolines proved to be relatively unsatisfactory with respect to convenience, yields, generality, and availability of starting materials. To avoid these difficulties, an alternate approach was conceived. The thought was to prepare 2-(nitromethyl)pyridines by aromatization of suitable tetrahydropyridine precursors (Eq. 2). In fact, the preparation of these precursors did not seem unreasonable based upon closely related processes patented by three different European companies including Bayer, Fratmann, and Pechiney-Saint-Goban (Eqs. 3-5; Petersen 1953; Fratmann 1971; Etienne and Correia 1969). However, when this plan was implemented at Shell several unexpected things happened (Eq. 6 and Table 1). First, the product arising from the reaction of nitromethane with the methyl imidate derived from 2-piperidone was not the expected 2-(nitromethyl)-3,4,5,6-tetrahydropyridine. Instead, it was the exocyclic double bond isomer, compound **3**, as evidenced by the H-NMR spectrum. It clearly lacked a methylene singlet but did reveal a single uncoupled olefinic proton and a broad low-field NH singlet that was exchangeable with D_2O.

More importantly, the biological activity against corn earworm had skyrocketed by an order of magnitude, and the nitroenamine moiety had been unveiled (Roman 1973a). It is interesting to note that Pechiney-Saint-Goban had also claimed insecticidal activity for their preparation (Eq. 5; Etienne and Corriera 1969), but described the product as a nitromethyl tautomer and apparently failed to follow up on their discovery.

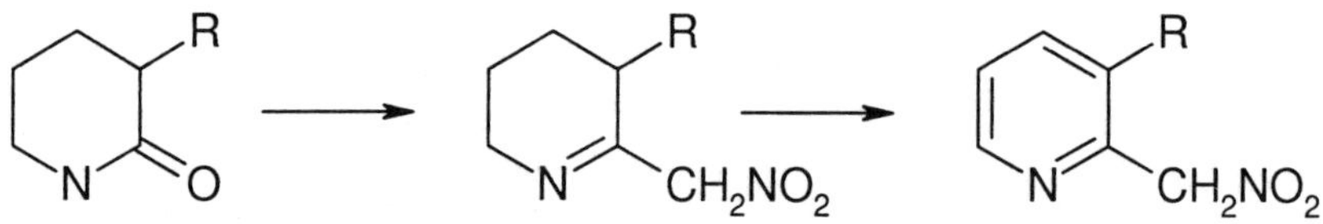
(2)

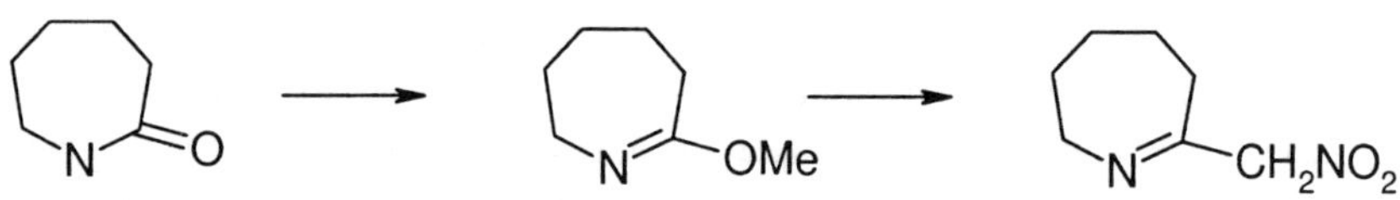
(3)

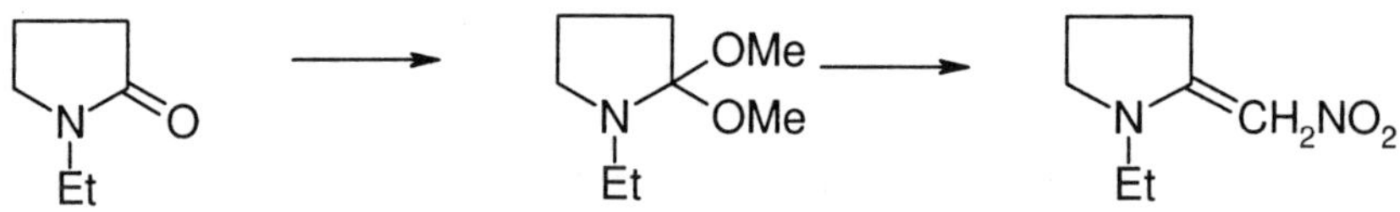
(4)

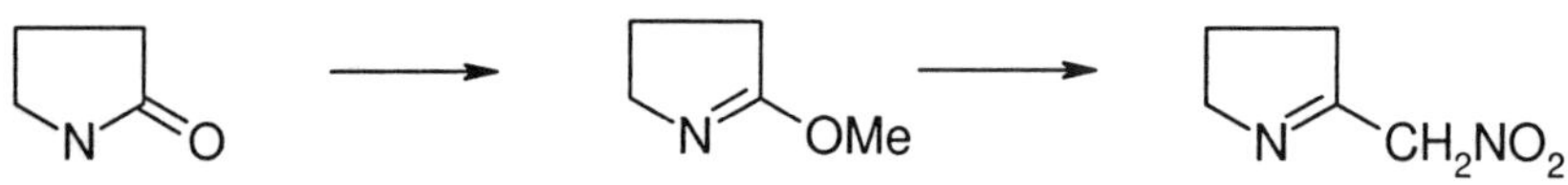
(5)

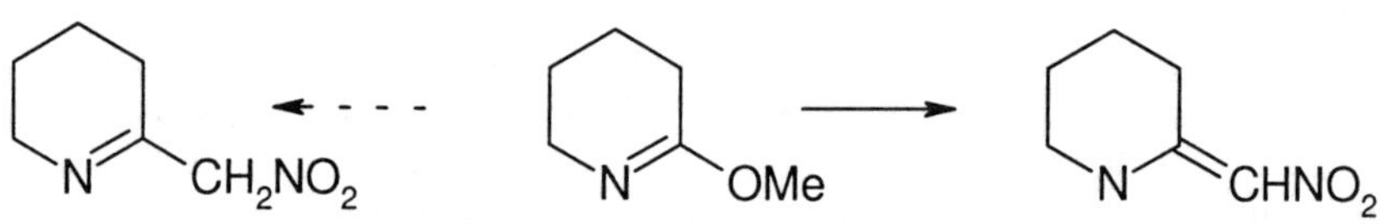
(6)

3

In short order several related analogs involving five-, six-, and seven-membered rings were synthesized and tested at Shell, e.g., compounds **4** through **8** in Table 1 (Roman 1973b and unpublished results). These materials quickly illustrated the vagaries of ring size and substitution at ring nitrogen on corn earworm activity. In any event it was now clear that a major new class of insecticide was at hand, and a vigorous synthesis program was laid out to explore the bounds of these observations.

Returning to the pyridines, the tactic of N-substitution was used to force adoption of a nitroenamine structural configuration, because the early lead compounds in this system tended to exist as the nitromethyl tautomers (Eq. 7). In fact, this approach did enhance the biological activity towards corn earworm as can be seen by comparing compounds **2** and **9** in Table 1. However, **9** and its congeners had zwitterionic characteristics, making them awkward to isolate because of their relative insolubility and inability to be extracted from water phases with the usual organic solvents (Eq. 8; Henry 1974). These compounds are described in *Chemical Abstracts* as pyridinium hydroxide inner salts. For example, compound **10**, whose registry number is 54436-77-0, is listed as 1-[(4-chlorophenyl)methyl]-2-(nitromethyl)pyridinium hydroxide, inner salt.

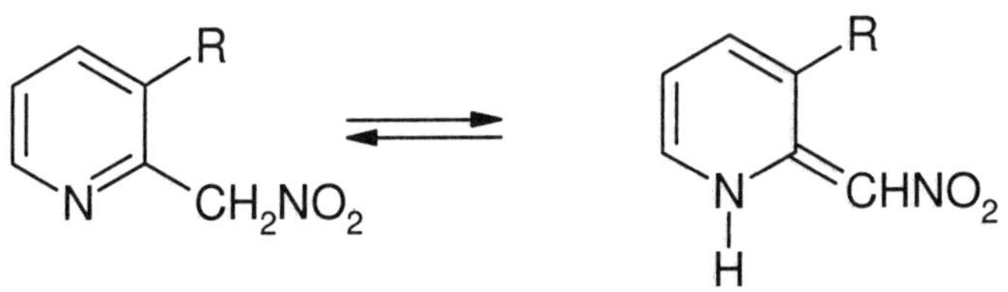

(7)

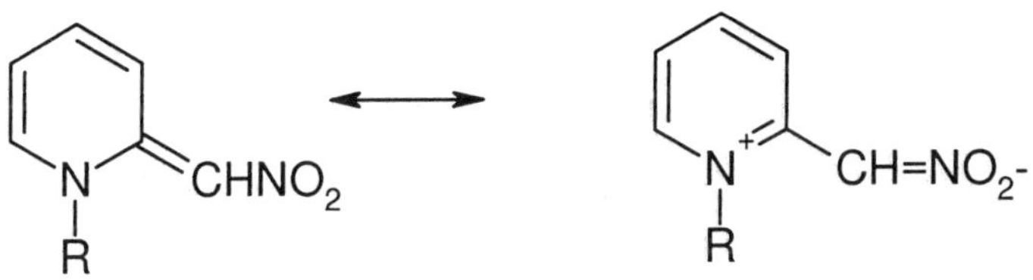

(8)

9 R = methyl
10 R = p-chlorobenzyl

With perfect hindsight it is interesting to note the early and encouraging activity of compound **10** towards pea aphid, a homopteran representative of important sucking insects towards which imidacloprid is so successful today (Elbert et al. 1991). Unfortunately, as mentioned previously, the Shell objective was a new cotton insecticide, which resulted in a focus on lepidopteran species such as *Heliothis zea*.

Returning to compound **3** as the newer and more potent lead to be explored, it quickly became apparent that incorporation of another heteroatom and ring size were very important (Scheme 1). Thus 1,3-diazacycloalkanes were the next heterocyclic subset to be exemplified. The first member to be prepared, compound **11**, was relatively uninteresting (see Table 1), but not so for the later members such as compounds **12** and **15** (Eq. 9; Tieman, Kollmeyer, and Roman 1975, 1976a). In this subset the effects of ring size and substitution at nitrogen on corn earworm activity were not predictable based upon the earlier results obtained with the pyridines and monoazacycloalkanes, particularly with respect to N-methylation.

Scheme 1. Structural Exploration Via Ring Size and Heteroatoms

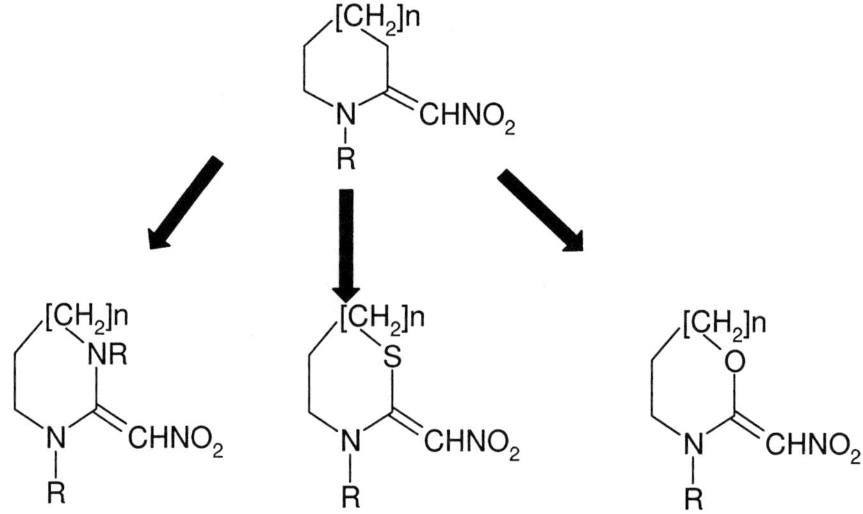

From a priority point of view, the preparation of the unsubstituted imidazolidine **11** and its similarly uninteresting N,N'-dimethyl analog **13** had already been disclosed in the literature (Gompper and Schaefer 1967), but the maximally insecticidal mono N-methyl analog compound **12** had not (Tieman, Kollmeyer, and Roman 1975). Although Gompper's original method was suitable for small-scale synthesis (Eq. 9), it was not used for large-scale, field trial preparations. As pointed out in the literature and subsequently observed in our own hands, the dipotassium 2-nitroethylene-1,1-dithiolate that precipitated from an ethanolic solution of carbon disulfide, nitromethane, and potassium hydroxide was a hazardous material. It could explode with rapid heating or upon impact (Freund 1919) or detonate when allowed to dry in air (Isaksson and Sandstrom 1973).

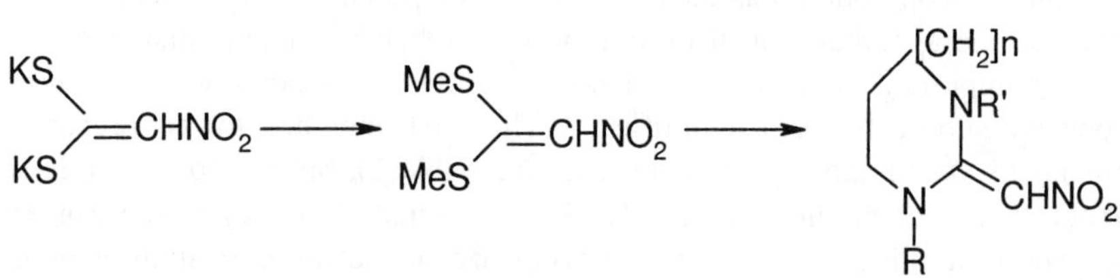

(9)

11, n=0, R=H, R[1]=H
12, n=0, R=Me, R[1]=H
15, n=1, R=Me, R[1]=H

Like the earlier 2-(nitromethyl)pyridinium hydroxide inner salts, compounds in the 1,3-diazacycloalkane subset also had unusual, nearly zwitterionic properties. Generally speaking, these materials could not be extracted from water phases with the usual organic solvents because of their extremely polar character arising from resonance interaction in the nitroenediamine system. In the extreme, this can be represented by resonance hybrid structures such as those shown in Eq. 10, and this type of phenomenum accounts for many of the remarkable chemical and perhaps biological properties of the NMHs in general.

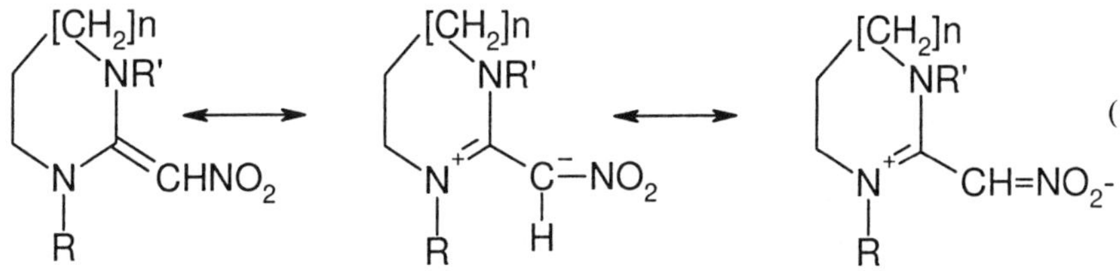

(10)

While the best 1,3-diazacycloalkane, compound **12**, represented a twofold activity increase over the lead compound **3**, a stunning discovery was soon to follow. Treatment of Gompper's versatile intermediate with 3-aminopropanethiol gave the first active example of the 1-aza-3-thiacycloalkane family (Eq. 11; Powell 1977). Biological evaluation of compound **19**, now known as nithiazine, revealed a toxicity index of approximately 1700 on corn earworm. Thus, an astounding one 1000-fold increase in whole organism activity had been achieved starting from the low-level lead originally obtained from Professor Feuer! However, when the N-methyl analog of nithiazine was prepared (compound **20**), it proved to be about 10 fold less active by comparison, which again illustrated the unpredictable effect of N-substitution in NMHs.

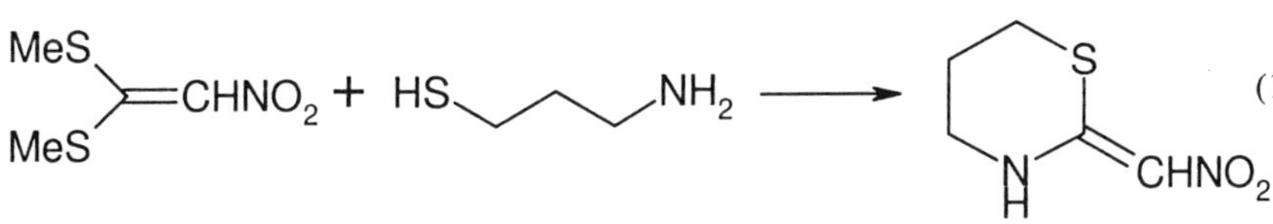

(11)

19

Subsequently, the five-membered thiazolidine analog was also synthesized, although by a different method. When freshly prepared in the Shell laboratories, this material (compound **21b**) was somewhat active, but only marginally so compared to its higher homolog nithiazine. It is interesting to recognize that Professor Feuer was actually the first to prepare this thiazolidine describing it as a zwitterionic species. He provided it to Shell for testing along with the original lead compound **1**. However, that first sample, **21a**, proved to be nearly inactive when tested in the insecticide screen. In retrospect it is reasonable to speculate that his sample had undergone some degree of degradation by the time it was biologically evaluated. Indeed, light-induced dissipation of active ingredient was soon recognized to occur with varying degrees of severity among the NMHs prepared at Shell, and it was extremely acute with nithiazine (Soloway et al. 1979). A related problem contributing to poor storage stability of many NMHs was sensitivity to Lewis acids, which caused technical material to slowly polymerize in situ resulting in insoluble oligomers. However, as we shall see below, the rich enamine chemistry of these NMHs offered many avenues to explore in an attempt to overcome the poor field stability of the parent compounds.

After observing the astounding effect brought about by introduction of sulfur, it was only logical to look at oxygen as indicated in Scheme 1. Unfortunately, with the exception of compound **23**, which had only moderate activity (Roman 1975), the 1-aza-3-oxacycloalkanes as well as their N-methyl derivatives proved to be relatively uninteresting (see compounds **24** through **26** in Table 1).

Although it was not obvious at that time, nithiazine represented the highwater mark for the NMH undertaking at Shell. In spite of the best intentions, plans and syntheses of more than 1,000 compounds, no more interesting ring systems or effective nitromethylene group replacements would be found. This included the preparation of several nitroimino versions of NMHs, all of which proved to be unremarkable in Shell's hands (compounds **29** through **33** in Table 2). The special virtues of this structural variation would remain hidden until Bayer scientists coupled it with the unique (6-chloro-3-pyridinyl)methyl N-substituent found in imidacloprid (Diehr et al. 1991).

Table 2. Nitroimino analogs

Compound number		
Text	Shell	Structure
29	SD-033387	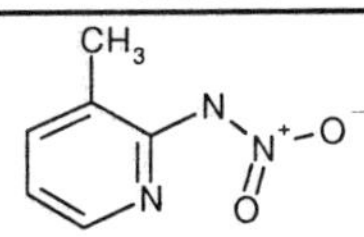
30	SD-034170	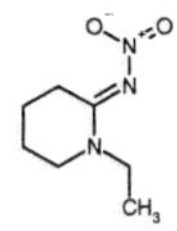
31	SD-008512	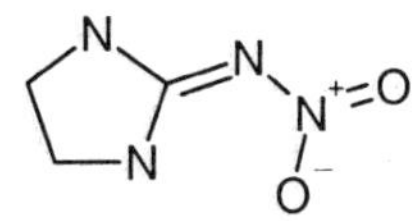
32	WL-082459	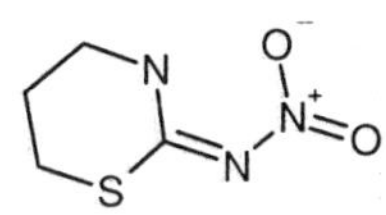
33	WL-082516	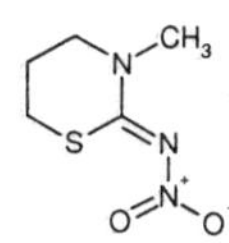

Nevertheless, a substantial amount of new chemistry was still to be discovered and exemplified for two compelling reasons. Practical methods were needed to prepare large quantities of field trial candidates such as compounds **3**, **12**, **15**, and **19** (nithiazine). And as the problem of poor field persistence of these materials became evident (Kleier et al. 1985), virtually no stone was left unturned in an attempt to prepare derivatives or prodrugs that might ameliorate the deficit of photolability. In this regard a substantial program was undertaken later on at Shell's Sittingbourne Research Center in England. That effort was directed towards development of a rice insecticide and resulted in the preparation and evaluation of the N-formyl derivative of nithiazine (Harris et al. 1986), which was widely field tested but never commercialized.

The significant Shell work directed towards large-scale synthesis and biologically active derivatives of various parent NMHs can be found in the patent literature associated with the principal investigators listed in the Acknowledgement at the end of this paper. For illustrative purposes, only brief excerpts are described here; a more complete presentation was given elsewhere (Powell 1990).

The cyclic imidate ester approach to synthesizing large quantities of NMHs was used, normally by taking advantage of analogous thioimidate-like structures that were more readily available. As shown in Table 3, the more basic of these intermediates were nitromethylenated upon direct treatment with nitromethane, i.e., compounds **34** and **35**. However, on one occasion a serious explosion resulted during a large-scale reaction of compound **34** and nitromethane, giving further credence to the reported hazards of mixing strong bases with nitromethane (Bretherick 1990). Fortunately, most of the less basic and less reactive intermediates such as compounds **37** through **40** did combine smoothly and uneventfully with nitroacetic acid esters, but in some cases requiring a Lewis acid catalyst as reported for compound **40** (Hirai et al. 1971). These observations can be explained, at least in part, by recognizing the substantially greater carbon acidity of nitroacetic acid esters versus nitromethane (Eq. 12). Subsequent conversion of the nitroacetic acid ester condensation products to the desired NMHs was straightforward by mild saponification and spontaneous decarboxylation upon neutralization.

Table 3. Basicity and reactivity of thioimidate-type intermediates

Compound number			Basicity	Condenses with	
Text	Shell	Structure	(pK_b)	CH_3NO_2	$O_2NCH_2CO_2R/ZnCl_2$
34	SD-036290	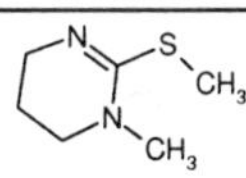	4.3	yes	yes (no $ZnCl_2$)
35	SD-036038	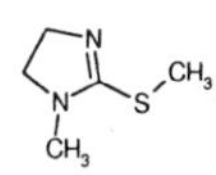	5.5	yes	yes (no $ZnCl_2$)
36	SD-037299	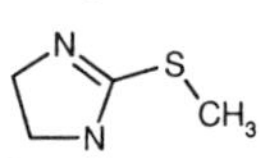	5.4	no	yes (no $ZnCl_2$)
37	SD-035798	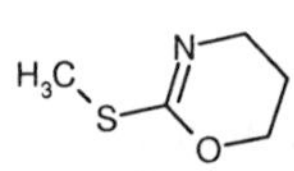	7.8	no	yes
38	SD-036236	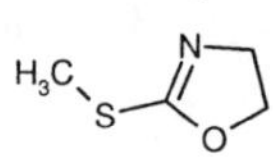	7.7	not determined	yes
39	SD-036015	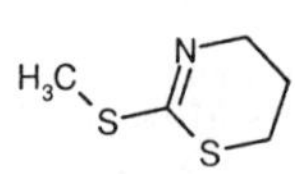	8.2	no	yes
40	WL-007989	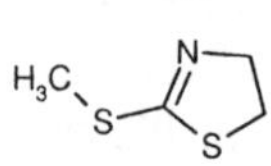	10.0	no	yes
41	SD-037202	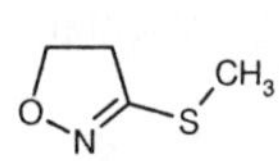	>11.3	no	no

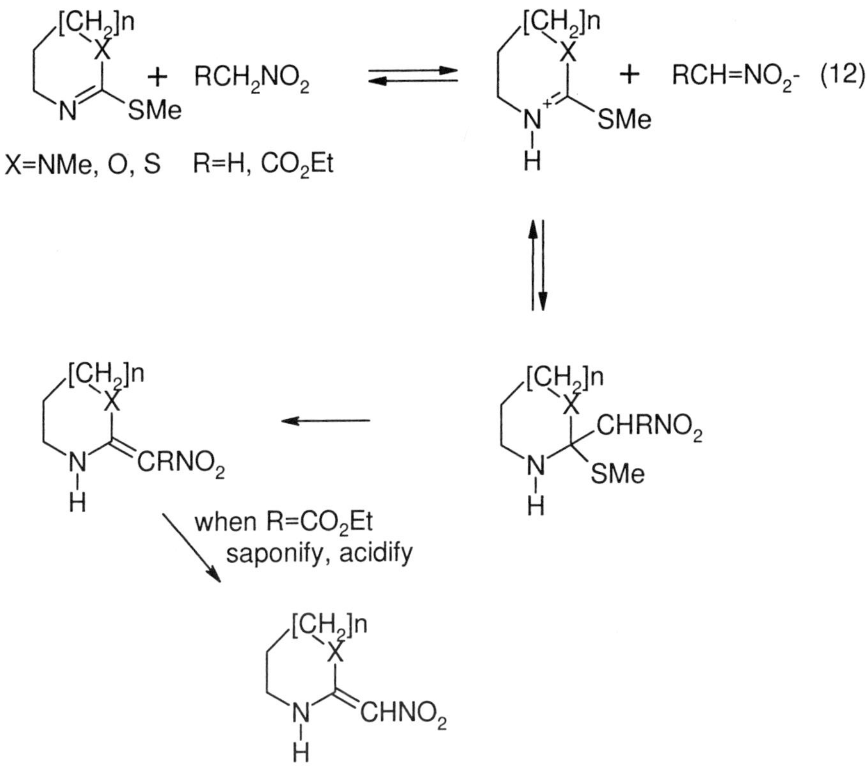

Besides providing field trial quantities of NMHs, another significant result of these process studies was finding substantial activity in the ester derivatives of nithiazine, e.g., compounds **27** and **28** in Table 1 (Roman 1976). This initiated a wide range of derivatization chemistry based upon the ambident nucleophilic character of the nitroenamine moiety residing in the NMHs. By this time it was quite obvious that photochemical reactivity leading to poor field persistence was a serious issue, especially with nithiazine, and the hope was that a suitable derivative or prodrug could be found to overcome this problem.

Nithiazine and to a lesser extent compound **12** were subjected to a broad range of electrophilic reagents including all manner of alkylating agents, acylating agents, halogenating agents, sulfonyl halides, sulfenyl halides, isocyanates, sulfonylisocyanates, aryldiazonium salts, benzyne precursors, aldehydes, nitrating agents, nitrosoating agents, and iminium ions (the Mannich reaction). This gave rise to a great range of products and structures, which could inevitably be explained by invoking an initial attack of the electrophilic reagent at one of the three nucleophilic sites in the nitroenamine moiety. In the simplest circumstance, this led to monosubstitution or adduct formation at either nitrogen or carbon or a mixture of substitution at both sites. With excess reagents, bis-substitution at

Scheme 2. Products Derived from NMHs by Treatment with Electrophilic Reagents

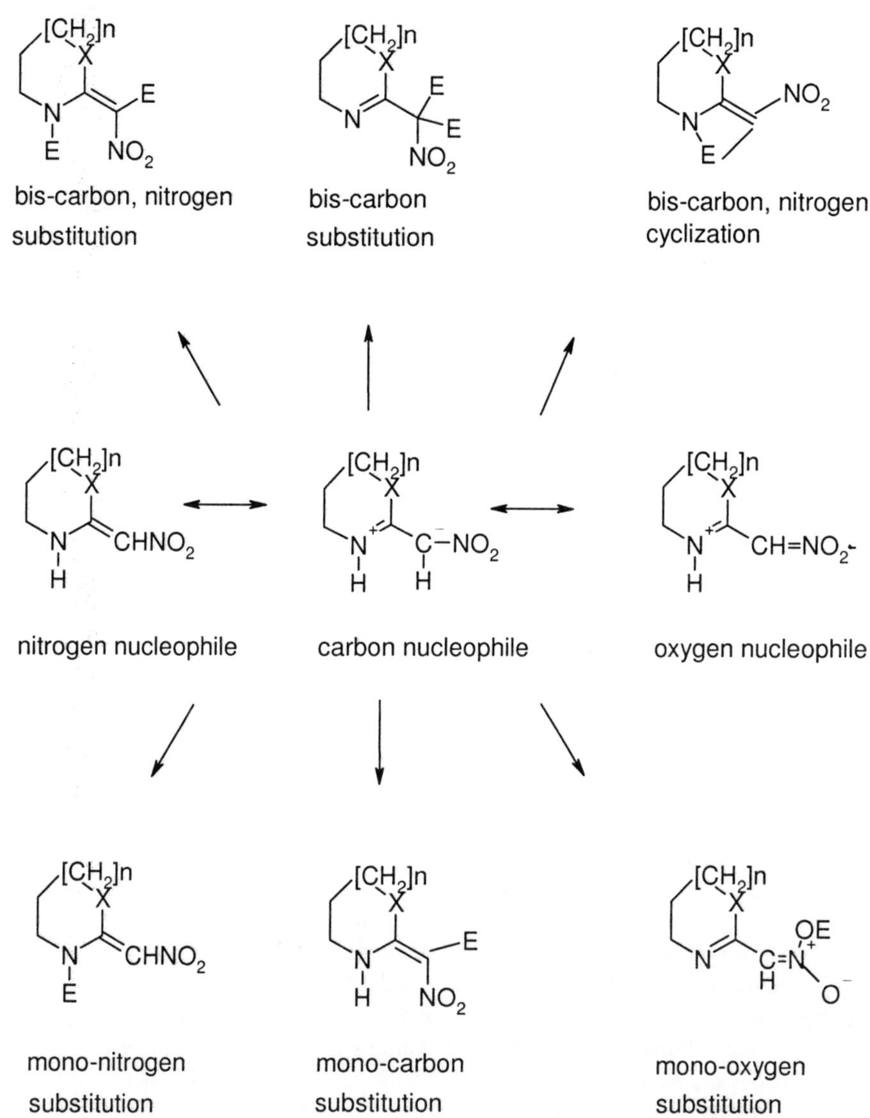

carbon or complete substitution at both carbon and nitrogen could be realized, depending upon the particular electrophile being used. Cyclization between carbon and nitrogen was also observed with appropriate bidentate reagents such as in the prolific Mannich reaction. To facilitate substitution at the ring nitrogen, it was often convenient to first form a sodium salt by treatment with sodium hydride in a polar solvent such as dimethylformamide or hexamethylphosphoramide and then to add the electrophile of choice, for example, an alkyl halide. A smattering of these many possibilities is shown generically in Scheme 2.

Although space does not permit a full delineation of the details of these hundreds of compounds, it suffices to say that many were active, but none had a suitable balance of activity and field persistence to qualify as a serious commercialization candidate.

4 Mode of Action Studies

As all this chemistry was unfolding, a major effort was being made to understand the mode of action of these new compounds. The extremely rapid and potent knockdown activity, and in some cases extraordinary levels of synergism, stimulated a great interest in how these compounds were achieving such effects. Because of its high neurochemical activity and early availability, compound **8** (see Table 1) became the research tool in the first significant studies on the mode of action in this new class (Schroeder and Flattum 1984). This work revealed that the NMHs represented not only a new structural class of insecticides but also a new mode of action. Specifically, it appeared that the NMHs were acting as cholinergic ligands at the postsynaptic receptor, and that this target site differed sufficiently between insects and other life forms to account for the wide margin of safety associated with the NMHs. Additional electrophysiological studies involving nithiazine and its N-formyl derivative were also carried out at Shell's Sittingbourne Research Center in England. The results confirmed that these newer NMHs were also cholinomimetics of exceptional potency and rapid action (Harris et al. 1986).

At the conceptual level, examination of Dreiding models confirmed that the cis-nitroenamine moiety, although not an atom for atom analog, could be viewed as a stereoelectronic mimic of acetylcholine (ACh), especially when the latter is in its preferred conformation in aqueous solution (Culvenor and Ham 1966; Cassidei and Sciacovelli 1981). Within a few tenths of an angstrom, the partially charged amine nitrogen of the nitroenamine maps to the fully charged nitrogen of ACh and the two nitro oxygens map to the carbonyl and ester oxygens of ACh (unpublished results, W.D. Kollmeyer, 1976). Unfortunately, this post facto insight did not lead to the design of any effective nitro group replacements. However, others had also used a stereoelectronic analogy to explain the neurochemical activity of a wide range of cholinergic ligands (Chothia 1970; Beers and Reich 1970; Baker et al 1971).

5 Nithiazine Fly Trap

It was perhaps ironic that Shell had at that time a viable business selling what was known as the No Pest Strip®. This consisted of a polymeric matrix plasticized with a vinyl phosphate known as Vapona Insecticide®, which was slowly released as a vapor. This simple device was very good at controlling houseflies and mosquitoes in confined areas where the insect continually took up the gaseous toxicant until a lethal dose was reached. So it was with great excitement that an unusual property of nithiazine was first observed. It, and to a lesser extent many of the earlier NMHs, caused extremely rapid knockdown when administered to houseflies by spray or injection methods. This prompted a series of experiments in a Peet-Grady chamber attempting to intoxicate flies in the manner of the No Pest Strip®, that is, by vapor application. However, when the exceptionally low vapor pressures of the NMHs became evident (nithiazine $\approx 4 \times 10^{-7}$ mm Hg), their relatively poor performance in this type of application was understandable. Even though aerosols showed some activity, spray deposits developed a yellow stain and were thus unacceptable to consumers.

However, it was observed that houseflies were almost instantaneously incapacitated after ingesting nithiazine. This led to a nithiazine/muscalure-doped sugar scatterbait that showed outstanding activity in this regard (unpublished results, J.P. Foster and W.T. Reed, 1973). A further improvement came from using water as an attractant in the dry California climate (unpublished results, R.A. Corey, 1973). For a practical demonstration, a simple wicking device was constructed and set up at the farm of one of the principal investigators who raised horses as a hobby (unpublished results, S.A. Roman, 1974). The results were astonishing - thousands of dead flies piled up around the device and the entomologists kept score by measuring the volume of dead flies rather than by counting them individually. Because of these promising results, device designs and compositions were patented (Foster and Reed 1982).

Unfortunately from a perverse marketing point of view the very efficacy of the nithiazine "fly trap" proved to be a liability for home and garden use. Because the original No Pest Strip® was predominantly sold for use as a household product, the nithiazine fly trap was viewed as a potential replacement. But, due to the rather slow speed of action of the No Pest Strip® , dead flies were widely and unobviously distributed throughout the room affected by the vapor-emitting device. In contrast, the nithiazine fly trap presented the user, assumed to be the housewife, with an unsightly pile of dead flies, and this was considered to be a definite liability from the consumer's point of view. In any event, the nithiazine fly trap never progressed to commercialization while in Shell's hands.

However, later work at DuPont led to new and effective trap designs for confined animal areas (Foster 1990; Foster and Jennings 1991, 1992), and an EPA registration for nithiazine was obtained for non-food use in traps. This paved the way for a nithiazine containing device that is currently marketed for control of flies in poultry and animal husbandry outlets under the brand of Quickstrike® fly-abatement strips (see the Internet at http://www.novartis.com and links therein to agribusiness and animal health products).

6 Summary

Agrochemical screening of α-nitroalkyl heterocycles from a university source revealed a low level of insecticidal activity for 2-(dibromonitromethyl)-3-methylpyridine. Chemical and biological exploration of this low-level lead resulted in the discovery of a new class of insecticides, the nitromethylene heterocycles (NMH). The most interesting members consisted of 2-(nitromethylene)-1-azacycloalkanes with an optional heteroatom at the 3-position as in nithiazine. These new materials typically caused extremely rapid knockdown of susceptible insects, and neurochemical studies indicated that they were acting as cholinergic ligands at the postsynaptic acetylcholine receptor. Unfortunately, the reactivity characteristics of the nitroenamine moiety caused most NMHs to have a very short lifetime under field conditions and in some cases poor storage stability. As a result, no NMH discovered by Shell was ever developed for crop protection purposes. However, the extraordinarily rapid action and potency of nithiazine against houseflies ultimately led to the development of a device for control of flies in poultry and animal husbandry outlets.

Acknowledgement

Finally, in a recounting of this sort, we would be remiss not to acknowledge all the principal investigators who contributed to the massive NMH undertaking. They are listed here along with their Shell affiliation, where BSRC denotes Shell Development Company, Biological Sciences Research Center, Modesto, California and SRC denotes Shell Research Ltd., Sittingbourne Research Center, Sittingbourne, Kent, England:

Principal Investigators - Biology: BSRC - R.A. Corey, C.A. Horne, R.F. Flattum, M.E. Schroeder, W.T. Reed; SRC - J.S. Badmin, J.F. Donallson, R.J. Dowson, T.E. May, J. Robinson, R.N. Price.

Principal Investigators - Chemical Synthesis: BSRC - H. Estreicher, A.C. Henry, W.D. Kollmeyer, W.M. Padgett, G.B. Payne, J.E. Powell, S.A. Roman, S.B. Soloway, C.H. Tieman; SRC - M. Anderson, M.J. Bull, J.H. Davies, R.H. Davis, M. Harris, R. Pettman, D.A. Wood, J. Wood.

Principal Investigators - Environmental Fate, Formulations and Physical Chemistry: BSRC - T. Brownscombe, T.K. Brunck, C.C. Chen, J. Cotton, J.P.Foster, D.A. Kleier, J. Morales, A.Y.S. Yang; SRC - T.G. Grayson.

References

Baker RW, Chothia CH, Pauling P, Petcher T (1971) Structure and activity of muscarinic stimulants. Nature 230:439-445.

Beers WH, Reich E (1970) Structure and activity of acetylcholine. Nature 228:917-922.

Bretherick L (1990) Bretherick's handbook of reactive chemical hazards, fourth edition. Butterworths, London , pp 161-166.

Cassidei L, Sciacovelli O (1981) Conformational analysis of the C(6)-O(1)-C(5)-C(4) fragment in acetylcholine by ^{13}C-NMR spectroscopy. J Am Chem Soc 103:933-934.

Chothia C (1970) Interaction of acetylcholine with different cholinergic nerve receptors. Nature 225:36-38.

Culvenor CCJ, Ham NS (1966) The proton magnetic resonance spectrum and conformation of acetylcholine. Chem Commun 15:537-539.

Diehr HJ, Gallenkamp B, Jelich K, Lantzsch R, Shiokawa K (1991) Synthesis and chemical-physical properties of the insecticide imidacloprid (NTN 33893). Pflanzenschutz-Nachrichten Bayer 44/2:107-112.

Elbert A, Becker B, Hartwig J, Erdelen C (1991) Imidacloprid - a new systemic insecticide. Pflanzenschutz-Nachrichten Bayer 44/2:113-136.

Etienne A, Correia Y (22 October 1969) Derivatives of 1-pyrroline. British Patent 1,167,809; Chemical Abstracts 72:31602p.

Feuer H, Lawrence JP (1969) The alkyl nitrate nitration of active methylene compounds. VI. A new synthesis of α-nitroalkyl heterocyclics. J Am Chem Soc 91:1856-1857.

Foster JP, Reed WT (19 January 1982) Device for combating flies. United States Patent 4,310,985.

Foster JP (20 March 1990) Device for killing arthropods. United States Patent 4,908,977.

Foster JP, Jennings PV (10 September 1991) Device for killing insects. United States Patent 5,046,280.

Foster JP, Jennings PV (29 September 1992) Device for killing insects. United States Patent 5,150,541.

Fratmann SA (15 October 1971) Process for the preparation of 1-alkyl-2-(aminomethyl)pyrrolidines. Belgian Patent 773,979.

Freund E (1919) Uber die einwirkung von schwefelkohlenstoff auf nitro-methan. Chem Ber 52:542-544.

Gompper R, Schaefer H (1967) Beitrage zur chemie der dithiocarbonsaureester und ketenmercaptale. Chem Ber 100:591-604.

Harris M, Price RN, Robinson J, May TE (1986) WL108477 - a novel neurotoxic insecticide. Proceedings of the 1986 British Crop Protection Conference. BCPC Publications, Surrey, UK, Vol 1, pp 115-121.

Henry AC (24 October 1974) Insecticidal 1-alkyl-2-(aci-nitromethyl)pyridinium compounds. German Patent 2,416,350; Chemical Abstracts 82:31263k.

Hirai K, Matsuda H, Kishida Y (1971) Syntheses and reactions of 2-substituted thiazolidines. Chem Pharm Bull 20:97-101.

Isaksson G, Sandstrom J (1973) Studies of polarized ethylenes. Part VI. Internal rotations, dipole moments, and ultraviolet spectra of nitroethylenes. Experimental results and PPP calculations. Acta Chem Scand 27:1183-1191.

Johnson ER, Reed WT, Tieman CH, Soloway SB (20 August 1974) Pyridine insecticides. US Patent 3,830,921; Chemical Abstracts 82:12301.

Kleier D, Holden I, Casida JE, Ruzo LO (1985) Novel photoreactions of an insecticidal nitromethylene heterocycle. J Agric Food Chem 33:998-1000.

Petersen S (15 January 1953) Process for the preparation of condensation products. German Patent 863,056.

Powell JE (27 December 1977) Tetrahydro-2-(nitromethylene)-2H-1,3-thiazine insect control agents. United States Patent 4,065,560.

Powell JE (1990) Chemistry of nitromethylene heterocycles. The 6th assembly for pesticide design research. Pesticide Science Society of Japan, pp 6-7.

Roman SA (15 November 1973a) Insecticidal 2-(nitromethylene)piperidines. German Patent 2,321,523; Chemical Abstracts 80:27110.

Roman SA (15 November 1973b) Insecticidal 2-(nitromethylene)pyrrolidines. German Patent 2,321,522; Chemical Abstracts 80:36988.

Roman SA (23 September 1975) Tetrahydro-2-(nitromethylene)-2H-1,3-oxazines. United States Patent 3,907,790; Chemical Abstracts 83:1975.

Roman SA (8 June 1976) Esters of nitro(tetrahydro-2H-1,3-thiazin-2-ylidene)acetic acids. US Patent 3,962,234; Chemical Abstracts 85:123955.

Schroeder ME, Flattum RF (1984) The mode of action and neurotoxic properties of the nitromethylene heterocycle insecticides. Pestic Biochem Physiol 22:148-160.

Soloway SB, Henry AC, Kollmeyer WD, Padgett WM, Powell JE, Roman SA, Tieman CH, Corey RA, Horne CA (1979) Nitromethylene heterocycles as insecticides. In: Geissbuehler H (Ed) Advances in Pesticide Science, 4th International Congress of Pesticide Chemistry. Pergamon, Oxford, Vol 2, pp 206-217.

Tieman CH, Kollmeyer WD and Roman SA (7 May 1975) 2-(Nitromethylene)-1,3-diazacycloalkanes. German Patent 2,445,421; Chemical Abstracts 83:97297.

Tieman CH, Kollmeyer WD and Roman SA (13 July 1976a) 2-(Nitromethylene)-hexahydropyrimidines. United States Patent 3,969,354; Chemical Abstracts 85:192727.

Tieman CH, Kollmeyer WD and Roman SA (27 July 1976b) 2-(Nitromethylene)-1,3-diazepines. United States Patent 3,971,774; Chemical Abstracts 86:5460.

4 Discovery of Chloronicotinyl Insecticides

Shinzo Kagabu
Department of Chemistry, Faculty of Education, Gifu University,
Yanagido 1-1, Gifu 501-1193, Japan

1 Introduction

The insecticide market has long been dominated, more than 80%, by organophosphates, carbamates, and synthetic pyrethroids. The consequence is the development of resistant strains to these three insecticide classes. For effective crop protection, development of a new insecticide class with a new mode of action has been urgently required. Future insecticides, as seen from the retreat of chlorinated hydrocarbons from the primary seat, are required to have not only high insecticidal potentials, but also low toxicity to vertebrates and no damage to the environment. Furthermore, these agents are expected to have sufficient stability under weathering conditions to reduce application times for the decreasing farmer population. The recently developed chloronicotinyl insecticides meet these requirements for the modern pesticides. This article describes how this new insecticide class was found.

2 Starting Point of the Research

Not a few insecticides have originated from natural active substances. Nicotine is an old insecticide extracted from tobacco leaves. Its action on the postsynaptic acetylcholine receptor is different from the modes of action of the present major classes. However, no major insecticidal class has been developed from this botanical insecticide, and this has been reasonably ascribed to its high mammalian toxicity and relatively low level of insecticidal activity. In view of this, Shell's new insecticide nithiazine (**1**, Table 1) caught our attention because it acts at the same receptor as nicotine even though there is no apparent structural similarity between them. More surprising and more interesting for us was that nithiazine is highly active against insects but of low toxicity to mammals (Soloway et al. 1979; Schroeder and Flattum 1984).

In 1979 we started to look at the chemistry and biology of nithiazine structure. We chose the green rice leafhopper (*Nephotettix cincticeps*), a major rice pest, as our target insect for screening and optimizing biological activity. At the beginning of the project, we devised a simple and reliable laboratory screening method for continuously observing the combined dermal and oral activity of test compounds against leafhoppers and planthoppers over 2 weeks. The procedures are done this way: an aqueous sample solution is sprayed onto young rice seedlings (freshly germinated from 25-30 seeds) placed on the cover of a plastic vessel (40 mm ϕ, 60 cm high). The ingredient solution is spread on leaves and stems, and many of the drops are pooled around the roots. After air-drying the foliar surface, 10 to 20 insect nymphs of are released into the vessel cage, and the mortality is evaluated. Since the systemic substances move continually to the upper parts of the plants from the roots, insecticidal activity through two routes can be observed simultaneously (Sone et al. 1995). We confirmed the systemic properties of nithiazine and its sufficient activity against the sucking insects by this method.

3 Structure-Activity Relationship Based on Laboratory Tests

The structure-activity relationships described in this section refer to published papers (Kagabu et al. 1992; Moriya et al. 1992; Shiokawa et al. 1992; Moriya et al. 1993a,b). Insecticidal activity was measured by exposing third-instar larvae of the organophosphates- and carbamates-resistant strains mixed with the sensitive ones to sprayed rice plants, and is expressed by the LC_{90}, the concentration in ppm to kill 90% of the insects after 4 days. By this assay, the LC_{90} value of nithiazine (**1**) was 40 ppm (Table 1).

We started our SAR study by modifying the hydrothiazine ring of nithiazine. First we constructed the hetero rings using commercially available α,ω-diaminoalkanes (Table 1). The activity of 1-benzyl-2-nitromethylene-1,3-diaminocycloalkanes depends on the ring size. As the ring size becomes larger, the activity decreases and the seven-membered ring nitromethylene compound (**5**) loses activity greatly. The open-ring compound (**6**), which corresponds to large-ring compounds in the sense of bioisosterism, is also inactive. We then had an idea that active molecules must be a ring, preferably a five- or six-membered plane ring. This idea was however a hasty conclusion. It became clear later that the real key to the activity was not the planarity of molecules, but the nature of the substituent R.

We continued the SAR study by introducing substituents into the imidazolidine ring, and recognized only benzyl-2-nitromethylene-imidazolidine (**2**) and the six-membered homologue (**4**) to possess some activity, even though the LC90 were far inferior to the lead (**1**) of 40 ppm (Table 2). The fact that substituents except benzyl did not show any activity prompted an assumption that the nitromethylene-imidazolidine itself would be responsible for the activity and that the benzyl residue contributes to the activity merely through metabolic scission.

Taking this result into account, we examined the activity of the benzyl imidazolidines where the phenyl ring is substituted (Table 3). The activity depends on the nature of the substituents and their position in the ring; the *ortho*-chloro

Table 1. Relationship between ring size and *N*-substituents and insecticidal activity ($R=CH_2C_6H_5$)

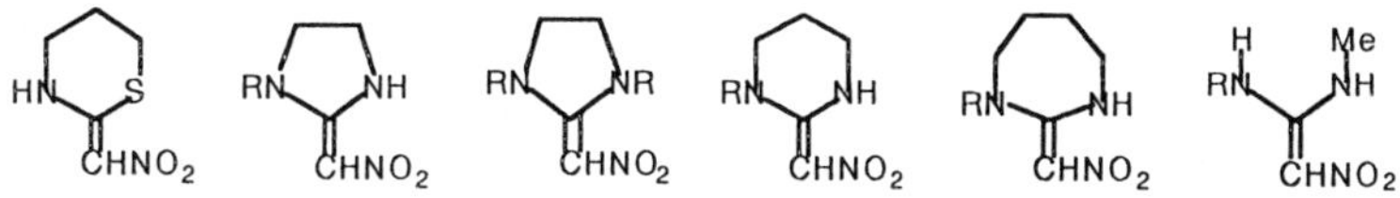

	1: nithiazine	2	3	4	5	6
LC_{90}:	40	200	200	~200	1000	Inactive

Insecticidal activity was measured by exposing third-instar larvae of organophosphate- and carbamate-resistant strains mixed with the sensitive ones of the green rice leafhopper (*Nephotettix cincticeps*) to sprayed rice plants, and expressed by LC90, the lowest concentration in ppm to kill over 90% of the insects after 4 days.

Table 2. Relationship between substituents and insecticidal activity in 2-nitromethyleneimidazolidine

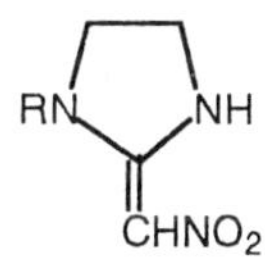

Compd.	7	8	9	10	2	11	12
R:	H	Me	cyclohexyl	Ph	$PhCH_2$	$PhCH_2CH_2$	PhCHMe
LC_{90}:	1000	1000	1000	1000	200	Inactive	Inactive

Table 3. Relationship between substituents on the benzene ring of 1-benzyl-2-nitromethyleneimidazolidine and insecticidal activity

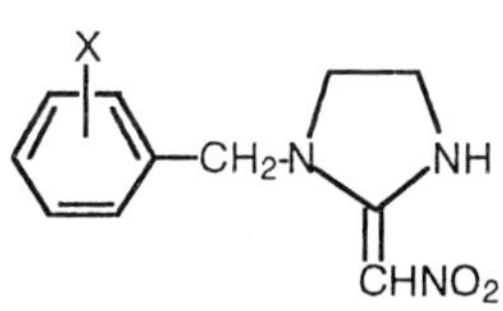

Compd.	2	13	14	15	16	17	18	19	20
X:	H	2-Cl	3-Cl	4-Cl	3-Me	4-Me	4-MeO	$4-NO_2$	4-CN
LC_{90}:	200	Inactive	200	40	1000	200	200	200	200

derivative (**13**) is utterly inactive, and chlorine in *meta*-position (**14**) leads to an LC_{90} of 200 ppm and in *para*-position (**15**) of 40 ppm. Derivatives carrying a methyl-, methoxy-, nitro-, and cyano group at this position (**17**-**20**) again are at 200 ppm. Considering and judging from these relationships, the aforementioned concept of the metabolic route must be open again. That is because it might be difficult to understand that the position and nature of the substituents in the phenyl ring should influence the metabolic displacement of the benzyl group so much. In any event it became clear that the *para*-chlorobenzyl group exerts a particularly advantageous effect on the nitromethyleneimidazolidine.

The heteroarylmethyl derivatives at the imidazolidine nitrogen suggested another aspect concerning the structure-activity relationships (Table 4). Furfuryl- (**21**), 2-thenyl- (**22**), and 2-pyrrolylmethyl (**23**) derivatives lead to the same activity level as the benzyl derivative (**2**). Based on the results that insecticidal efficacy of the nitromethylenes can be increased by these five-membered heteroaromatics at the imidazolidine nitrogen, it would be logical that not only the nitromethylene parent skeleton itself is responsible for the activity, but the molecule as the whole including the *N*-substituents must be considered.

Table 4. Heteroarylmethylnitromethylene imidazolidines and the insecticidal activity

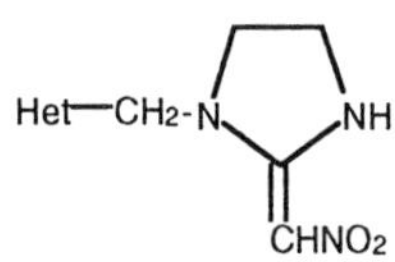

Compd.	2	21	22	23	24	
Het:	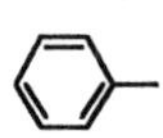	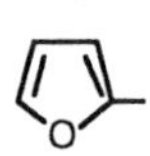	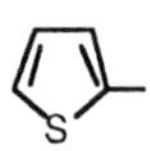	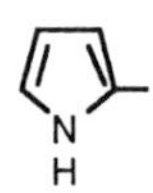	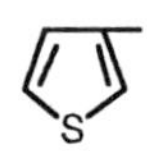	
LC_{90}:	200	200	200	200	200	
Compd.	**25**	**26**	**27**	**28**		
Het:	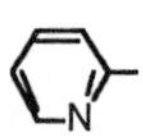	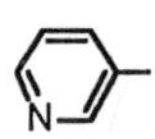	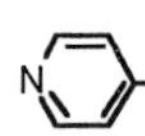	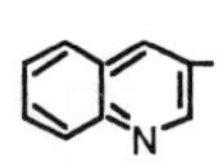		
LC_{90}:	200	8	8	Inactive		
Compd.	**29**	**30**	**31**	**32**	**33**	**34**
Het:	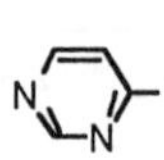	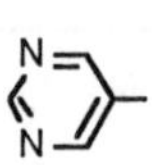	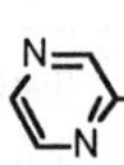	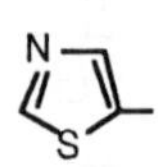	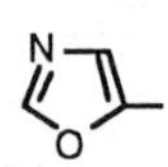	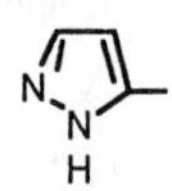
LC_{90}:	40	8	8	8	40	20

Until this stage we were not so optimistic as to expect to develop a new insecticide from this class, thinking of the modest insecticidal potencies of the compounds prepared. However, the situation changed in the next study with pyridylmethyl (picolyl)series, where we could find a clue of the following strategy. These were compounds, 3-picolyl (nicotinyl **26**) and 4-picolyl (**27**) nitromethylenes, whose activity is better than the benzyl prototype (**2**) by a factor as great as 25. The activity of the compounds of this class depends greatly on the position of the nitrogen atom in the ring, as seen by the 2-picolyl derivative (**25**) showing a LC_{90} of 200 ppm at the same level as that of **2**. This positional selectivity was also corroborated in the diazine (**29**-**31**) and the azole rings (**32**-**34**) with two heteroatoms, showing the kind of influence the position of the nitrogen atoms exerts on the activity: nitrogen in *meta*-position increases the activity, nitrogen in *ortho*-position decreases it, and the annellated pyridine system, quinoline (**28**), is totally inactive. Thus, we arrived at a compounds of LC_{90} of 8 ppm, five times better than our lead compound.

In parallel to these studies, we had examined the activity of substituted picolyl derivatives, and found an exceptionally active compound (Table 5), 6-chloronicotinyl-2-nitromethylene imdazolidine (**35**). The LC_{90} was as low as less than 0.32 ppm, more than 125 times better than **1**, and this was counted as one of the most active insecticides known so far. Further, the insecticidal spectrum was considerably widened by introducing a 6-chloronicotinyl group. Compound **35** controls not only the hemiptera species but also thisanoptera, and some of the lepidoptera, coleoptera, diptera, and isoptera species.

Introduction of a chlorine atom to the other positions on the pyridine ring afforded no such activity enhancement (**36**, **37**). Other substituents in this specific ring position (**39**-**45**) did not induce such a remarkable effect either, excepting fluorine (**38**).

Now we extended the SAR study to the heteroaromatic rings with two hetero atoms (Table 6). Here also the rule from the pyridine series was confirmed. As seen in compound **46**, chlorinating in the *quasi-para* position in thiazol (**32**) increases the activity 25 times and the introduction of a methyl group 5 times (**47**). Oxazol (**48**), pyrazol (**49**), and isoxazol derivatives (**50**, **51**) with the nitrogen at *quasi meta*-position all show high activity. This rule holds also for the diazine ring series with pyridazines, pyrimidines and pyrazines (**52**-**56**). Also, the false position of the ring nitrogen and the false sites of substituents decrease the activity drastically, as *ortho*-methyl pyrazine (**53**),. the chloro atom exerts generally a more favorable influence than the methyl group. Of the examined compounds the chlorothiazolyl (**46**) and chloropyrimidinyl derivatives (**56**) showed the highest activity, although the latter is yet too unstable.

Table 5. Substituent variation of the pyridine ring of 1-nicotinyl-2-nitromethylene-imidazolidine and insecticidal activity

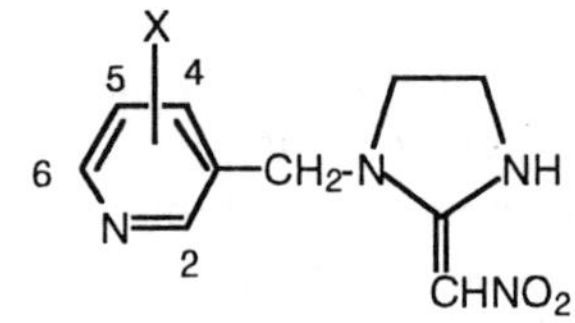

Compd.	**26**	**35**	**36**	**37**	**38**	**39**
X:	H	6-Cl	5-Cl	2-Cl	6-F	6-Br
LC_{90}:	8	0.32	40	200	0.32	1.6
Compd.	**40**	**41**	**42**	**43**	**44**	**45**
X:	6-Me	6-CF_3	6-CN	6-MeS	6-MeO	6-AcNH
LC_{90}:	1.6	8	200	1000	200	1000

Table 6. Chloro- and methyl-substituted heteroarylmethyl nitromethyleneimidazolidines and the insecticidal activity

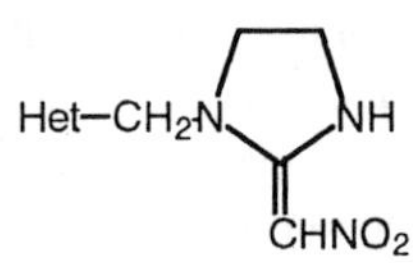

Compd.	**46**	**47**	**48**	**49**	**50**	**51**
Het:	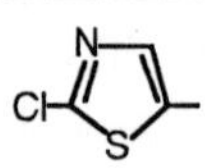	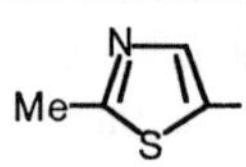	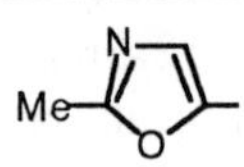	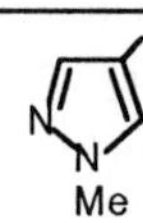	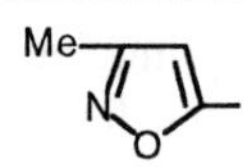	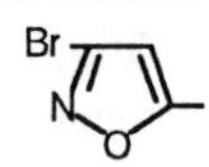
LC_{90}:	0.32	1.6	1.6	1.6	1.6	1.6
Compd.	**52**	**53**	**54**	**55**	**56**	
Het:	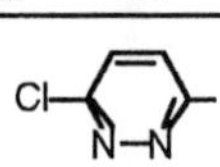	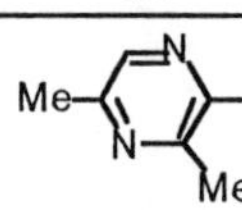	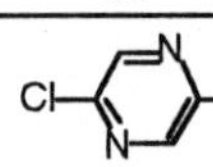	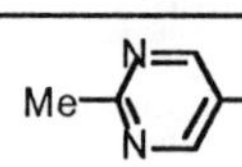	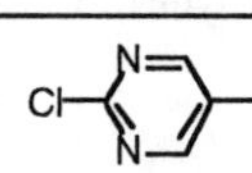	
LC_{90}:	40	200	1.6	1.6	0.32	

Table 7. Variation of chain length of the space between both rings

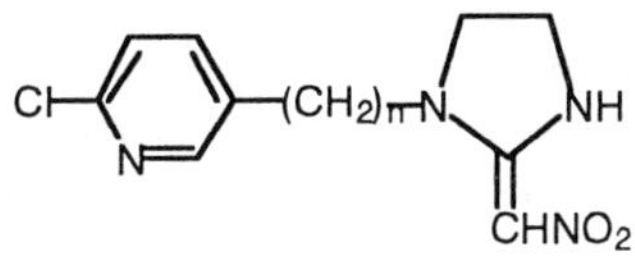

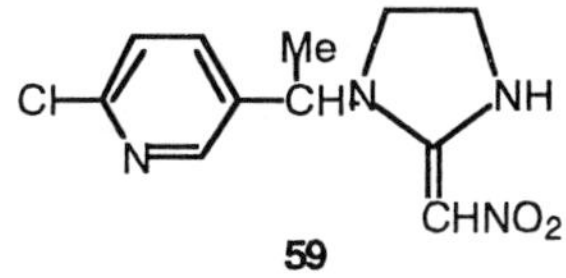

59

Compd.	**57**	**35**	**58**	**59**
n :	0	1	2	1
LC_{90} :	8	0.32	40	1.6

Table 8. Variation of imidazolidine ring

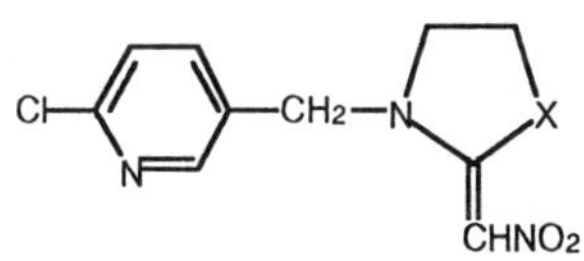

Compd.	**35**	**60**	**61**	**62**	**63**
X:	NH	NMe	S	CH_2	O
LC_{90}:	0.32	8	0.32	1.6	8

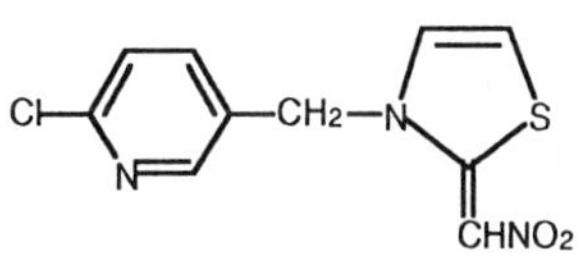

64, LC_{90}: 0.32

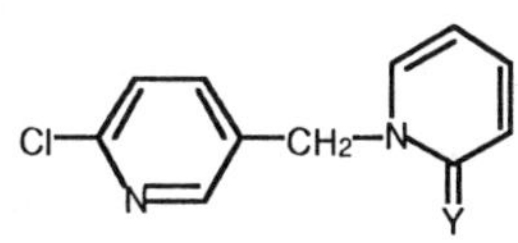

65 (Y=NCN), LC_{90}: 0.32
66 (Y=NNO_2), LC_{90}: 0.32

As we see in Table 7, concerning the joint length between the pyridine ring and the imidazolidine moiety, methylene (*n*=1, **35**) is the best spacer. With a longer or shorter chain length, the activity is lowered significantly, while methylation at the joint reduces the activity only slightly (**59**). If we associate this with the positional specificity of the nitrogen atom of the pyridyl ring, we could assume that the distance between the pyridyl and the imidazolidine nitrogen atoms should be important for the activity. This means the 1,5-atom length should be the suitable distance to the recognition site on the receptor.

Modification of the central imidazolidine ring gave us another aspect as for the SAR (Table 8). In place of the imidazolidine ring, hetero five-membered rings impart a high level of activity, and among them the activity of the thiazolidine ring (**61**) is pronounced. The level is comparable to the imidazolidine derivative. It is noteworthy that this thiazolidine derivative shows fairly highly insecticidal efficacy also against lepidoptera species, in contrast to the prepared other heterocyclic compounds, which have generally rather weak activity against this kind. Another remarkable SAR is that even if a double bond was introduced into the central heterocyclic ring (like **64**), the insecticidal efficacies remained at a high level. This phenomenon was also observed in full-conjugated pyridone derivatives (**65** and **66**). This suggests that for the central ring system not the electronic effect of the ring system but the exact arrangement of the ring atoms seems to govern the activity.

As for the functional group Y on Table 9, we had long believed the nitromethylene to be the best since we started with nithiazine as our lead compound. In fact, insufficient activity of urea (**67**), guanidine (**68**), as well as cyanomethylene (**69**) and ethoxycarbonylmethylene (**70**) substituted compounds agreed with this prediction. The dicyanomethylene (**71**) had an activity of LC_{90} of 40 ppm at most. However, unexpectedly cyanoimine (**72**) was highly insecticidal at 0.32 ppm. Other imino derivatives like carbamate (**73**) and sulfonylimine (**74**) showed again essentially weaker activity. We continued the SAR study with the functional groups and finally discovered a nitroimino derivative, as imidacloprid (**75**), with the highest activity (Shiokawa et al. 1986).

The study with the derivatives of imidacloprid, where another imidazolidine nitrogen is substituted, provided us information that in general the longer alkyl chain reduces the insecticidal activity as seen in the compounds in Table 10. There is however an exception: the linear chain derivatives with 3 C atoms like propyl (**78**), allyl (**80**), and propargyl (**81**) show interestingly the same level of efficacy as the H-compound (**75**). Admitting the activity tends to decrease to some degree by alkyl substitution in the imidazolidine ring, it should be mentioned that the nitroimino derivatives hold still sufficient potency, in contrast to the corresponding nitromethylene derivatives where the activity decreases greatly by alkylation (not shown). This suggests that nitroimino group must have inherently a specific activating power in this system. It is self-evident that acetyl, benzoyl, and cyano derivatives (**84-86**) exert activity through the metabolic dissociation of the substituents.

Table 9. Variation of functional group on the imidazolidine ring and the insecticidal activity

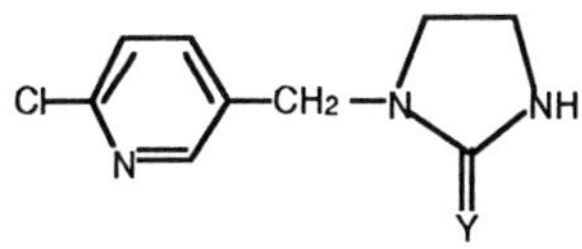

Compd.	**67**	**68**	**69**	**70**	**71**
Y:	O	NH	CHCN	$CHCO_2Et$	$C(CN)_2$
LC_{90}:	1000	1000	Inactive	Inactive	40
Compd.	**72**	**73**	**74**	**75** (imidacloprid)	
Y:	NCN	NCO_2R	NSO_2R	NNO_2	
LC_{90}:	0.32	40	40	0.32	

Table 10. Introduction of substituents to the imidazolidine ring and insecticidal activity

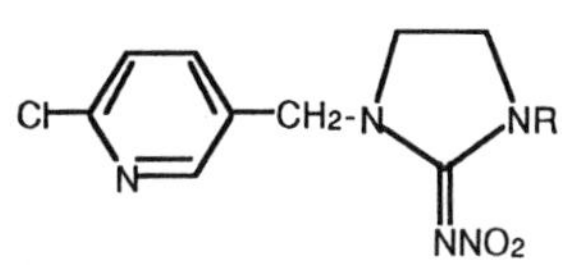

Compd.	**75**	**76**	**77**	**78**	**79**	**80**
R:	H	CH_3	CH_2CH_3	n-C_3H_7	$CH(CH_3)_2$	$CH_2CH=CH_2$
LC_{90}:	0.32	1.6	8	0.32	40	0.32
Compd.	**81**	**82**	**83**	**84**	**85**	**86**
R:	$CH_2C\equiv CH$	n-C_4H_9	n-C_5H_{11}	$COCH_3$	COC_6H_5	CN
LC_{90}:	0.32	1.6	40	0.32	0.32	0.32

4 Comparison of Physicochemical Properties of Imidacloprid and Related Compounds Under Practical Conditions

About 2000 compounds were prepared in the laboratory during the project, and dozens of these compounds possessed high insecticidal activity in laboratory tests. We then narrowed down to approximately ten compounds for further intensive tests under practical conditions and finally selected imidacloprid for commercial development, considering the field results as well as taking the toxicological and environmental aspects into account.

Laboratory and greenhouse results are not always confirmed in the field. Stable reliable activity under weathering conditions is an important criterion for a practical insecticide. Pesticides exposed to nature forces are deactivated by several factors. The representative ones are photolysis, hydrolysis and washing away by rain, vaporization, oxidation by air, carboxylation by carbon dioxide, degradation by soil microorganisms, and conjugation by plant tissue components.

First we tested stability toward oxidation and found that chloronicotinyl compounds tested were fairly stable in an oxygen-saturated aqueous solution (Kagabu and Medej 1995). Examination of their stability towards hydrolysis showed that chloronicotinyls tested and nithiazine are quite stable in neutral regions, *i.e.*, under physiological conditions, although these compounds tend to be decomposed in an alkaline milieu (Kagabu and Medej 1995).

Potential field candidates were then studied under photolysis conditions using a simulated sunlamp. Their half-lives are given in Table 11 together with their electronic absorptions, maximum absorptions, and absorptions at 290 nm. There are remarkable differences among these compounds. Nitromethylene compounds, **35**, **78**, **61**, and **1**, independent of the heterocyclic structure and the ring size, decompose rapidly under the sunlamp, whereas the half-lives of the nitroimines, compounds **71**, **77**, and **76**, and of cyanoimine **69** are significantly longer. Due to absorption in the upper atmosphere, only sunlight of wavelengths between 290 and 400 nm is relevant at ground level. The nitromethylene chromophore absorbs strongly in this range and rapidly decomposes, losing intrinsic activity in the field. On the other hand, nitroimines hold sufficient efficacy for a properly long time, and will decompose gradually in the environment owing to the significant end-tailing absorption at 290 nm. Cyanoimines like **69**, having an essentially shorter wavelength absorption are expected to have greater photostability (Kagabu and Medej 1995).

The difference in photostability between nitromethylene and nitroimine compounds was considered in the light of quantum chemistry. The AM1(Austin Model 1) and *ab initio* calculations were performed for the *N*-methyl derivatives (abbreviated respectively as 1-Me-$CHNO_2$ and 1-Me-NNO_2; Figure 1) instead of the compounds **35** and **75** for calculation economy. The frontier orbital maps displayed that the HOMOs (Highest Occupied Molecular Orbitals) reside mostly in the amidino and guanidino parts and the LUMOs (Lowest Unoccupied Molecular

Table 11. Electronic absorption of chloronicotinyl compounds in water and the half-life times ($t_{1/2}$) by photolysis

Compd.	R: 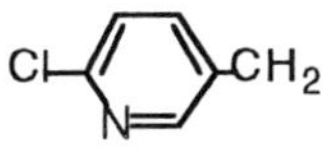	λ_{max} (nm)/log ε (log ε at 290 nm)	$t_{1/2}$ (hr)
71	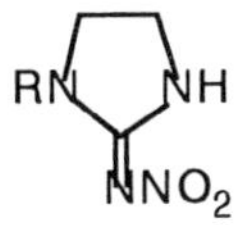	269 / 4.17 (3.73)	3
77	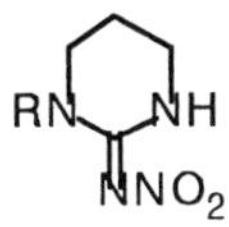	269 / 3.61 (3.21)	3.5
76	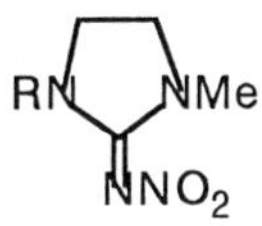	255 / 4.10 (2.85)	3.5
35	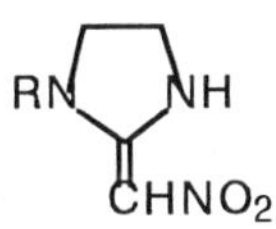	323 / 4.09	1
78	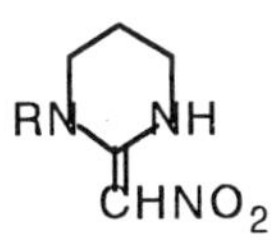	313 / 4.17	2
61	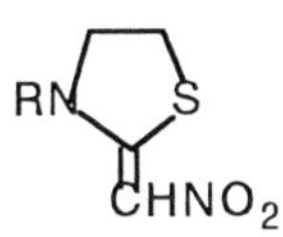	357 / 4.23	<0.5
69	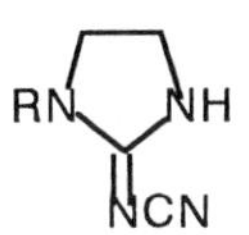	268 / 3.61 (< 0.01)	>24
1	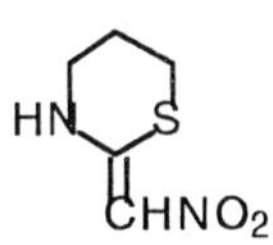	348 / 4.07	0.5

The $t_{1/2}$ data were obtained by irradiation of the compound in acetonitrile-water (8:2, v/v) solution by a 250-W sun lamp at 30°C.

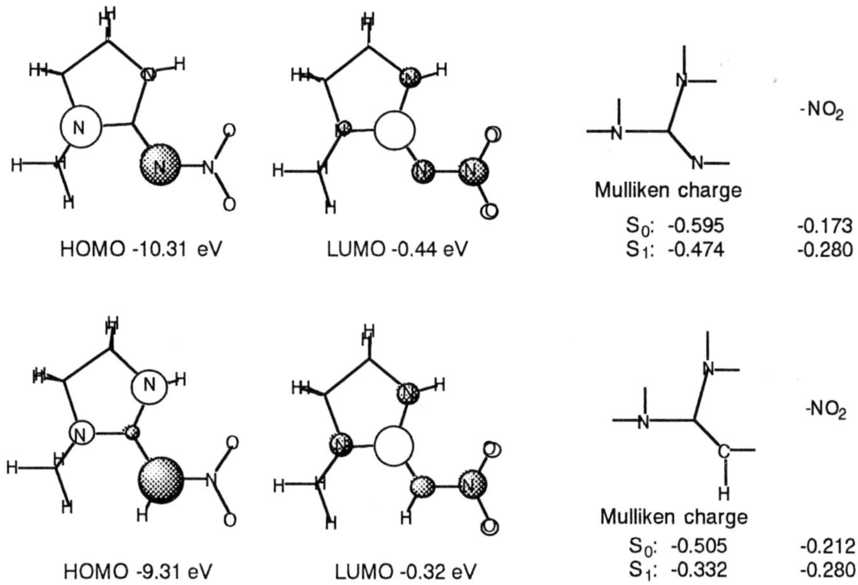

Fig. 1. Frontier orbitals by AM1 method and the Mulliken charge of nitroimino- and nitromethylene-imidazolidines, 1-Me-NNO_2 (upper) and 1-Me-$CHNO_2$ (lower), respectively.

Orbitals) stretch toward the nitro groups. The primary excitation by UV absorption is classified as the intramolecular donor-acceptor electron transfer from the Mulliken charge differences between the S_0 and S_1 states of the both groups. The energy gaps from the ground state to the singly excited state is calculated to be 8.99 eV and 9.87 eV for nitromethylene and nitroimine compounds, respectively. Interestingly, the nitro group of 1-Me-$CHNO_2$ in the triplet excited states becomes twisted from the amidine plane; on the other hand in 1-Me-NNO_2 the distortion degree is so small that resonance stability can still be expected. The AM1 and *ab initio* calculations ground for the better photostability of the nitroimino derivatives than nitromethylene compounds by the greater excitation energy for the former and the still expected bonding nature in the double bonds of C=NNO_2 and C=NCN in the excited states (Kagabu and Akagi 1997).

Lipophilicity and hydrophilicity of insecticidal ingredients are connected with the adsorption of molecules on leaves and soil, systemicity in plants, permeability into and movement within the insect body, and accumulation in internal organs of animals. The water solubility and the P_{ow} of imidacloprid and related compounds are given in Table 12. Chloronicotinyl insecticides are push-pull olefins made up of

Table 12. Partition coefficients (log Pow) and water solubility of chloronicotinyl compounds

Compd.		log Pow	Water solubility (g/L)
1[a]	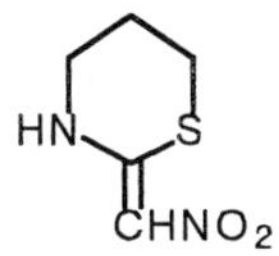	-0.4	200 (23 °C)
35	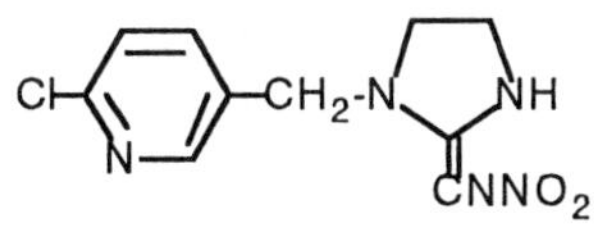	-0.19	3.4 (20 °C)
71 (Imidacloprid)	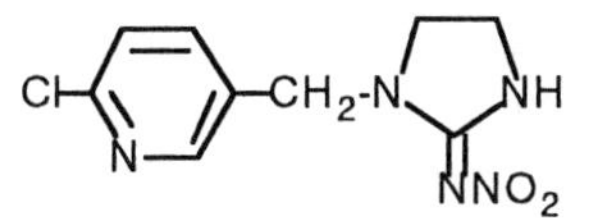	0.57	0.61 (20 °C)
61	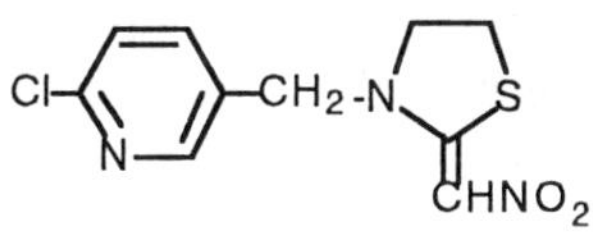	0.65	0.046 (20 °C)
79[b] (Nitenpyram)	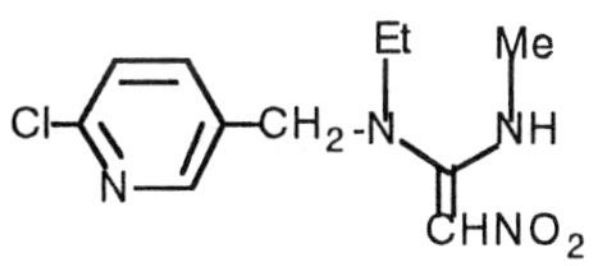	-0.64	840 (20 °C)
80[c] (Acetamiprid)	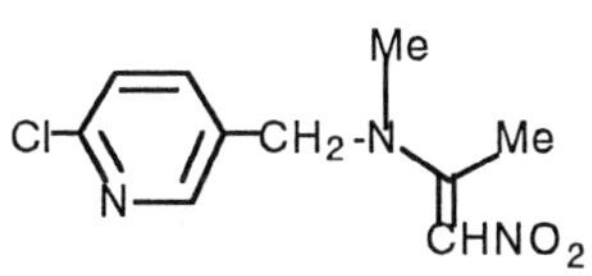	0.8	4.2 (25 °C)
Fenitrothion[d]		3.44	21 (20 °C)
Cypermethrin[d]		6.6	0.004
Nicotine[e]		1.17	∞

[a] Soloway et al. 1979.
[b] Kashiwada 1996.
[c] Matsuda and Takahashi 1996.
[d] Tomlin 1995.
[e] Hansch and Elkins 1971.

conjugated electron-donating and -accepting parts, and such polar molecules have as a whole great water solubility and eventually low P_{ow} values compared with other nonpolar insecticidal classes. We have studied the relationships between the structure and lipophilicity of a series of chloronicotinyl insecticides based on the log k values on HPLC. A general tendency is as follows: first, chain molecules are less lipophilic than the corresponding cyclic molecules; second, concerning the functional group of Y, the water solubility is in the order of $CHNO_2$ > NCN > NNO_2; and third, concerning the atom of Z, the lipophilic order is S > C > O > NH (refer to the structures in Figure 2) (Kagabu 1996).

The hydrophilicity is well correlated with the systemicity of the compounds, and also the hydrophilic properties of a molecule will contribute greatly to low fish toxicity and low bioaccumulation. However, excessive water solubility often has adverse effects on the practical application in some cases. For water-soluble insecticides, water-surface application to a paddy field will be limited and seed-coating treatments are rarely applied because constant release of a specified quantity of ingredient is often difficult. Imidacloprid owes its good systemicity and low fish toxicity to the proper lipophilicity/hydrophilicity. Imidacloprid is quite stable to air oxidation, fairly stable under hydrolytic conditions, and sufficiently stable under sunlight. Such favorable properties connected with the environmental conditions certainly contribute to its highly estimated utility in nursery-box treatment and seed-coating applications (Ishii et al. 1994; Kagabu and Medej 1995).

In looking back now at the chain of events that led us to the new insecticide, finding the chloronicotinyl group was the literally the key to the new domain. In fact, compounds developed from this class by ourselves and other groups following the discovery of imidacloprid carry this activating group or the bioisosteric groups. From this background, we are calling the new class of insecticides chloronicotinyl insecticides. The general structural requirements for chloronicotinyl insecticides are summarized in Figure 2. The basic molecule is made up of a chloro-substituted heteroaromatic ring (A) and a saturated nitrogen hetero ring (B) jointed at the 1 position of a five- or six-membered heteroaromatic ring. Concerning ring (A): (1) a chlorine atom should be in the 4- position, fluorine atom and methyl group may be also possible; (2) the 2-position of the ring should be unsubstituted; and (3) the methylene is the best spacer. Concerning ring (B): (1) the preferred ring size is five- or six membered, and open-ring structures composed of the corresponding atom alignment are also possible; (2) the ring atom Z should be NH, NR or S, and may be CH_2. The functional group Y is strongly electron withdrawing and has a hydrogen-accepting head like NNO_2, $CHNO_2$, or NCN (Kagabu and Matsuno 1997).

The discoveries of the chloronicotinyl class of insecticides represent important advances in technology for insect control. The attributes of the high levels of insecticidal activity, application flexibility, excellent crop tolerance, and low levels of toxicity to mammals exhibited by chloronicotinyls are important characteristics for modern agrochemicals. The performance characteristics of these materials are linked partly but strongly to the unique mechanism of action of the selective agonistic binding to the postsynaptic ACh receptor of insects and the appropriate systemic properties and to environmental stability.

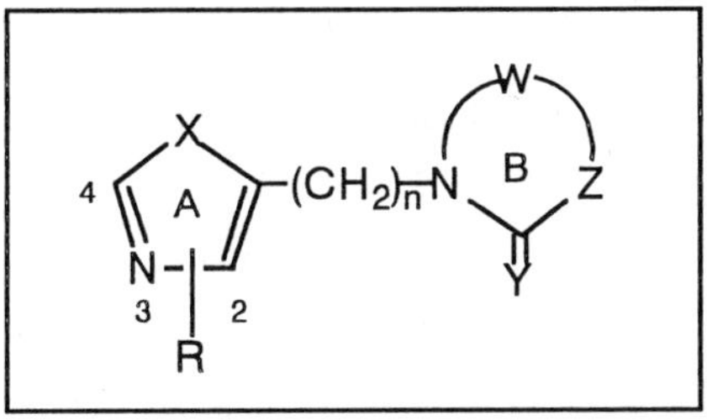

A ring
R: 4-Cl >> 2-Cl, Cl > Me >> OMe > H
N position: 3-N > 4-N >> 2-N
X: CH=CH, CH=N, O, S
Joint
n: 1 >> 0 > 2
B ring
W: CH_2CH_2, CH=CH, $CH_2(Z)CH_2$ or open-ring
Y: NNO_2, $CHNO_2$, NCN
Z: NH, NR, S, O, CH_2

Fig. 2. Structural requirements for chloronicotinyl insecticides.

Acknowledgment

I am particularly indebted to my colleagues of the research division of Nihon Bayer Agrochem Co., Ltd. and the plant protection institute of Bayer AG for co-working and cooperation for this project. I am especially grateful to Dr. Shimpei Kuyama, the former institute director of NBA, for his enduring encouragement from the start of this research until the finding of this new insecticide class.

References

Hansch LA, Elkins CD (1971) Partition coefficients and their uses. Chem Rev 71:525-616

Ishii Y, Kobori I, Araki, Y, Kurogochi S, Iwaya K, Kagabu S (1994) HPLC determination of the new insecticide imidacloprid and its behavior in rice and cucumber. J Agric Food Chem 42:2917-2921

Kagabu S, Moriya K, Shibuya K, Hattori Y, Tsuboi S, Shiokawa K (1992) 1-(6-Halonicotinyl)-2-nitromethylene-imidazolidines as potential new insecticides. Biosci Biotechnol Biochem 56:362-363

Kagabu S, Medej S (1995) Chloronicotinyl insecticides - stability comparison of imidacloprid and related compounds under simulated sunlight, hydrolysis conditions, and to oxygen. Biosci Biotechnol Biochem 59:980-985

Kagabu S (1996) Studies on the synthesis and insecticidal activity of neonicotinoid compounds. J Pestic Sci 21:231-239

Kagabu S, Akagi T (1997) Quantum chemical consideration of photostability of imidacloprid and related compounds. J Pestic Sci 21:231-239.

Kagabu S, Matsuno H (1997) Crystal and molecular structures of imidacloprid and analogous compounds. J Agric Food Chem 45:276-281

Kashiwada Y (1996) Bestguard (nitenpyram, TI-304)- New systemic insecticide. Agrochemicals Japan 68:18-19.

Matsuda M, Takahashi H (1996) Mospilan (acetamiprid, NI-25)- New systemic insecticide. Agrochem Jpn 68:20-21

Moriya K, Shibuya K, Hattori Y, Tsuboi S, Shiokawa K, Kagabu S (1992) 1-(6-Chloronicotinyl)-2-nitroimino-imidazolidines and related compounds as potential new insecticides. Biosci Biotechnol Biochem 56:364-365

Moriya K, Shibuya K, Hattori Y, Tsuboi S, Shiokawa K, Kagabu S (1993a) Structural modification of the 6-chloropyridyl moiety in the imidacloprid skeleton: introduction of a five-membered heteroaromatic ring and the resulting insecticidal activity. Biosci Biotechnol Biochem 57:127-128

Moriya K, Shibuya K, Hattori Y, Tsuboi S, Shiokawa K, Kagabu S (1993b) 1-Diazinyl-2-nitromethylene-imidazolidines as new potential insecticides. J Pestic Sci 18:119-123.

Schroeder ME, Flattum RF (1984) The mode of action and neurotoxic properties of the nitromethylene heterocycle insecticides. Pestic Biol Physiol 22:148-160

Shiokawa K, Tsuboi S, Kagabu S, Moriya K (1986) Jpn Pat Showa 61-267575

Shiokawa K, Moriya K, Shibuya K, Hattori Y, Tsuboi S, Kagabu S (1992) 3-(6-Chloronicotinyl)-2-nitromethylene-thiazolidine as a new class of insecticide acting against lepidoptera species. Biosci Biotechnol Biochem 56:1364-1365.

Soloway SB, Henry AC, Kollmeyer WD, Padgent WM, Powell JE, Roman SA, Tieman CH, Corey RA, Horne CA (1979) Nitromethylene insecticides. In: Geissbuhler H, Brooks G T and Kearney P C (Eds) Advances in pesticide science, part 2. Pergamon, Oxford, pp 206-227

Sone S, Yamada Y, Tsuboi S (1995) New pesticide screening method for leaf hoppers and plant hoppers using rice seedlings. Jpn J Appl Entomol Zool 39:171-173

Part 2

Synthetic Nicotinoid Insecticides

5 Chloronicotinyl Insecticides: A Success of the New Chemistry

Detlef Wollweber * and Klaus Tietjen **
Bayer AG, Crop Protection Business Group, Molecular Target Research and Biotechnology, Monheim, Building 6240, D-51368 Leverkusen, Germany
Key words: insecticides, chloronicotinyl insecticides, nitromethylene insecticides, neonicotinoid insecticides, nicotinic acetylcholine receptor, nAChR, binding studies, electrophysiology, imidacloprid, acetamiprid, nitenpyram

1 Introduction

The insecticide world market, worth about 7 billion at the beginning of the 1990s, has long been dominated by old and well-established products belonging to the organophosphate, carbamate, and organochlorine classes and the newer class of pyrethroids. All these commercially important insecticides act at targets in the central nervous system and were responsible for more than 85% of worldwide insecticide sales. Inhibitors of the nicotinic acetylcholine receptor, the target of chloronicotinyl insecticides, were represented by earlier products like nicotine itself or nereistoxin analogs without any real commercial relevance at that time (Figure 1).

Main Chemical Classes	Mechanisms
• Organochlorines	GABA/Na-Channel
• Organophosphates	ACE
• Carbamates	ACE
• Pyrethroids	Na-Channel

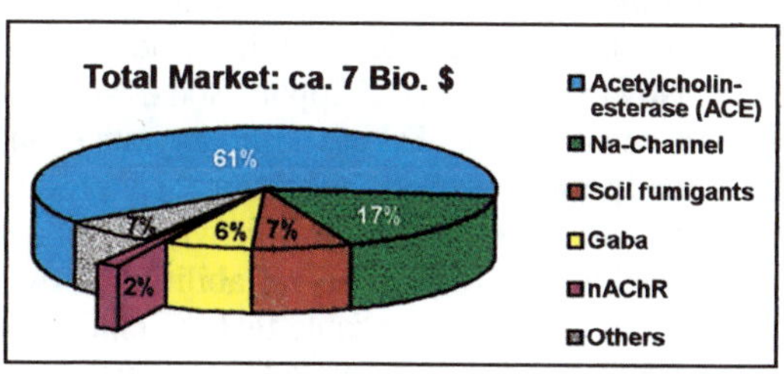

Figure 1. Insecticide market 1991 according to mechanisms

There was a strong demand in the market for new broad-spectrum insecticides with new modes of action and favorable toxicological and environmental properties. The introduction of the first chloronicotinyl insecticide, imidacloprid, by Bayer in the beginning of the 1990s met that demand, culminating in a tremendous economic success.

In this review, the following aspects of chloronicotinyl insecticides are summarized and discussed:

- Chemical synthesis
- Biological performance and environmental aspects
- Modern application methods
- Economic success
- Biochemical research

The history and discovery of chloronicotinyl insecticides are the topic of detailed preceding articles in this volume. Nevertheless, it is worthwhile to mention again that the key to the market launch of chloronicotinyl insecticides was the discovery by Bayer of the chloropyridin moiety. Compared to the first product, nithiazine, from Shell, the biological efficacy against sucking pests could be enhanced by a factor of more than 125 by introducing the chloropyridin moiety (Table 1).

Compounds	LC_{90}
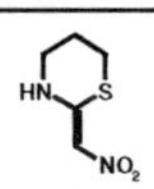 Nithiazin	40 ppm
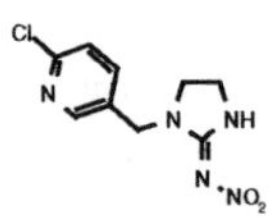 Imidacloprid	< 0,32 ppm

LC_{90} = Concentration at which > 90% of insects are killed.

Table 1. Efficacy of nithiazin and imidacloprid against green rice leafhopper (Kagabu 1996)

First patent applications covering chloronicotinyl insecticides were published in 1985 and triggered substantial activities within the research-based agchem industry (Figure 2). Novartis, Takeda, Nippon Soda, and others immediately entered this new research area. Takeda (nitenpyram) and Nippon Soda (acetamiprid) have already succeeded in launching their products and it is to be expected that there are more to come. Their entry was facilitated by the fact that the new chemistry showed a relatively broad variability. An overview of possible variations is given in Figure 3. It is remarkable that the chloropyridine moiety can be replaced by other aromatic and even saturated heterocyclic systems that possess a heteroatom in the position corresponding to the pyridine nitrogen in imidacloprid. Heteroannelations as in Bay T 9992, open-chain guanidine substructures, and heterocyclic bridged nitroguanidines also exhibit high insecticidal activities.

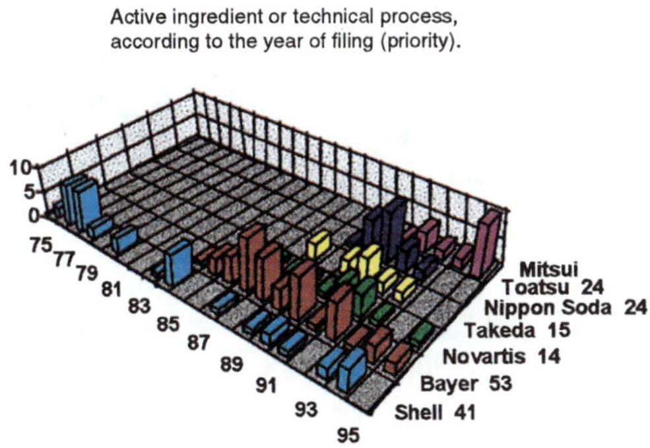

Figure 2. Patent applications – Bayer/main competitors

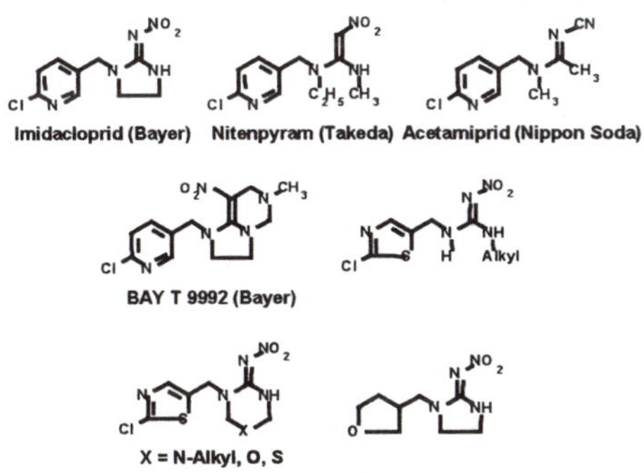

Figure 3. Chloronicotinyl insecticides: structural diversity

2 Chemical Synthesis

2.1 Synthesis of Chloropyridine Intermediates

The key intermediate for chloronicotinyl insecticides is 2-chloro-5-methylpyridine (CMP) or its derivative 2-chloro-5-chloromethylpyridine (CCMP). In the first laboratory synthesis of imidacloprid, chloronicotinic acid was used as the starting material for the synthesis of CCMP, which can be obtained in moderate yield by reduction of the acid and subsequent chlorination of the resulting alcohol (Figure 4).

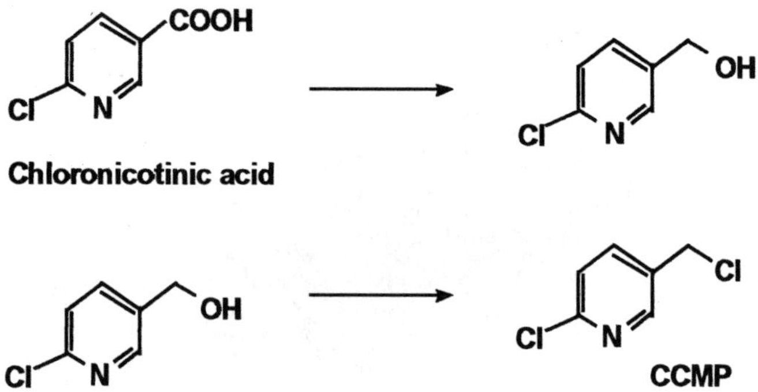

Figure 4. Laboratory synthesis of CCMP

This sequence was intensively studied to develop a viable technical process based on these commercially available starting materials. However, the reduction of chloronicotinic acid turned out to be not as straightforward as one might think (R. Lantzsch, personal communication). In catalytic reduction the elimination of the chlorosubstituent of the pyridyl moiety always leads to undesired by-products. Reduction of the acid chloride with one equivalent of sodium borohydride proceeds slowly and esterification with formed alcohol already occurs as a side reaction. Using a large excess of borohydride led to a much better result, but at the same time to an unacceptable increase in the manufacturing costs. This situation prompted some work on possible alternative routes to CMP and CCMP (Diehr et al. 1991), e.g., based on ß-picoline, nicotinic acid derivatives, or starting with aliphatic precursors (Figure 5). The amination of ß-picoline for example, using the well-known Tschitschibabin reaction, is carried out on a technical scale, giving 2-amino-5-methylpyridine (AMP). CMP is obtained via formation of the diazonium salt and subsequent Sandmeyer reaction.

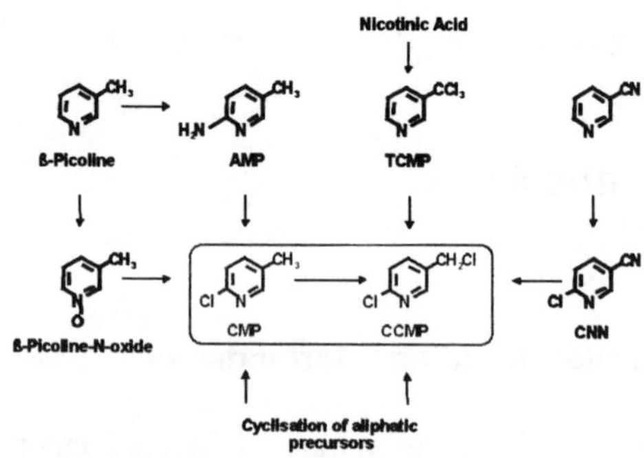

Figure 5. CMP/CCMP processes

2.2 Synthesis of Imidacloprid

For the synthesis of nitenpyram and acetamiprid starting from chloropyridine intermediates, see the chapters by A. Akayama and T. Yamada, this volume. Imidacloprid can be synthesized according to Figure 6 (Moriya et al. 1992), starting from 2-chloro-5-chloromethylpyridine, by reaction with 2-nitroimino-1,3-pyrrolidine in acetonitrile with potassium carbonate as base. 2-nitroimino-1,3-pyrrolidine is available by cyclizing nitroguanidine with diaminoethane. Nitroguanidine is used on a manufacturing scale for the manufacture of explosives and propellants.

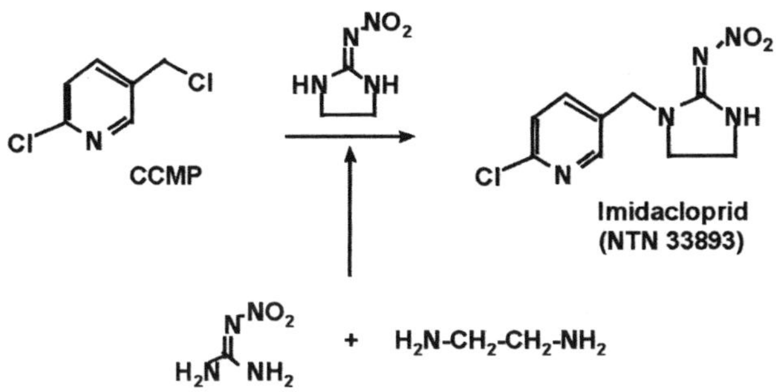

Figure 6. Synthesis of imidacloprid

3 Biological Performance and Application Methods

3.1 Biological Performance

The biological properties for chloronicotinyl insecticides in general can be summarized as follows:

- Different mode of action compared to established insecticides
- High efficacy against sucking and some biting insect species
- Good efficacy against resistant populations
- Systemic properties, important for soil and seed treatment
- Long-lasting efficacy and good plant compatibility
- Antifeedant effects at sublethal doses
- Control of various virus vectors in sugar beet, cereals, and vegetables

These favorable properties are, of course, present to different degrees in the products now available on the market.

The following overview, illustrating these properties, is mainly focused on imidacloprid (detailed summary of the biological activity and agricultural importance of imidacloprid is given by Elbert et al. in press). Nitenpyram of

Takeda and acetamiprid of Nippon Soda are discussed separately in other chapters of this volume.

The chloronicotinyl insecticides are highly effective against homopteran pests, e.g., aphids, whiteflies, leafhoppers and planthoppers, as well as thrips. Sideeffects are to be mentioned against some species of coleopteran, lepidopteran, and dipteran species. Because their systemic properties result in good xylem mobility, these products can be used in foliar and soil applications as well as in seed treatments.

3.1.1 Foliar Application of Imidacloprid

Imidacloprid is used in foliar applications under the trade names Confidor™ and Provado™. Its efficacy in foliar application against various sucking pests in greenhouse trials is shown in Figure 7. The LC_{95}, that is, the concentration at which 95% of the insects were killed, varies from 30 to 0.3 ppm, which is the magnitude of efficacy known from established insecticides. Imidacloprid can be used against sucking pests in such crops as cereals, corn, rice, vegetables, sugarbeet, cotton, etc. Its plant compatibility is highly satisfactory (Elbert et al. 1991).

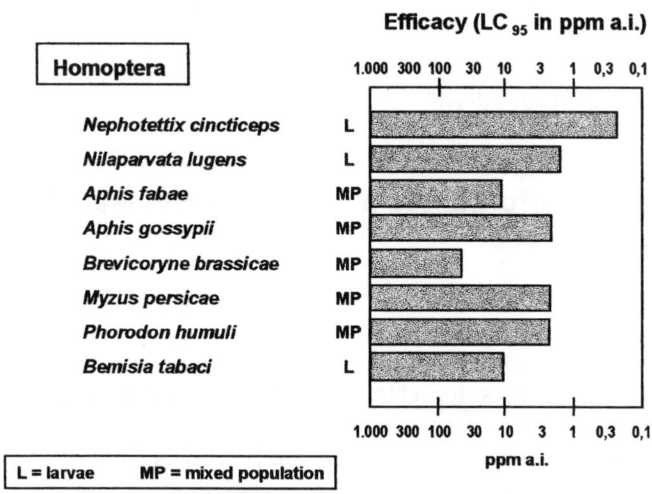

Figure 7. Imidacloprid: foliar application

3.1.2 Soil and Seed Treatment

One major strength of imidacloprid is based on its high efficacy in soil application. In Figure 8 the efficacy against typical soil insects is summarized. *Agriotes* spp. and *Diabrotica balteata* are controlled by incorporation of 2.5 to 5 ppm a.i. into the soil. It is also very effective against termites (*Reticulitermes flavipes*) and has become a new standard in the marketplace, especially in the United States under the trade name Premise™. Due to its systemic properties, imidacloprid not only acts against soil insects but is also taken up by the roots and translocated within the plants. Imidacloprid is highly effective against early-season sucking pests such as *Myzus persicae* and *Aphis fabae* at soil concentrations as low as 0.15 ppm a.i.. Combined with long residual activity, this gives young plants protection during the first growth stages. Comparable effects can also be recognized on seed treatments, and the residual efficacy of imidacloprid is clearly superior against aphids compared to standard insecticides following soil or seed treatments. It is marketed under the trade names Admire™ and Gaucho™.

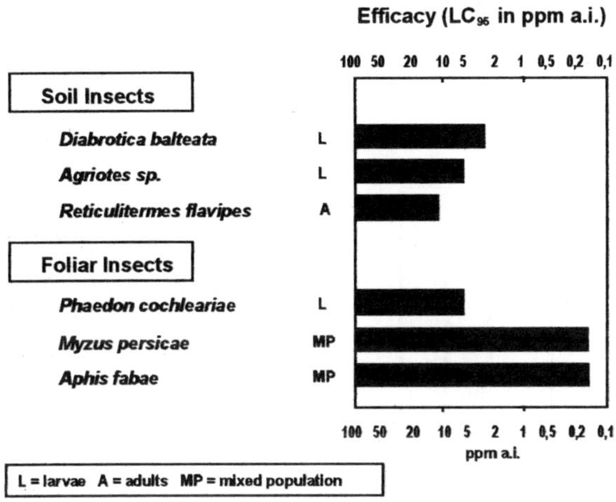

Figure 8. Imidacloprid: soil application

3.1.3 Antifeedant Effects

When applied at sublethal doses (<10 ppb) imidacloprid dramatically affects the behavior of aphids, resulting in depression of honeydew excretion, wandering, and finally death from starvation (Nauen 1995; Nauen and Elbert 1997). These antifeedant effects are also described for other insect species (Lösel and Goodman 1993; Drinkwater 1994; Drinkwater and Groenewald 1994; Nauen et al. 1998). The viability of nymphs deposited by aphids is also significantly reduced at very low concentrations (Devine et al. 1996). These effects may be responsible for preventing the rapid buildup of populations in the field, even later in the season when the concentration of imidacloprid in seed-treated plants has declined.

3.1.4 Activity on Resistant Insect Species

Resistance to synthetic insecticides is a serious problem in the field because of the long and continuous use of older commercially available products, especially from classes of compounds such as organophosphates, carbamates, and pyrethroids. It is very important for new insecticides to have a good efficacy against such pest species. The efficacy of imidacloprid against these so-called multiresistant pest insects was already investigated in its early developmental phase. Its efficacy against multiresistant *M. persicae*, for example, is comparable to its efficacy against a susceptible reference strain (Figure 9). In contrast these aphids show high resistance factors to organophosphates such as parathion-ethyl, carbamates such as pirimicarb, and pyrethroids such as cypermethrin. Cypermethrin, for example, is more than 125 times less active against this multiresistant aphid strain compared to the susceptible strain (Elbert et al. 1991).

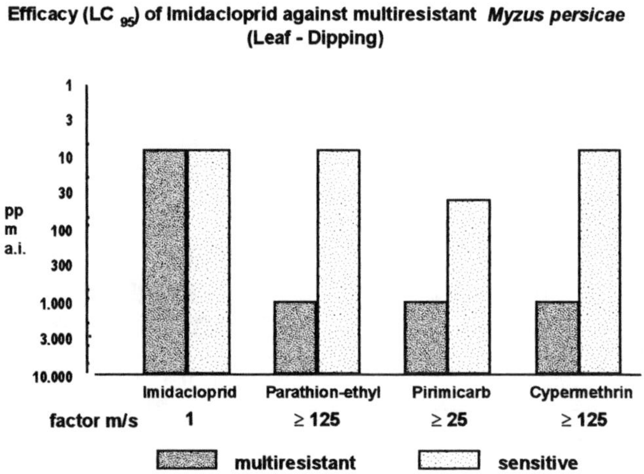

Figure 9. Imidacloprid: activity on resistant insect species

Several investigations were performed in the past and are still continuing to monitor imidacloprid susceptibility in certain target pests from a wide geographic area. Extensive resistance monitoring programs have already been implemented before or just after launch of imidacloprid, in particular to establish and recommend resistance management guidelines to prevent early development of resistance (Elbert et al. 1996). Work was focused on pest species already known for their resistance to other chemical classes of insecticides, such as certain aphid species, whiteflies, and Colorado potato beetles.

European monitoring of imidacloprid susceptibility in *Myzus* spp. revealed a homogeneous baseline response (Elbert et al. 1996), although some strains of the tobacco aphid, *Myzus nicotianae*, often showed a 5- to 10fold lower tolerance level probably not associated with any known mechanism of resistance in that species (Nauen et al. 1996; Devine et al. 1996; Nauen et al. 1997). Speculations on reduced acetylcholine receptor affinity from nicotine cross-tolerance of tobacco-feeding strains of *Myzus* spp. could not be confirmed in binding bioassays (Nauen et al. 1996). Other aphid pests resistant to most conventional insecticides, such as the damson hop aphid, *Phorodon humuli*, and the cotton aphid, *Aphis gossypii*, are still excellently controlled with imidacloprid and no cases of resistance has been reported so far (R. Nauen, personal communication).

The tobacco whitefly, *Bemisia tabaci*, and the recently described B-subtype, *B. argentifolii*, are, because they can transmit more than 60 different Geminiviruses, deleterious pests of growing importance in a vast range of crops worldwide (Perring et al. 1993; Bedford et al. 1994). Whiteflies have developed resistance against all classes of control agents, but imidacloprid is active on several highly resistant strains from different locations worldwide (Cahill et al. 1996a). Some investigations on Spanish strains from Almeria revealed a lower susceptibility to imidacloprid in laboratory bioassays (Cahill et al. 1996b). Electrophoretically derived esterase patterns showed that the Spanish subtype of *B. tabaci* was not comparable to the common B-type and some other strains (Nauen et al. in press). However, the field performance in Spain is still very good, even 6 years after market introduction (Elbert and Nauen 1996). Whiteflies however are still considered to be the pest with the highest likelihood of developing resistance insecticides, and therefore research is especially focused on that species. Apart from the homopteran pests, some coleopteran species, e.g., the Colorado potato beetles (CPB), *Leptinotarsa decemlineata*, are also target organisms for imidacloprid. Monitoring of CPB baseline susceptibility to imidacloprid revealed a certain degree of heterogeneity supposed to be natural variation in response rather than resistance (Elbert et al. 1996; Olson et al. 1996). The findings of the worldwide studies on pests controlled by imidacloprid have been incorporated into resistance management guidelines published recently (Elbert et al. 1996).

3.2 Environmental and Toxicological Aspects

3.2.1 Activity of Imidacloprid on Beneficial Arthropods

The profile of imidacloprid against beneficial arthropods was investigated under greenhouse and field conditions. In soil applications most beneficials like honeybees, beneficial beetles, and predatory bugs and mites are unaffected up to 250 g/ha, i.e., above the normal application rates (H.W. Schmidt, private communications). Therefore imidacloprid can be used in systemic soil and seed treatment applications in integrated pest management systems without reservation.

In foliar applications, correct timing of the spray enables the farmer to avoid most problems; e.g., imidacloprid is harmful to bees and should not be applied during the flowering period (Pflüger and Schmuck 1991).

3.2.2 Ecobiological Profile of Imidacloprid

The ecobiological profile of imidacloprid was established by Pflüger and Schmuck (1991). The product does not impact soil microorganisms, earthworms, algae, or fish species at the relevant application rates. The low fish toxicity is important for use in the paddy rice markets in Asia. If used as recommended, relevant impacts are also not anticipated for aquatic and terrestrial arthropods. Although acutely toxic to birds, imidacloprid poses no significant risk to these organisms due to its strong antifeedant effects.

3.2.3 Toxicological Profile of Imidacloprid

For a summary of the toxicological profile of imidacloprid see the chapter by J.H. Thyssen, this volume.

4 Agricultural Importance of Imidacloprid

The excellent biological profile of imidacloprid combined with its highly systemic activity is the basis for new and innovative agricultural application methods, e.g., in seed treatments, rice cultivation, stem painting and ornamental greenhouse cultivation. The advantages of the first two points are briefly discussed.

4.1 Seed Treatment with Imidacloprid

Compared to older techniques, seed treatment with imidacloprid reduces application rates and treated areas significantly. As shown in Figure 10, in old conventional treatments insect pests were controlled with 1.3 kg a.i. of lindane, for example, sprayed on an area of 1 ha. Newer in-furrow applications, for example with carbofuran (Curaterr^{TM}), reduced the application rate to 600g/ha and the area actually treated to about 500 square meters. In seed treatments with imidacloprid (Gaucho^{TM}), only up to 125 g/ha are used and the area actually treated is further reduced to roughly 60 square meters. Moreover, due to the long-lasting efficacy of imidacloprid, the first spray applications of a conventional spray program to protect the crop against early-season pests can even become unnecessary. The advantage for the farmer is the saving in time and labor. The advantage for the environment is that the seedlings and young plants are treated effectively at the point of attack of the pests. Today, seed treatments for most of the important crops, e.g., cotton, cereals, corn, and sugar beet, are commercialized.

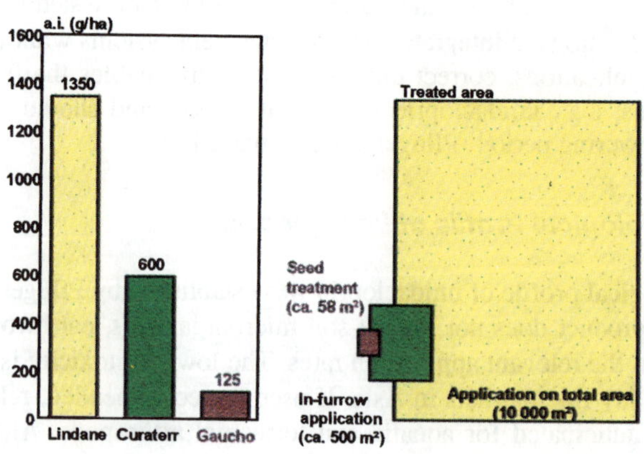

Figure 10. Development of application methods

4.2 Rice Cultivation with Imidacloprid

Rice growing in Japan has been developed by generations of farmers into a sophisticated technique starting with sowing of rice in seedling boxes and transplanting of the young seedlings to the rice fields. During this growing phase the seedlings were protected against various insect pests by products such as carbosulfan, benfuracarb, and propoxur, but these could not control all commercially relevant pests. Imidacloprid, applied as a granular formulation in seedling boxes, on the one hand broadened the spectrum of insects that could be controlled. On the other hand, due to its systemic and long residual activity, it is transferred with the seedlings in the field and continues to control important sucking pests further. up to 50 to 90 days after transplantation. Again, subsequent spray applications can even become unnecessary. In addition imidacloprid can also be used in conventional applications in rice cultivation, which make it a very effective and flexible tool for the rice farmer.

5 Economic Success

Because of its outstanding biological and technical properties, imidacloprid is now on the market in more than 76 countries. The chloronicotinyl insecticides in general are achieving rapid market growth all over the world and are probably the fastest growing chemical group in the 1990s. The market shares of chloronicotinyl insecticides are estimated to increase from 2% for nAChR insecticides at the beginning of the 1990s to 10% to 15% of the insecticide world market at the beginning of the new century (Figure 11). Imidacloprid will be the leading compound, but other chloronicotinyls will also gain significant market shares, providing a full range of chloronicotinyl products for many different technical problems. Of course, this represents a tremendous success for this new chemistry.

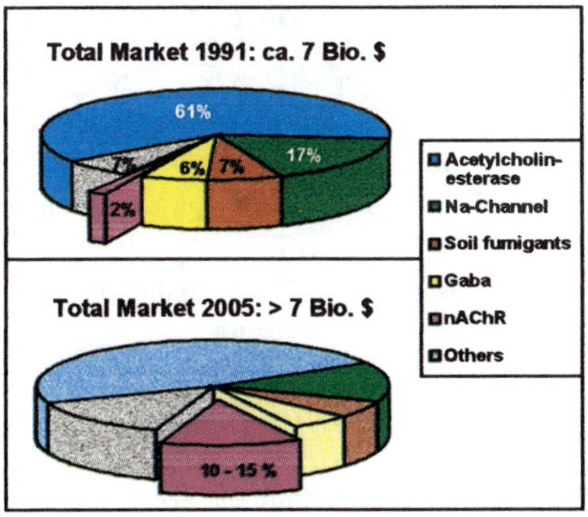

Figure 11. Chloronicotinyl insecticides: market development

6. Biochemistry of Imidacloprid Receptors

It was only after the discovery of the chloronicotinyl insecticides that their target became known. In 1984 it was first shown in electrophysiological studies by Schroeder and Flattum (1984) of Shell that their nitromethylene compound nithiazin acts on nicotinergic acetylcholine receptors from insects. At that time we started to measure the binding of compounds to insect acetylcholine receptors using competition with tritiated α-bungarotoxin in insect brain preparations. This test system in fact existed before the synthesis of imidacloprid itself (Mansour et al. 1980). Figure 12 illustrates pI_{50} values obtained for different compounds with housefly head membrane preparations and [^{3}H]-α-bungarotoxin. Chloronicotinyl insecticides exhibit pI_{50} values ranging from 6 to 9, but there are some questions concerning the relevance of α-bungarotoxin competition. On the one hand nithiazin, without the chloropyridinyl moiety, has a rather low pI_{50} value of only 4.5, which contrasts markedly with its remarkable insecticidal properties. On the other hand, methyllycaconitine exhibits a high pI_{50} value of 8.1 but is insecticidally ineffective. Moreover, pI_{50} values measured with [^{3}H]-α-bungarotoxin did not give rise to satisfactory structure-activity relationships for chloronicotinyls.

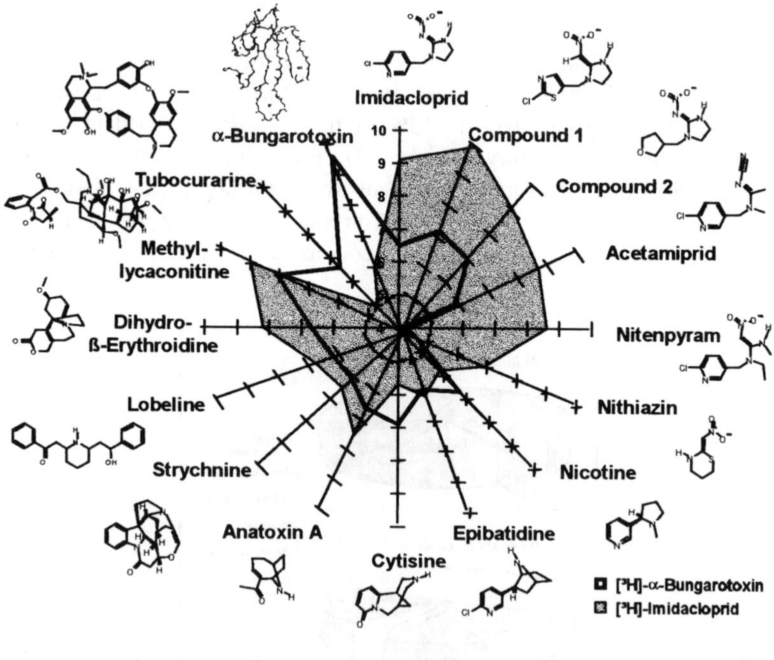

Figure 12. pI_{50} Values for housefly head membrane acetylcholine receptors, using [^{3}H] α-bungarotoxin or [^{3}H]-imidacloprid

One explanation for this failure could be the existence of other receptors apart from those binding [^{3}H]-α-bungarotoxin with high affinity. The presence of such receptors was already suggested by Sattelle et al. in 1980. Using tritiated imidacloprid, such receptors were identified by Liu and Casida (1993). Figure 12 compares the results obtained with [^{3}H]imidacloprid to those obtained with [^{3}H]-α-bungarotoxin. Clearly, imidacloprid and α-bungarotoxin appear to bind to distinct sites. Imidacloprid and other chloronicotinyls exhibit high affinities in the range between 9 and 10 in the [^{3}H]imidacloprid assay. Structure-activity relationships for chloronicotinyls are more satisfactory using these results.

The existence of imidacloprid receptors different from α-bungarotoxin receptors fits well with the fact that the chloronicotinyl insecticides do not act on vertebrate muscle receptors (Methfessel 1992). Tubocurarine, a vertebrate muscle receptor ligand, binds only weakly to imidacloprid receptors. Moreover, the affinity of imidacloprid for vertebrate brain receptors (using [^{3}H]cytisine) is 3 orders of magnitude lower than in the insect brain (K. Tietjen, unpublished). Conversely, epibatidine, which is extremly active in vertebrate brains, is only moderately active in insect brains (Figure 12).

Not all compounds binding to acetylcholine receptors are activators of the sodium channel. Electrophysiology can distinguish between agonists, opening the channels like the natural ligand acetylcholine does, and antagonists, closing the channel in some way. Our electrophysiological results (U. Ebbinghaus-Kintscher, unpublished) show that the chloronicotinyls are clear agonists of insect acetylcholine receptors, as are nicotine, epibatidine, cytisine, and anatoxin A (Figure 13). Other natural products are clear antagonists.

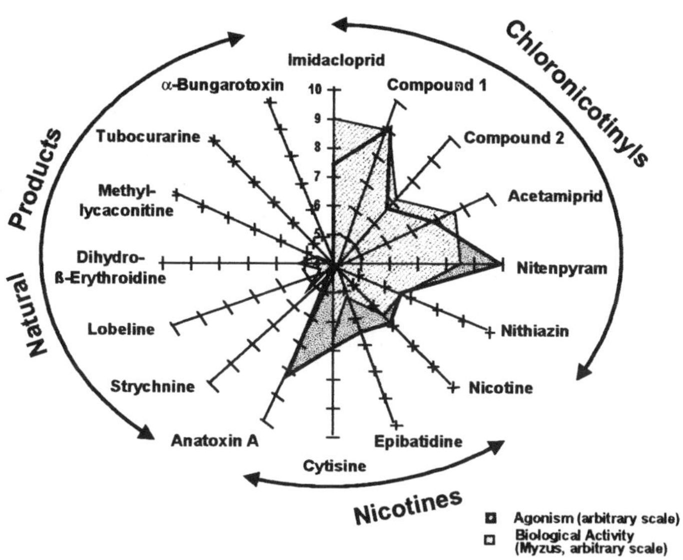

Figure 13. Electrophysiological activity at housefly head membrane acetylcholine receptors (Ebbinghaus, unpublished) and biological activity against *M. persicae* (Nauen, unpublished)

How does binding to the acetylcholine receptor and opening or closing the sodium channel correlate with biological efficacy? Figure 13 combines electrophysiological data with biological data. Obviously, binding and agonistic effect have to coincide to give a good insecticide. The nicotines are somewhat weaker in insecticidal activity although they are strong agonists. This might be explained by the poor bioavailabilty of such compounds at their target site. Anatoxin is even totally inactive. Antagonists only occasionally have some biological efficacy.

The nicotinic acetylcholine receptor is one of the best known members of the ligand-gated ion channel superfamily (McLane et al. 1996). Insect receptors are less well characterized than mammalian receptors (Eldefrawi and Eldefrawi 1997). Figure 14 shows a sketch of a single subunit of the pentameric acetylcholine receptor protein. Ion flux is from outside the cell to the inside. The C-terminal part of the protein, being integrated into the nervous membrane with the regions M1 to M4, does not need to be considered here. Imidacloprid binds to the acetylcholine binding site in the extracellular N-terminal domain, as shown by the competition experiments. This region is known to contain the vicinal pair of cysteines, typical for all α-subunits of acetylcholine receptors, and several aromatic residues from different parts of the protein.

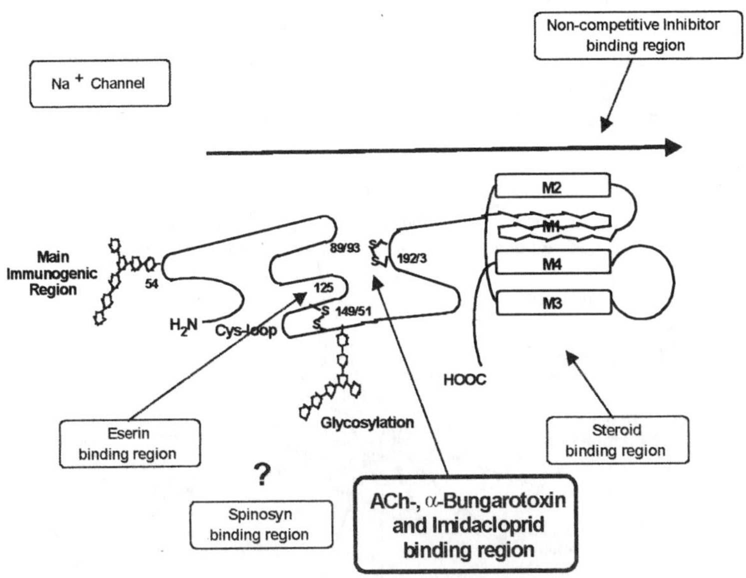

Figure 14. Sketch of an α-subunit of insect acetylcholine receptors

Despite many attempts, nothing is known about the real 3-dimensional structure of this site, but a detailed 3-dimensional model has recently been published (Tsigelny et al. 1997). Although it is the natural ligand binding site, the sequence in the region around the cysteine pair is somewhat variable between different receptor subtypes and animals. The insect receptor sequences published to date are all of the neuronal type, but significantly different from vertebrate neuronal receptors. Interestingly, it has been found by Salgado et al. (1997) from DowElanco that the natural products of the spinosyn type also act on insect nicotinic acetylcholine receptors. But spinosyn does not compete with imidacloprid or with α–bungarotoxin binding, so a new binding site has to be sought. This site may be related to the avermectin-binding site on insect glutamate (Cl^-) and $GABA_A$ receptors.

The spinosyn-binding site could be located on another subunit in the pentameric receptor away from the imidacloprid-binding site, but to date it is still unknown which α-subunits of insect acetylcholine receptors bind imidacloprid or α-bungarotoxin. Nor is it known how many different acetylcholine receptors or how many different imidacloprid binding subunits there are in insects. Results from Casida's group indicate that the imidacloprid and α-bungarotoxin-binding sites might be located within the same pentameric receptor, but on different α-subunits (Tomizawa et al. 1996; also see the chapter by J.E. Casida, this volume).

Even less is known about the physiological function of the imidacloprid receptor than about the precise roles of nicotinic acetylcholine receptor in the insect brain in general (Homberg 1994). Imidacloprid should therefore be useful as a scientific tool for future insect nervous system research.

Acknowledgment

The work covered by this review and the success of imidacloprid are based on the efforts and support of almost all our colleagues in Nihon Bayer and Bayer's insecticide research and development group worldwide.

References

Bedford ID, Pinner M, Liu S, Markham P G (1994) *Bemisia tabaci* - potential infestation, phytotoxicity and virus transmission within European agriculture. Proceedings of the Brighton Crop Protection Conference - Pests and Diseases 7C-4, pp 911-916

Cahill M, Denholm I, Byrne F, Devonshire A L (1996a) Insecticide resistance in *Bemisia tabaci* - current status and implications for management. Proceedings of the Brighton Crop Protection Conference - Pests and Diseases 2B-3, pp 75-80

Cahill M, Gorman K, Day S, Denholm I, Elbert A, Nauen R (1996b) Baseline determination and detection of resistance to imidacloprid in *Bemisia tabaci* (Homoptera: Aleyrodidae). Bull Entomol Res 86:343-349.

Devine GJ, Harling ZK, Scarr AW, Devonshire AL (1996) Lethal and sublethal effects of imidacloprid on nicotine-tolerant *Myzus nicotianae* and *Myzus persicae*. Pestic Sci 48:57-62

Diehr H-J, Gallenkamp B, Jelich K, Lantzsch R, Shiokawa K (1991) Synthesis and chemical properties of the insecticide imidacloprid (NTN 33893) PflanzNachr Bayer (Engl Ed) 44:107-112

Drinkwater TW (1994) Comparison of imidaclopird with carbamate insecticides, and the role of planting depth in the control of false wireworms, *Somaticus* species, in maize. Crop Prot 13:341-345

Drinkwater TW, Groenewald LH (1994) Comparison of imidacloprid and furathiocarb seed dressing insecticides for the control of the black maize beetle, *Heteronychus arator* Fabricius (Coleoptera: Scarabaeidae), in maize. Crop Prot 13:421-424

Elbert A, Becker B, Hartwig J, Erdelen C (1991) Imidacloprid - a new systemic insecticide. PflanzNachr Bayer 44:113-136

Elbert A, Nauen R, Cahill M, Devonshire A, Scarr A, Sone S, Steffens R (1996) Resistance management for chloronicotinyl insecticides using imidacloprid as an example. PflanzNachr Bayer 49:5-54

Elbert A, Nauen R (1996) Bioassays for imidacloprid for a resistance monitoring against the whitefly *Bemisia tabaci*. Proceedings Brighton Crop Prot Conf, Pests and Diseases 6D-8, pp 731-738

Elbert A, Nauen R, Leicht W (1998) Imidacloprid, a novel chloronicotinyl insecticide: biological activity and agricultural importance. In: Ishaaya I, Degheele D (Eds) Insecticides with novel modes of action, mechanism and application. Springer, Berlin Heidelberg New York, pp 50-73

Eldefrawi ME, Eldefrawi AT (1997) Comparative molecular and pharmacological properties of cholinergic receptors in insects and mammals. ACS Symp Ser 658:327-338

Homberg U (1994) Distribution of neurotransmitters in the insect brain. Gustav Fischer, Stuttgart Jena New York, pp 47-49

Kagabu S (1996) Studies on the synthesis and insecticidal activity of neonicotinoid compounds. Pestic Sci 21:231-239

Liu MY, Casida JE (1993) High affinity binding of [^{3}H]imidacloprid in the insect acetylcholine receptor. Pestic Biochem Physiol 46:40-46

Lösel PM, Goodman LJ (1993) Effects on the feeding behaviour of *Nilaparvata lugens* (Stål) of sublethal concentrations of the foliarly applied nitromethylene heterocycle 2-nitromethylene-1,3-thiazinan-3-yl-carbamaldehyde. Physiol Entomol 18:67-74

Mansour NA, Pessah IN, Eldefrawi AT (1980) Binding of [^{125}I]α-bungarotoxin and reversible cholinergic ligands to proteins in housefly brains. In: Insect neurobiological pesticide action. Society of Chemical Industry, London, pp 201-207

McLane KE, Dunn SJM, Manfredi AA, Conti-Tronconi BM, Raftery MA (1996) The nicotinic acetylcholine receptor as a model for a superfamily of ligand-gated ion channel proteins. In: Carey PR (Ed), Protein engineerung Des, Academic Press, San Diego, pp 289-352

Methfessel C (1992) Action of imidacloprid on the nicotinergic acetylcholine receptors in rat muscle. PflanzNachr Bayer (Germ. Ed), 45:369-380

Moriya K, Shibuya K, Hattori Y, Tsuboi S, Shiokawa K, Kagabu S (1992) 1-(6-Chloronicotinyl)-2-nitroimino-imidazolidines and related compounds as potential new insecticides. Biosci Biotechnol Biochem 56:364-365

Nauen R (1995) Behaviour modifying effects of low systemic concentrations of imidacloprid on *Myzus persicae* with special reference to antifeeding response. Pestic Sci 44:145-153

Nauen R, Strobel J, Otsu K, Tietjen K, Erdelen C, Elbert A (1996) aphicidal activity of imidacloprid against a carbamate and organophosphate resistant japanese strain of the tobacco feeding form of *Myzus persicae* (Homptera: Aphididae) closely related to *Myzus nicotianae*. Bull Entomol Res 86:165-171

Nauen R, Elbert A (1997) Apparent tolerance of a field-collected strain of *Myzus nicotianae* to imidacloprid due to strong antifeeding response. Pestic Sci 49:252-258

Nauen R, Hungenberg H, Tollo B, Tietjen K, Elbert A (1998) Antifeedant effect, biological efficacy and high affinity binding of imidacloprid to acetylcholine receptors in tobacco-associated *Myzus persicae* (Sulzer) and *Myzus nicotianae* Blackman (Homoptera: Aphididae). Pestic Sci 53:133-140

Nauen R, Koob B, Klüver T, Elbert A (1997) Biochemical characterization of insecticide resistant strains of the tobacco whitefly *Bemisia tabaci* (Homoptera: Aleyrodidae). MittDtschGes Allgemeine Angewandte Entomol 11:217-221

Nauen R, Koob B, Elbert A (in press) Behavioural effects of sublethal concentrations of imidacloprid to *Bemisia tabaci* (Homoptera: Aleyrodidae). Entomol Exp Appl

Olson ER, Dively GP, Nelson JO (1996) Survey of susceptibility to imidacloprid (Admire) in Colorado potato beetle (Coleoptera: Chrysomelidae). Resis Pest Manage 8:39-41

Perring TM, Cooper AD, Rodriguez RJ, Farrar CA, Bellows TS Jr (1993) Identification of a whitefly species by genomic and behavioural studies. Science (Washington DC) 259:74-77.

Pflüger W, Schmuck R (1991) Ecotoxicological profile of imidacloprid. PflanzNachr Bayer 44:145-158

Salgado VL, Watson GB, Sheets JJ (1997) Studies on the mode of action of spinosad, the active ingredient in tracer insect control. Proceedings Beltwide Cotton Conferences (Vol 2). National Cotton Council, Memphis, pp 1082-1084

Sattelle DB, David JA, Harrow ID, Hue B (1980) Actions of α-bungarotoxin on identified insect central neurones. In: Sattelle DB, Hall LM, Hildebrand JG (Eds) Receptors for neurotransmitters, hormones and pheromones in insects. Elsevier, Amsterdam, pp 125-139

Schroeder ME, Flattum RF (1984) The mode of action of neurotoxic properties of the nitromethylene heterocycle insecticides. Pestic Biochem Physiol 22:148-160

Tomizawa M, Latli B, Casida JE (1996) Novel neonicotinoid-agarose affinity column for *Drosophila* and *Musca* nicotinic acetylcholine receptors. J Neurochem 67:1669-1676

Tsigelny I, Sugiyama N, Sine SM, Taylor P (1997) A model of the nicotinic receptor extracellular domain based on sequence identity and residue location. Biophys J 73:52-66

6 Discovery of a New Systemic Insecticide, Nitenpyram and Its Insecticidal Properties

Atsuo Akayama and Isao Minamida
Agricultural Research Laboratories, Agro Company, Takeda Chemical Industries, Ltd., 10 Wadai, Tsukuba, Ibaraki 300-42, Japan

1 Introduction

Rice is cultivated on about 2,000,000 ha of land, and its production is by far the largest among all crops in Japan. One of the most serious rice pests in Japan was the rice stem borer, *Chilo suppressalis*, whose population density has begun to decrease rapidly since the early 1970s (Hirai 1994). Consequently the insect has become a minor and local pest. On the other hand, outbreaks of the rice leaffolder, *Cnaphalocrocis medinalis*, have increased in frequency since the late 1960s, and therefore this insect has now become one of the major rice pests (Wada and Kobayashi 1980). The brown planthopper, *Nilaparvata lugens*, is also one of the major rice pests in Japan. Since ancient times there have been numerous records of brown planthopper outbreaks, which sometimes have caused severe famine (Dyck and Thomas 1979).

Cartap hydrochloride was discovered by modifying the natural insecticidal substance, nereistoxin, which was extracted from the marine annelid, *Lumbriconereis heteropoda*. This unique insecticide was developed by Takeda Chemical Industries, Ltd. and has been widely used to control rice stem borers and rice leaffolders since 1967. However, the brown planthopper, which usually occurs simultaneously with the rice leaffolder, cannot be controlled by cartap, and the insect recently developed a resistance against carbamate and organophosphorous insecticides (Nagata et al. 1979; Kilin et al. 1981; Endo et al. 1988). Therefore, in 1986, we started a project to discover a brown planthopper control agent to meet the demand for a novel insecticide.

2 Discovery

Insecticidal activities of nitromethylene heterocyclic compounds (**1**) were described by Soloway et al. in 1978. Following this, its *N*-formyl derivative (**2**) by Harris et al. (1986) and its *N*-pyridylmethyl derivatives (**3**,**4**) by Shiokawa et al. (1985, 1986) were presented (Fig. 1).

We found acyclic nitromethylene compounds (Fig. 2) that were active against brown planthoppers in 1986. Soloway et al. (1978) reported that the heterocyclic part of the nitromethylene compounds was essential for the insecticidal activity, while we found these acyclic nitromethylene compounds have insecticidal activity. Therefore, replacing the heterocyclic skeleton with the corresponding acyclic moiety was considered to be very interesting, and the research study was started by focusing on the modification of compounds **3** and **4**. The basic structure of the compounds studied is shown in Fig. 3.

2.1 Synthesis of Compounds

The compounds, having amino groups as R^1, were synthesized by the methods shown in Fig. 4. The details are described in three separate papers (Minamida et al. 1993a, 1993b; Tabuchi et al. 1994).

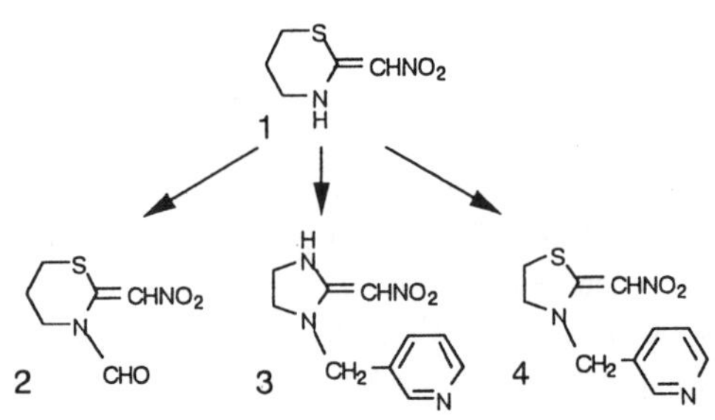

Fig. 1. Heterocyclic nitromethylene compounds

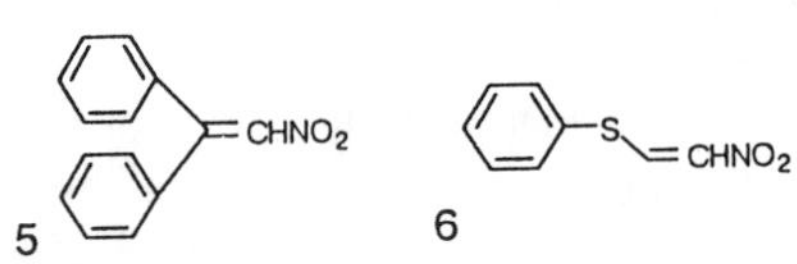

Fig. 2. Acyclic nitroethene compounds found to be active by Takeda Chemical Industries, Ltd.

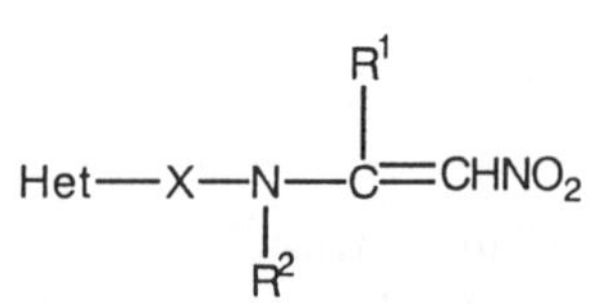

Fig. 3. Basic structure of acyclic nitroethene compounds studied

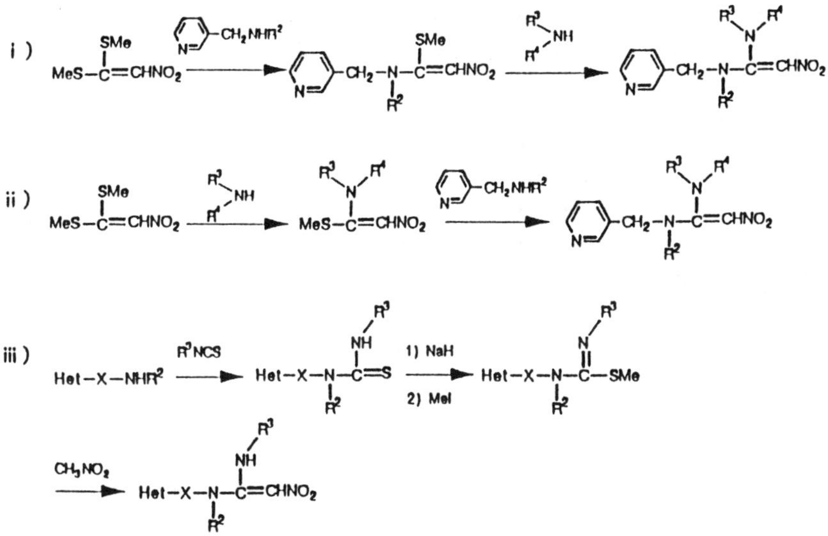

Fig. 4. Methods of synthesizing acyclic nitroethene compounds

2.2 Structure-Activity Relationship

2.2.1 Biological Tests

The test chemical solutions were prepared by dissolving 20 mg of the compounds in 0.5 mL of *N,N*-dimethylformamide containing 5% Tween-20 and then diluting to 500, 200, 40, 2.5, and 0.5 ppm with water containing 67 ppm of polyoxyethylene-nonylphenylether and 40 ppm of ligninsulfonic acid calcium salt.

Six rice seedlings of the two-leaf stage, cultivated in a paper pot (D: 1.5 cm, W: 1.5 cm, H: 2.0 cm), were treated by foliar spray with 10 mL of a test solution until runoff using a handheld airbrush sprayer. After drying, the plants were placed in a glass tube with 5 mL of water, and 10 third instar nymphs of the brown planthopper or the green rice leafhopper were added. Mortality was assessed at 7 days after treatment (DAT).

2.2.2 The Effect of the Substituents on the Insecticidal Activity of 1-(3-Pyridylmethylamino)-2-Nitroethene

The insecticidal activity of 1-(3-pyridylmethylamino)-2-nitroethene compounds (Fig. 3) against the brown planthopper and the green rice leafhopper are summarized in Tables 1-3. Using 3-pyridylmethyl as Het-X- and hydrogen to R^2, the effect of R^1 was studied (Table 1) for the first step. The compound **10** substituted with a methylamino group was more active than **7**, **8**, and **9**. The introduction of sterically large amino groups (compounds **11-15**, **17-21**) decreased the activity against the brown planthopper. Among these compounds, **10** and **16** were the most active against the brown planthopper, but ineffective against the green rice leafhopper. Then the effect of R^2 was studied, using the methylamino group as R^1 (Table 2). The result was that the introduction of methyl (**22**) and ethyl (**23**) increased the activity against the green rice leafhopper. Therefore, the effect of R^1 by using methyl as R^2 was examined again (Table 3). The introduction of the larger alkylamino (**31**, **33**, **34**, and **35**) and secondary amino (**30**) decreased activity against the green rice leafhopper.

Table 1. The effect of the substituents (R^1) on the insecticidal activity of 1-(3-pyridylmethylamino)-2-nitroethene against the brown planthopper (BPH) and the green rice leafhopper (GLH)

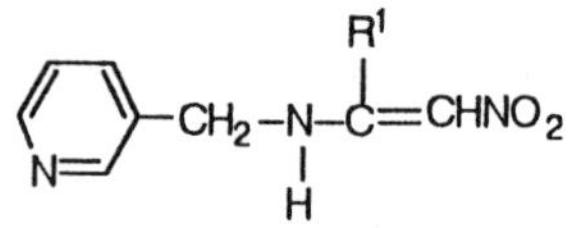

Compounds	R^1	Method[a]	LC_{95} (ppm) [b]	
			BPH	GLH
7	Et	-[c]	>500	NT[d]
8	OMe	-	>500	NT
9	SMe	i	>500	NT
10	NHMe	i	40	>200
11	NHEt	i	500	>200
12	NHPri	i	>500	NT
13	NHBu^n	i	>500	NT
14	NHPenn	i	>500	NT
15	NHPh	i	>500	NT
16	NMe2	i	40	>200
17	NHNH2	i	500	>200
18	NHNMe2	ii	500	>200
19	Pyrrolidino	i	500	>200
20	Piperidino	ii	>500	NT
21	Morpholino	i	>500	NT

[a]Roman numerals indicate the methods for synthesis described in the text.
[b]The lowest concentration to achieve 95% mortality. The concentrations of the test chemical solutions were 40, 200, and 500 ppm for the BPH, and 40 and 200 ppm for the GLH.
[c]Method other than i-ii was employed.
[d]Not tested.

Table 2. The effect of the substituents (R^2) on the insecticidal activity of 1-methylamino-1-(3-pyridylmethylamino)-2-nitroethene against the BPH and the GLH

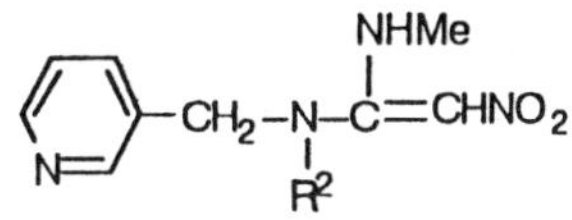

Compounds	R^2	Method[a]	LC_{95} (ppm)[b]	
			BPH	GLH
10	H	i	40	>200
22	Me	iii	40	40
23	Et	iii	40	40
24	Pr^n	iii	40	>200
25	Bu^n	iii	40	>200
26	Benzyl	iii	40	200
27	OMe	iii	200	>200
28	NMe_2	iii	40	>200

[a]Roman numerals indicate the methods for synthesis described in the text.
[b]The lowest concentration to achieve 95% mortality. The concentrations of the test chemical solutions were 40, 200 and 500 ppm for the BPH, and 40 and 200 ppm for the GLH.

Table 3. The effect of the substituents (R^1) on the insecticidal activity of 1-[N-methyl-(3-pyridylmethyl)amino]-2-nitroethene against the BPH and the GLH

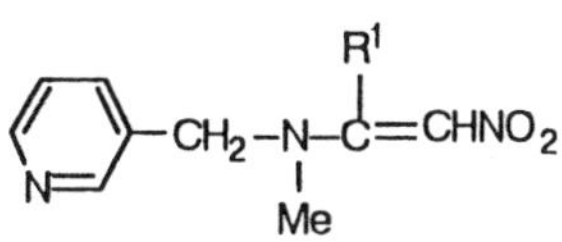

Compounds	R^1	Method[a]	LC_{95} (ppm)[b]	
			BPH	GLH
22	NHMe	iii	40	40
29	NH_2	i	40	40
30	NMe_2	ii	40	200
31	NHEt	iii	40	>200
32	N(Me)CHO	-[c]	40	40
33	$NHPr^i$	iii	40	>200
34	$NHBu^n$	iii	40	>200
35	$NHCH_2Ph$	iii	40	>200

[a]Roman numerals indicate the methods for synthesis described in the text.
[b]The lowest concentration to achieve 95% mortality. The concentrations of the test chemical solutions were 40, 200, and 500 ppm for the BPH, and 40 and 200 ppm for the GLH.
[c]Other method than i-iii was employed.

2.2.3 Effect of the Linkage Between the 3-Pyridyl and the Nitrogen

The methylene part between the 3-pyridyl and the nitrogen was replaced with other linkages (Table 4). All attempts of shortening and lengthening the linkages failed to increase the activities against the brown planthopper and the green rice leafhopper.

2.2.4 Effect of the Heteroaryl Moiety

Heterocyclic aromatic substituents, other than 3-pyridyl, were introduced into the nitroethene compounds (Table 5). The replacement of 3-pyridyl (**22**) with 6-fluoro-3-pyridyl (**46**), 6-chloro-3-pyridyl (**47**), 6-bromo-3-pyridyl (**48**), and 2-chloro-5-thiazolyl (**49**) enhanced the activity against the brown planthopper.

2.2.5 Residual Activity

Good residual activity is favored for new insecticides, because most of the conventional insecticides for the rice hoppers control do not have sufficient residual effects. Therefore, the residual activity of the compounds with combination of the favorable substituents was evaluated: R^1 = methylamino, R^2 = methyl and ethyl, X = methylene, Het = 6-fluoro-3-pyridyl, 6-chloro-3-pyridyl, 6-bromo-3-pyridyl, and 2-chloro-5-thiazolyl.

The residual activity was evaluated by a pot test. Ten mature adult females were allowed to lay eggs for 1 day on the rice plants. One day after removing the adult females, the rice plants were treated with an aqueous solution at the concentration of 2.7 ppm. The eggs of the brown planthopper were laid in the tissue of the leaf sheath. All the compounds tested showed no ovicidal activity; however, most of them exhibited insecticidal activity about 10 DAT when eggs began to hatch. The number of surviving nymphs was counted when untreated nymphs developed from third to fourth instar. The result indicated that the ethyl group is favorable for R^2 and 6-chloro-3-pyridyl is favorable for the heterocycle (Table 6).

Intensive field studies were conducted to evaluate the field performance of some compounds described in Table 6, which revealed that the compound with these favorable substituents, 1-[*N*-(6-chloro-3-pyridylmethyl)-*N*-ethyl]amino-1-methylamino-2-nitroethene (Fig. 5), was the best candidate for development. After 6 years of official field tests under the code number of TI-304, the compound was launched in 1995 with the name of nitenpyram and the trade name Bestguard.®

Table 4. The effect of the linkage between 3-pyridyl and nitrogen on the insecticidal activity of 1-methylamino-2-nitroethene against the BPH and the GLH

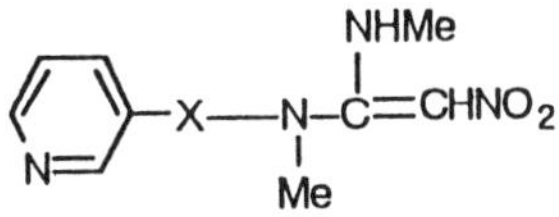

Compounds	X	Method[a]	LC_{95} (ppm)[b]	
			BPH	GLH
36	-	iii	40	>200
22	CH_2	iii	40	40
37	CH_2CH_2	iii	200	>200
38	CHMe	iii	500	200

[a]Roman numerals indicate the methods for synthesis described in the text.
[b]The lowest concentration to achieve 95% mortality. The concentrations of the test chemical solutions were 40, 200, and 500 ppm for the BPH and 40 and 200 ppm for the GLH.

Table 5. The effect of the heteroaryl group on the insecticidal activity of 1-[*N*-(heteroaryl-methyl)-*N*-methyl]amino-1-methylamino-2-nitroethenes against the BPH

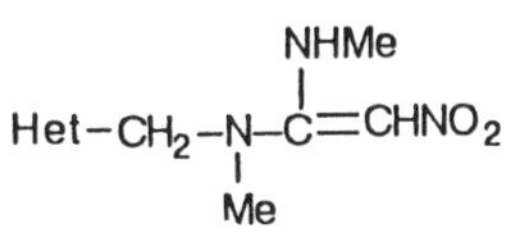

Compounds	Het	Method[a]	LC_{95} (ppm)[b]
			GLH
39	4-pyridyl	iii	40
40	pyradinyl	iii	>40
41	4-thiazolyl	iii	>40
42	4-chlorophenyl	iii	>40
43	5-bromo-3-pyridyl	iii	>40
44	6-methyl-3-pyridyl	iii	2.5
45	6-methoxy-3-pyridyl	-[c]	>40
46	6-fluoro-3-pyridyl	iii	0.5
47	6-chloro-3-pyridyl	iii	0.5
48	6-bromo-3-pyridyl	iii	0.5
49	2-chloro-5-thiazolyl	iii	0.5

[a]Roman numerals indicate the methods for synthesis described in the text.
[b]The lowest concentration to achieve 95% mortality. The concentrations of the test chemical solutions were 0.5, 2.5, and 40 ppm.
[c]Other synthetic method than iii was employed.

Table 6. The effect of the heterocycles (Het) and substituents (R^2) on the residual activity of 1-[*N*-(heteroaromatic-methyl)]amino-1-methylamino-2-nitroethenes against the BPH

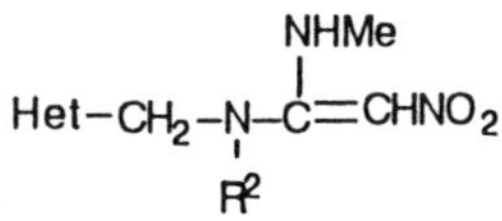

R^2	Het			
	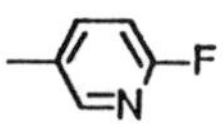	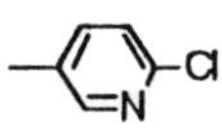	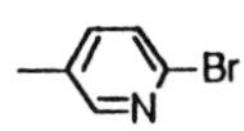	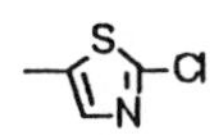
Me	0[a]	77	13	NT[b]
Et	61	100	82	16

[a]Values in the table indicate the control ratio. Control ratio=100×(1−number of insects in the treated plot / number of insects in the untreated plot).
[b]Not tested.

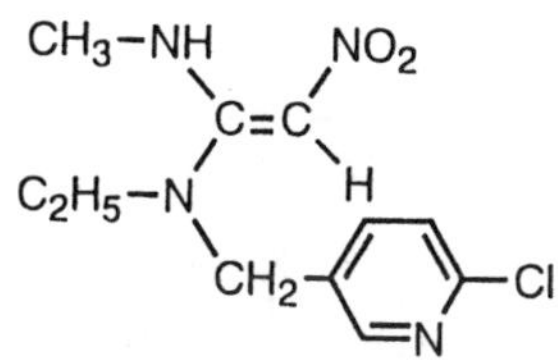

Fig. 5. Chemical structure of nitenpyram

3 The Properties of Nitenpyram

3.1 Chemical and Physical Properties

Nitenpyram is characterized by its high water solubility, 840 g/L, and low partition coefficient, -0.64 (Table 7).

3.2 Toxicological Properties

Nitenpyram has minimal adverse effect against mammals (Table 8), and low toxicity against birds (Table 9) and aquatic organisms (Table 10).

Table 7. Chemical and physical properties of nitenpyram

Molecular formula	$C_{11}H_{15}ClN_4O_2$
Molecular weight	270.7
Appearance	Pale yellow crystals
Odor	Odorless
Melting point	83-84℃
Specific gravity	1.40 (26℃)
Vapor pressure	8.3×10^{-12} mmHg
Solubility (g/L, 20℃)	Water 840, chloroform 700, methanol 670, acetonitorile 430, acetone 290, ethanol 89, ethyl acetate 33, xylene 4.5
Partition coefficent	-0.64 (Log $P_{O/W}$) (25℃)

Table 8. Mammalian toxicities of nitenpyram

Acute oral (LD_{50})	Rat (male/female)	1680/1575 mg/kg
	Mouse (male/female)	867/1281 mg/kg
Acute dermal(LD_{50})	Rat (male,female)	>2000 mg/kg
Acute inhalation(LD_{50})	Rat (male,female)	>5.8 mg/L
Primary eye irritation	Rabbit	Minimal irritant
Primary dermal irritation	Rabbit	Nonirritant
Dermal sensitization	Guinea pig	Nonsensitiser

Table 9. Avian toxicities of nitenpyram

Bird	Acute oral toxicity (LD_{50}, LC_{50})	
Mallard duck	1124 mg/kg	(capsule)
	>5620 ppm	(dietary)
Bobwhite quail	>2250 mg/kg	(capsule)
	>5620 ppm	(dietary)

3.3 Animal Metabolism

Animal metabolism was investigated by the oral administration of nitenpyram labeled with ^{14}C to rats. As a result, 95-98% of ^{14}C administered was excreted mainly into the urine within 2 DAT and no accumulation of ^{14}C in the internal organs was observed. The results indicate that the high water solubility and the low partition coefficient of nitenpyram may contribute to its low toxicity against mammals.

Table 10. Aquatic toxicities of nitenpyram

Organism		LC_{50}(48 h)
Fish	Carp, medaka	>1,000 mg/L
	Goldfish, asian pond loach	>10 mg/L
	Japanese eel, guppy	>10 mg/L
	Rainbow trout, mullet	>10 mg/L
Crustacean	Water flea	>10,000 mg/L[a]
	Crawfish, striped prawn	>10 mg/L
Shellfish	Pond snail, sakamaki snail	>10 mg/L
	Corbicula, shortnecked clam	>10 mg/L
Amphibian	Tadpole	>10 mg/L

[a]LC_{50}(24 h).

Table 11. Half-life of nitenpyram in water and soil

Aqueous photolysis	pH 5	18-21 min
	Purified water	16 min
Soil half-lives under the aerobic condition:		
	Floaded	<1 day
	Upland	1-3 days

3.4 Environmental Fate

Nithiazine (**1**), a lead compound, has not been developed because of its poor field efficacy, which was considered to be a result principally of its short photochemical half-life (Harris et al. 1986). Nitromethylene compounds, including nithiazine (**1**), are considered to decompose far more rapidly than nitroimines (Kagabu and Medej 1995). Nitenpyram rapidly decomposed in water under the condition of 12,000 lux irradiation intensity using a 500 W xenon lamp, and in soils under aerobic conditions (Table 11). The photostability of nitenpyram did not improve compared to its lead compound. Poor photostability is usually considered to be a defect of a pesticide. However, rapid decomposition of pesticides is a favorable characteristic for environmental safety.

3.5 Insecticidal Properties

3.5.1 Contact and Oral Toxicity

The fourth-instar nymphs of brown planthoppers were anaesthetized with carbon dioxide, and were topically treated with 0.012 µL of acetone solution of the test chemicals. The treated insects were transferred to a petri dish with rice seedlings. Oral toxicity against brown planthoppers was determined by applying the test chemicals in 5% sucrose solutions. The third instar nymphs were placed in a small glass jar with its top covered with an expanded film membrane (Parafilm). The chemicals in the diet were placed on the film to allow the insects to feed. Mortality was assessed at 6 DAT. The insecticidal activity of nitenpyram against the brown planthopper by topical and oral applications was much higher than those of conventional insecticides (Table 12).

3.5.2 Systemic Action

The translaminar activity of nitenpyram against the cotton aphid, *Aphis gossypii*, was examined. The test method was to treat the upper surface of the cucumber leaf with the chemical solution after placing the adult females of the apterous vivipara onto the undersurface of the leaf. The LC_{50} of nitenpyram was as low as 0.59 ppm. The result indicates that nitenpyram is effective against aphids on the undersurfaces of leaves where little deposit of chemical solution is expected.

The insecticidal activity of nitenpyram by translocation from the treated leaf to the untreated leaves was examined against the cotton aphid. The second leaf of the cucumber was sprayed with the chemical solution of the practical concentration, with the other parts of the plant covered with a polyethylene bag to keep them free from chemical deposits. The adult females of the apterous vivipara were placed onto the untreated first and third leaves on the day of treatment, and 1 week after and 2 weeks after treatment. The numbers of aphids were counted 5 days after the inoculation. Nitenpyram highly suppressed the aphid densities, not only on the first leaf, but on the third leaf for 2 weeks after treatment (Table 13).

The excellent insecticidal activity by translaminar action and translocation was the outstanding characteristic of nitenpyram, and its good systemicity will minimize the adverse effect of biased deposit distribution of the sprayed chemical solution.

Table 12. Contact and oral toxicities against brown planthopper

Compound	Contact LD_{50} (ug/g)	Oral LC_{50} (ppm)
Nitenpyram	0.016	0.001
Ethofenprox	0.62	45
Buprofezin	2.9	0.69
Fenobucarb	6.6	1.8

Table 13. Insecticidal activities against cotton aphid of chemicals translocated from the treated leaf to the untreated leaves

Chemicals[a]	Concentration (ppm)	Leaf position[c]	Modified aphid density[b]		
			0DAT	7DAT	14DAT
Nitenpyram WSG	100	1st	0	0	10
		3rd	0	0	8
Pirimicarb WP	240	1st	75	66	-[d]
		3rd	50	129	-
Methomyl WP	450	1st	80	106	-
		3rd	70	122	-
Fenitrothion EC	500	1st	97	80	-
		3rd	57	132	-

[a]WSG, water-soluble granule; WP, wettable powder; EC, emulsifiable concentrate.
[b]Modified aphid density = 100×(number of aphids on the treated plant / number of aphids on the untreated plant).
[c]Second leaf was treated with chemical solution and the aphids were placed on the first leaf and the third leaf. The leaf position was counted from the basal true leaf.
[d]Not tested.

3.6 Field Performance

3.6.1 Foliar Spray

Foliar spray of the dust formulation of pesticides is the common method to control rice pests in Japan. Figure 6 demonstrates control efficacy against the brown planthopper by a spray of a dust formulation. Nitenpyram applied at the rate of 75 g a.i./ha as a 0.25% dust formulation strongly suppressed brown planthopper density for more than 3 weeks. Application of the dust formulation of nitenpyram at the rate of 100 g a.i./ha successfully controlled stink bugs (mixed population of *Nezara viridula, Leptocorisa chinensis*, and *Cletus punctiger*) in the rice field (Fig. 7). A mixed formulation of nitenpyram and cartap hydrochloride was developed for the simultaneous control of these hemipterous pests and the lepidopterous pests, which are the major rice pests in the middle and late cultivation period of rice in Japan.

Nitenpyram showed good control of the cotton aphid for 3 weeks by a foliar spray of chemical solutions at the rate of 100 g and 200 g a.i./ha of the 10% WSG (water-soluble granule) formulation (Fig. 8). Nitenpyram is considered to rapidly decompose in the environment; however, it shows good residual effect by foliar spray. Thus, nitenpyram has the desirable properties of a foliar spray pesticide.

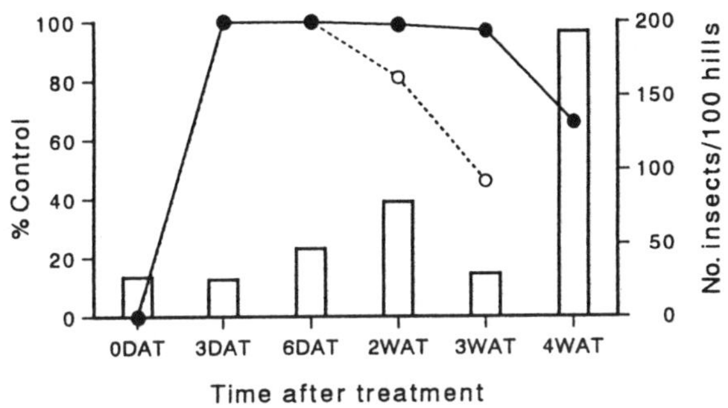

Fig. 6. Control of the brown planthopper by a foliar spray. *Solid line*, 0.25% dust formulation of nitenpyram, 75 g a.i./ha. *Dotted line*, 0.5% dust formulation of ethofenprox, 150 g a.i./ha. *Open bars*, the insect density in the untreated control. *DAT*, days after treatment; *WAT*, weeks after treatment

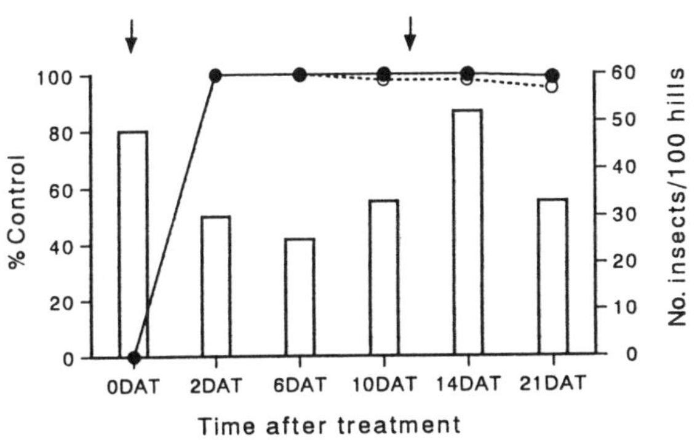

Fig. 7. Control of the stink bug on rice plants by a foliar spray. Chemicals were applied two times (*arrows*) at an interval of 11 days. The rates of injured rice grain, examined at 25 days after initial treatment, were 0.2%, 0.1%, and 1.4% for nitenpyram, fenthion, and untreated check, respectively. *Solid line*, 0.25% dust formulation of nitenpyram, 100 g a.i./ha. *Dotted line*, 2.0% dust formulation of fenthion, 800 g a.i./ha. *Open bars* represent the insect density in the untreated control

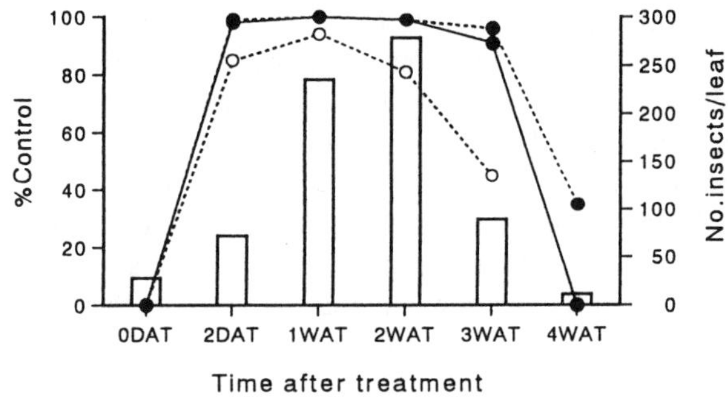

Fig. 8. Control of the cotton aphid on eggplants by foliar spray. *Solid line,* 10% WSG (water-soluble granule) of nitenpyram, 100 g a.i./ha. *Dotted line with solid circles,* 10% WSG of nitenpyram, 200 g a.i./ha. *Dotted line with open circles,* 50% WP (water-soluble powder) of acephate, 1000 g a.i./ha. *Open bars,* the insect density in the untreated control

3.6.2 Granular Broadcast on Paddyfield

Application of granules requires no special equipment, which saves the application cost. Nitenpyram of 1% granule formulation applied at the rate of 200 g a.i./ha showed excellent control against the brown planthopper (Fig. 9).

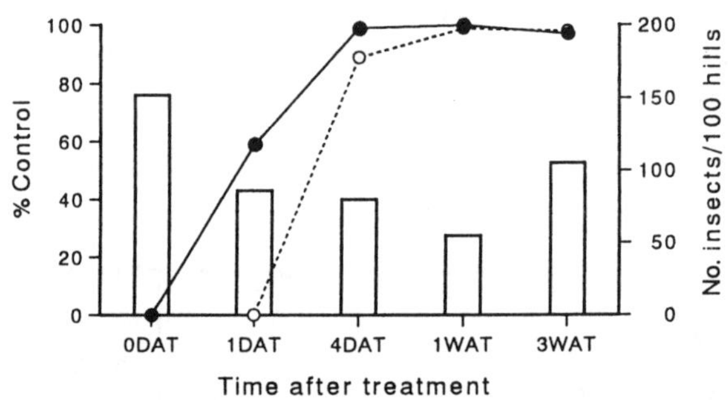

Fig. 9. Control of the brown planthopper by a granular broadcast. *Solid line,* 1.0% granule formulation of nitenpyram, 200 g a.i./ha. *Dotted line,* 2.0% granule formulation of buprofezin, 800 g a.i./ha. *Open bars,* the insect density in the untreated control

3.6.3 Soil Treatment

Nitenpyram was demonstrated to easily translocate from roots to leaves, which may be due to its high water solubility. The application of systemic pesticides to the root zone of crops is sometimes restricted by the phytotoxicities of the product used. However, in tests of nitenpyram, we have been able to confirm the absence of harmful side effects to 20 different crop types, and no adverse effect on crops has been reported since nitenpyram reached the market. The excellent systemic property of nitenpyram, combined with its high safety against crops, enable the compound to control pests by some soil treatment methods: planting hole application, plant foot treatment before/after transplanting, and soil incorporation, etc. Nitenpyram can also be applied combined with fertilizer using irrigation systems. Nitenpyram, by these application methods, controlled the cotton aphid for 6 weeks (Fig. 10), the silverleaf whitefly, *Bemisia argentifolii*, for 8 weeks (Fig. 11), the melon thrips, *Thrips palmi* (Fig. 12), and the western flower thrips, *Frankliniella occidentalis* (Fig. 13), for 4 weeks, and the serpentine leafminer, *Liriomyza trifolii*, for 3 weeks (Fig. 14), at the rate of 10-20 mg/plant. These pest insects are the major greenhouse pests that are difficult to control by conventional insecticides. Nitenpyram can control these greenhouse major pests simultaneously by a single application.

3.7 Effect Against Nontarget Arthropods

Recently, integrated pest management (IPM) has been prevailing, and minimal adverse effects of pesticides against beneficial insects are desired. IPM compatibility is now an important characteristic of the pesticide.

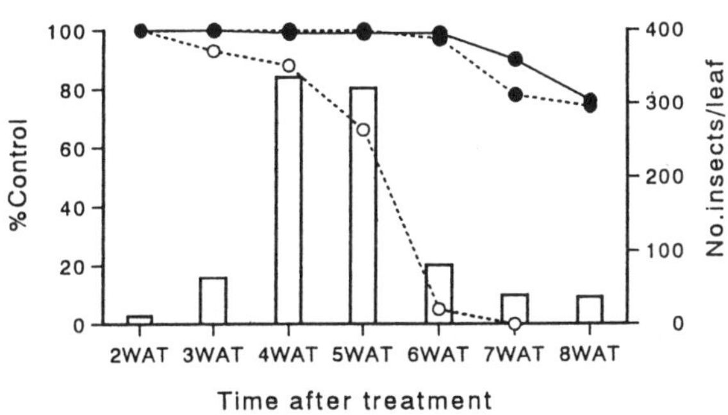

Fig. 10. Control of the cotton aphid on eggplants by soil application methods. *Solid line with solid circles*, planting hole application of 1.0% granule formulation of nitenpyram, 10 mg a.i./plant. *Dotted line with solid circles*, plant foot treatment with a chemical solution of nitenpyram before transplanting on potted seedling, 10 mg a.i./plant. *Dotted line with open circles*, planting hole application of 5.0% granule formulation of acephate, 50 mg a.i./plant. *Open bars*, the insect density in the untreated control

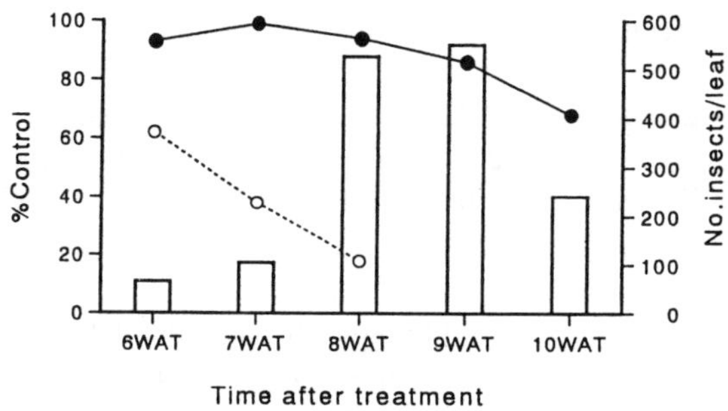

Fig. 11. Control of the silverleaf whitefly on tomatoes by a soil application. *Solid line*, plant foot treatment before transplanting on potted seedlings with 1.0% granule formulation of nitenpyram, 10 mg a.i./plant. *Dotted line*, planting hole application of 5.0% granule formulation of acephate, 50 mg a.i./plant. *Open bars*, the insect density in the untreated control

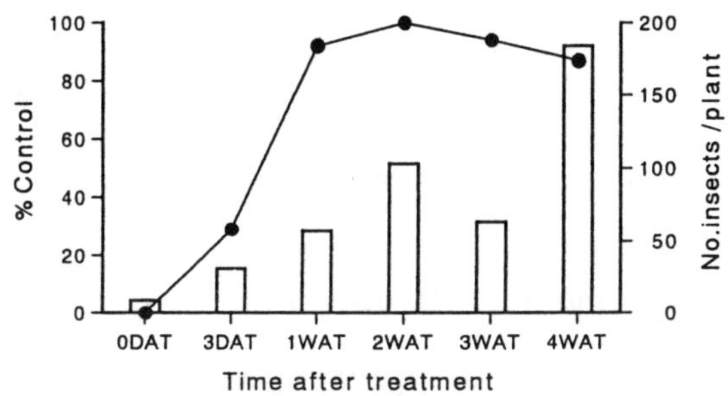

Fig. 12. Control of the melon thrips on eggplants by a soil application. *Solid line*, planting hole application with 1.0% granule formulation of nitenpyram, 10 mg a.i./plant. *Open bars*, the insect density in the untreated control

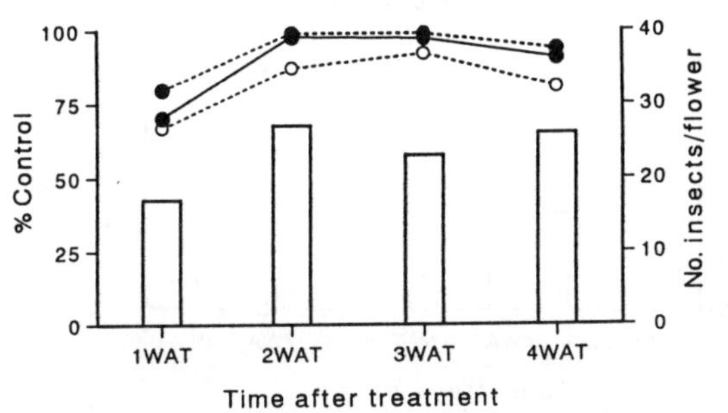

Fig. 13. Control of the western flower thrips on chrysanthemums by a soil application. *Solid line*, plant foot treatment with 1.0% granule formulation of nitenpyram, 10 mg a.i./plant. *Dotted line with solid circles*, plant foot treatment with 1.0% granule formulation of nitenpyram, 20 mg a.i./plant. *Dotted line with open circles*, plant foot treatment with 5.0% granule formulation of acephate, 50 mg a.i./plant. *Open bars*, the insect density in the untreated control

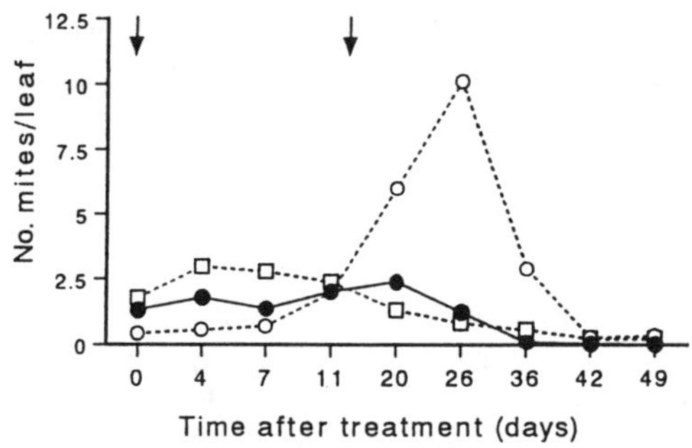

Fig. 14. Control of the serpentine leafminer on tomatoes by a soil application. *Solid line*, plant foot treatment before transplanting on potted seedling with 1.0% granule formulation of nitenpyram, 20 mg a.i./plant. *Dotted line*, plant foot treatment after transplanting with 1.0% granule formulation of nitenpyram, 20 mg a.i./plant. *Open bars*, the insect density in the untreated control

3.7.1 Effect Against Pollinators

The employment of some kinds of bees as pollinators, combined with natural enemies as biological control agents, has recently increased especially in greenhouses. The insecticidal activity of nitenpyram was not low against the honeybee, *Apis mellifera*, and the bumblebee, *Bombus terrestris*; however, the residual periods are estimated to be 5 days against the honeybee and 10 days against the bumblebee after nitenpyram application with a 100 ppm solution. The effects against these bees were investigated by introducing the bees to the greenhouse at the practical timing, and no adverse effect was observed for the planting hole application of nitenpyram at the rate of 20 mg/plant. The instability of nitenpyram in the environment may contribute to minimize its adverse effects against beneficial insects.

3.7.2 Resurgence of the Phytophagous Mite

Foliar spray of some insecticides sometimes induces the outbreak of phytophagous mites (Hoyt 1969; McCoy 1977; Hall 1979), which restricts their application. Nitenpyram showed miticidal activity against the predatory mite, *Phytoseiulus persimilis*, but the activity was not high and decreased rapidly (Table 14). As shown in Fig. 15, nitenpyram has little risk of inducing the resurgence of phytophagous mites.

3.8 Mode of Action

Nitromethylene derivatives are considered to act as an agonist on the nicotinic acetylcholine receptors in postsynaptic membranes (Harris et al. 1986; Bai et al.

1991; Tomizawa and Yamamoto 1992). The nicotinic acetylcholine receptor is also the target site of cartap (Sakai 1966). The mode of action of nitenpyram was compared with that of cartap hydrochloride.

3.8.1 Increase of Insecticidal Efficacy with the Lapse of Time

Nitenpyram and cartap hydrochloride were applied to the brown planthopper and the mortality was assessed every day. Nitenpyram at high concentrations rapidly killed nymphs of the brown planthopper, while at the low concentrations it took a few days to show its insecticidal activity. The LC_{50} of nitenpyram on 5 DAT was less than 1/100 of that on 1 DAT, while that of cartap hydrochloride was 1/3 (Fig. 16). The result suggests that there is a difference in the mode of action between the two compounds.

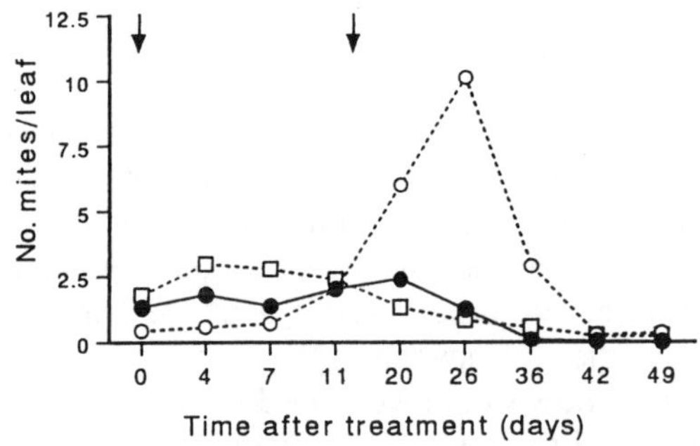

Fig. 15. Resurgence of the phytophagous mite induced by insecticide applications by foliar spray. Chemicals were applied two times (*arrows*) at an interval of 12 days. *Solid line*, 100 ppm solution of nitenpyram 10% WSG. *Dotted line with open circles*, 7 ppm solution of tralomethrin 1.4% SC (soluble concentrate). *Dotted line with open squares*, untreated control

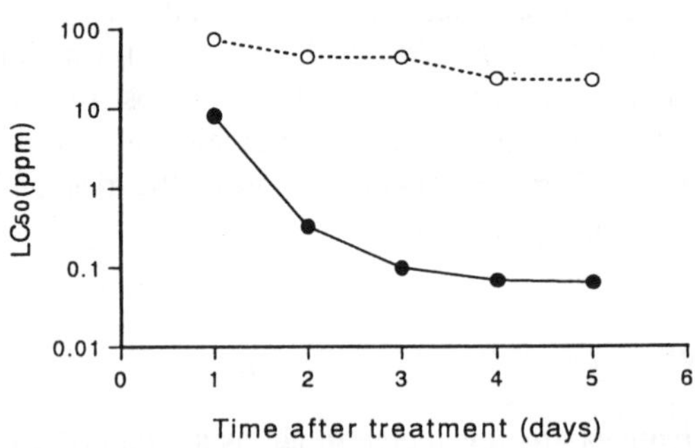

Fig. 16. Change of LC_{50} with the lapse of time. *Solid line*, nitenpyram. *Dotted line*, cartap hydrochloride

Table 14. Miticidal activities against the predatory mite, *Phytoseiulus persimilis*

Chemicals	LC_{50}(ppm)	
	0 DAT	5 DAT
Nitenpyram 10WSG	24	>100
Fenitrothion 50EC	230	250-500
Trichlorfon 50EC	6.1	33

Miticidal activities were determined by foliar spray of the chemical solutions on the predatory mites placed on the leaves of kidney bean. Mortality was assessed 2 days after inoculation.

Table 15. Joint action of nitenpyram and cartap mixed in the ratio of LC_{50} of each compound

Insects	Mixture ratio[a]	LC_{50} (ppm)	Co-toxicity coefficient[b]
Brown planthopper	1 : 0	0.017	-
	1 : 1500	17	67
	0 : 1	22	-
Rice stem borer	1 : 0	53	-
	7.5 : 1	52	48
	0 : 1	5.1	-

[a]Mixure ratio of nitenpyram and cartap hytdrochloride (w:w).
[b]Co-toxicity coefficient was calculated by the equation proposed by Sun and Johnson (1960).

Table 16. Joint action of nitenpyram and cartap mixed in the ratio of the content of each compound in the mixed dust formulation

Insect	Mixture ratio[a]	LC_{50} (ppm)	Co-toxicity coefficient[b]
Brown planthopper	1 : 0	0.035	-
	1 : 8	0.31	102
	0 : 1	47	-

[a]Mixure ratio of nitenpyram and cartap hytdrochloride (w:w).
[b]Co-toxicity coefficient was calculated by the equation proposed by Sun and Johnson (1960).

3.8.2 Joint Action with Cartap

The mixed chemical solutions of nitenpyram and cartap hydrochloride were applied to the brown planthopper and the rice stem borer, and the co-toxicity coefficients were calculated by the equation proposed by Sun and Johnson (1960). The values of the co-toxicity coefficients were 67 for the brown planthopper and 48 for the rice stem borer, when the two compounds were mixed at the ratio of each LC_{50} (Table 15). Nevertheless, the value was 102 for the brown planthopper when nitenpyram and cartap hydrochloride were mixed at the same ratio as that in the mixed dust formulation on the market (Table 16). The results indicate that these two compounds act antagonisticly to each other; however, in the mixed dust formulation, the insecticidal activity of nitenpyram may not be influenced by cartap hydrochloride in practical use.

3.8.3 Effect on the Nerve Activity of the American Cockroach

The electrophysiological study of the effect of nitenpyram on neurotoxic action against the American cockroach demonstrated that nitenpyram was characterized by an initial increase in the frequency of spontaneous giant fiber discharges, followed by the complete block of nerve impulse transmission. The result agreed with those obtained for the other nitromethylene derivatives (Schroeder and Flattum 1984; Harris et al. 1986; Sone et al. 1994) and nicotine (Sakai 1967), and disagreed with that obtained for cartap hydrochloride (Sakai 1967; Bettini et al. 1973).

3.8.4 Effect on the Threshold Stimulus to Induce Action Potential

The sixth abdominal ganglion of the American cockroach was treated with 10 micromolar of the test chemical, and the threshold stimulus to induce action potential was measured. The threshold stimulus was neither affected by nitenpyram nor nicotine, while cartap hydrochloride gradually increased it (Fig. 17). These results suggest there may be differences between nitenpyram and cartap in their binding affinity to the nicotinic acetylcholine receptors or in their binding sites, although the two compounds may act on the same receptor.

4 Conclusion

Nitenpyram was discovered in the process of the optimization of the substituents of an acyclic nitroetene (see Fig. 3). The compound maintains poor photostability as its lead compound, nithiazine (**1**); however, it shows excellent residual effect against rice hoppers and aphids by foliar spray. The poor photostability of nitenpyram is considered to be advantageous for environmental safety and for minimization of the resurgence of the phytophagous mite and adverse effects against beneficial insects.

Nitenpyram is highly soluble in water, and shows excellent systemic action and no phytotoxicity, which characteristics enable the various application methods of the compound. The compound has been developed for controlling the brown planthopper, however, it is highly active against some other pest species, and a single soil application of nitenpyram effectively controls aphids, whiteflies, thrips, and the serpentine leafminer, which are the major greenhouse pests difficult to control by conventional insecticides.

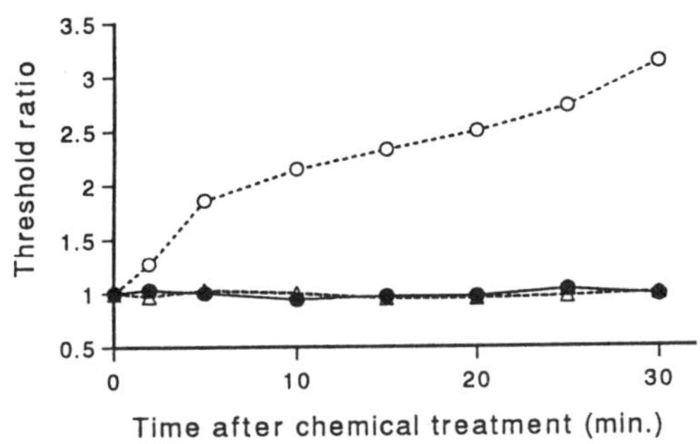

Fig. 17. Effect of time lapse upon the action induction threshold for nicotinoid insecticides and cartap hydrochloride. *Solid line*, nitenpyram. *Dotted line with open triangles*, nicotine. *Dotted line with open circles*, cartap hydrocloride

References

Bai D, Lummis SCR, Leight W, Breer H, Sattelle DB (1991) Actions of imidacloprid and a related nitromethylene on cholinergic receptors of an identified insect motor neurone. Pestic Sci 33:197-204

Bettini S, D'Ajello V, Maroli M (1973) Cartap activity on the cockroach nervous and neuromuscular transmission. Pestic Biochem Physiol 3:199-205

Dyck VA, Thomas B (1979) The brown planthopper problem. In: Brown planthopper: threat to rice production in Asia, International Rice Research Institute, pp 3-17

Endo S, Nagata T, Kawabe S, Kazano H (1988) Changes of insecticide susceptibility of the white backed planthopper *Sogatella furcifera* HORVATH (Homoptera: Delphacidae) and the brown planthopper *Nilaparvata lugens* STAL (Homoptera: Delphacidae). Appl Entomol Zool 23:417-421

Hall FR (1979) Effects of synthetic pyrethroids on major insect and mite pests of apple. J Econ Entomol 72:441-446

Harris M, Prece RN, Robinson J, May TE (1986) WL108477 - a novel neurotoxic insecticide. British Crop Protection Conference-Pests and Diseases 2B-4:115-122

Hirai K (1994) Abundance of *Chilo suppressalis* (WALKER) for the latest 30 years in Japan (in Japanese). Plant Prot 48:51-52

Hoyt SC (1969) Integrated chemical control of insects and biological control of mites on apple in Washington. J Econ Entomol 62:74-86

Kagabu S, Medej S (1995) Stability comparison of imidacloprid and related compounds under simulated sunlight, hydrolysis, and to oxygen. Biosci Biotechnol Biochem 59:980-985

Kilin D, Nagata T, Masuda T (1981) Development of carbamate resistance in the brown planthopper, *Nilaparvata lugens* STAL (Homoptera: Delphacidae). Appl Entomol Zool 16:1-6

McCoy CW (1977) Resurgence of citrus rust mite populations following application of methidathion. J Econ Entomol 70:748-752

Minamida I, Iwanaga K, Tabuchi T, Uneme H, Dantsuji H, Okauchi T (1993a) Synthesis and insecticidal activity of acyclic nitroethene compounds containing a 3-pyridylmethylamino group. J Pestic Sci 18:31-40

Minamida I, Iwanaga K, Tabuchi T, Aoki I, Fusaka T, Ishizuka H, Okauchi T (1993b) Synthesis and insecticidal activity of acyclic nitroethene compounds containing a heteroarylmethylamino group. J Pestic Sci 18:41-48

Nagata T, Masuda T, Moriya S (1979) Development of insecticide resistance in the brown planthopper, *Nilaparvata lugens* STAL (Homoptera: Delphacidae). Appl Entomol Zool 14:264-269

Sakai M (1966) Studies on the insecticidal action of nereistoxin, 4-*N,N*-dimethylamino-1,2-dithiolane III. Scientific pest control (Bochu Kagaku) 31:61-67

Sakai M (1967) Studies on the insecticidal action of nereistoxin, 4-*N,N*-dimethylamino-1,2-dithiolane V. Scientific Pest Control (Bochu Kagaku) 32:21-33

Schroeder ME, Flattum RF (1984) The mode of action and neurotoxic properties of the nitromethylene heterocycle insecticides. Pestic Biochem Physiol 22:148-160

Shiokawa K, Tsuboi S, Kagabu S, Seike K (1985) New heterocyclic nitromethylene derivatives - useful as insecticides, acaricides and nematocides. Japanese Patent Application Laid Open 60-172976

Shiokawa K, Tsuboi S, Kagabu S, Seike K (1986) New nitromethylene or nitroimino heterocyclic compounds - useful as insecticides. Japanese Patent Application Laid Open 61-178982

Soloway SB, Henry AC, Kollmeyer WD, Padgett WM, Powell JE, Roman SA, Tieman CH, Corey RA, Horne CA (1978) In: Geissbuehler H, Brooks GT, Kearney PC (eds) Advances in pesticide science, Part 2. Pergamon Press, Oxford, pp 206-217

Sone S, Nagata K, Tsuboi S, Shono T (1994) Toxic symptoms and neural effect of a new class of insecticide, imidacloprid, on the American cockroach, *Periplaneta americana* (L.). J Pestic Sci 19:69-72

Sun YP, Johnson ER (1960) Analysis of joint action of insecticides against house flies. J Econ Entomol 53:887-892

Tabuchi T, Fusaka T, Iwanaga K, Minamida I, Okauchi T (1994) Synthesis and insecticidal activity of acyclic nitroethene compounds containing (6-substituted)-3-pyridylamino group. J Pestic Sci 19:119-125

Tomizawa M, Yamamoto I (1992) Binding of nicotinoids and the related compounds to the insect nicotinic acetylcholine receptor. J Pestic Sci 17:231-236

Wada T, Kobayashi M (1980) Outbreaks of rice leaffolder in 1980 and its ecological properties (in Japanese). Plant Prot 34:528-532

7 A Novel Insecticide, Acetamiprid

Tomio Yamada, Hidemitsu Takahashi, and Renpei Hatano
Odawara Research Center, Nippon Soda Co., Ltd.,
345 Takada, Odawara, Kanagawa, Japan

1 Introduction

Acetamiprid, (E)-N^1-[(6-chloro-3-pyridyl)methyl]-N^2-cyano-N^1-methylacetamidine, is a novel insecticide developed by Nippon Soda Co., Ltd. Although the compound belongs to the neonicotinoids, it possesses characteristic insecticidal properties different from others in the same category of chemical structure. Acetamiprid shows excellent activities against Hemiptera and Thysanoptera, as do other neonicotinoids; it exhibits excellent activity against Lepidoptera as well, and the insecticide is applicable for controlling pests of vegetables, fruit trees, the tea tree, and so on. There are various kinds of insect pests damaging agricultural crops, and development of resistance to insecticides in many insect pests such as the diamondback moth and aphids has become a serious problem in recent years. Especially in the diamondback moth, the speed of resistance development is relatively fast; therefore, a compound that possesses a mode of action different from conventional insecticides needs to be developed. Also, pest control strategy that is safe for the environment is essential. Under these circumstances, we attempted to find a compound that possessed excellent efficacy against insects which are difficult to control, showed no cross-resistance to conventional insecticides, and was benign to the environment.

2 Physical and Biological Properties

2.1 Chemical Structure and Physical Properties

Acetamiprid has a cyanoamidine structure, which contains a 6-chloro-3-pyridylmethyl moiety. The compound was invented in the search for nitromethylene derivatives by Nippon Soda Co., Ltd., in 1989 and was registered in 1995 in Japan. The insecticide is being developed all over the world under the experimental code number NI-25 and was commercialized with the trade name Mospilan® in Japan.

Its chemical structure and physical properties are shown in Fig. 1. The insecticide was registered as formulations of 20% soluble powder and 2% granules in Japan (Takahashi et al. 1992).

Common name : acetamiprid (ISO)
Trade name : MOSPILAN®
Code name : NI-25
Chemical name : (E)-N^1 [(6-chloro-3-pyridyl) methyl] N^2-cyano-N^1-methylacetamidine
Molecular formula : $C_{10}H_{11}Cl\ N_4$
Structure formula :

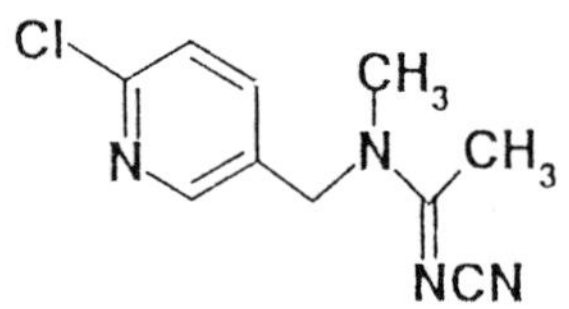

Molecular weight : 222.68
Physical appearance : White crystal
Melting point : 98.9℃
Vapor pressure : $<1\times10^{-6}$Pa at 25℃
Solubility at 25℃ : 4250mg/*l* in water. Soluble in acetone, methanol, ethanol, dichloromethane, chloroform, acetonitrile, tetrahydrofuran
Partition coefficient (Log P_{ow}) : 0.8
Hydrolytic stability : Degraded gradually at pH9 and 45℃
Photostability : Stable under sunlight
Formulation : 20%SP(W/W), 2%G(W/W)

Fig. 1. Chemical structure and physical properties of acetamiprid.

2.2 Toxicological Studies

Acetamiprid is of relatively low toxicity to mammals. The acute oral toxicity of the compound to male rat and mouse is about 200 mg/kg by LD_{50} value. Toxicological tests for acetamiprid, e.g., irritation to skin and eyes, sensitization to skin, and mutagenecity, as evaluated by Ames test, were all negative.

Ecotoxicities of the compound to fish and *Daphnia* are very low, showing LC_{50} values for carp and *Daphnia* $>$ 100 mg/l and $>$ 1000 mg/l, respectively. Furthermore, the sub-acute and chronic toxicity studies proved that acetamiprid was safe in common agricultural usage (Table 1) (Takahashi et al. 1992; Mastuda and Takahashi 1996a,b; Matsuda 1995).

Table 1. Toxicity and ecotoxicity of acetamiprid

Tests	Results
Toxicity	
Acute oral LD_{50}	Rat: male,217 mg/kg; female,146 mg/kg
	Mouse: male,198 mg/kg; female,184 mg/kg
Acute dermal LD_{50}	Rat: male and female,>2,000 mg/kg
Skin and eye irritation	Nonirritating (rabbit)
Skin sensitization	Negative (guinea pig)
Mutagenicity	Negative (Ames test)
Ecotoxicity	
Carp, LC_{50}	>100 mg/ l
Daphnia , LC_{50}	>1,000 mg/ l

Table 2. Adverse effect of acetamiprid on beneficials

Beneficials	Results
Predaceous mite	
Amblyseius womersleyi	LC_{20}: 20~25 ppm (adult female)
Agritemus sp.	LC_{20}: 25 ppm (adult female)
Phytoseiulus persimilis	LC_{20}: 25 ppm (egg)
Honeybee	
Apis mellifera	LD_{50}: >200 (adult)
Bumblebee	
Bumbus terrestris	LD_{50}: >200 (adult)

2.3 Adverse Effect on Beneficials

Although acetamiprid possesses a broad insecticidal spectrum as described later, it was ascertained that the compound showed less adverse effect on both natural enemies, such as predaceous mites, and beneficial insects, such as the honeybee and bumblebee, in laboratory tests (Table 2) (Takahashi et al. 1998a).

3 Discovery of Acetamiprid

3.1 Lead Generation

As a lead compound for discovering a new insecticide, nithiazine (reported by Shell Company) attracted our attention as the compound possessing a mode of action different from most conventional insecticides. Nithiazine (Soloway et al. 1978) showed good activity, but its insufficient persistency was not improved by the analogues (Harris et al. 1986). Later, the patents of Nihon Bayer compounds (Shiokawa et al. 1985) were disclosed. These compounds aroused our keen interest because they possessed excellent systemic activity as well as high activity against hemipterous insect pests such as aphids, planthoppers, and leafhoppers. It was recognized that these compounds, however, showed rather weak activity against Lepidoptera compared to that against Hemiptera, and our research started targeting compounds with the highest activity against Lepidoptera. In screening tests, it was noted that a cyanoimino compound, belonging to Nihon Bayer's compounds shown in Fig. 2, induced severe excitation to the German cockroach, *Blattella germanica*, by injection, regardless of its less lethal activity against the armyworm, *Pseudaletia separata*, and the cotton aphid, *Aphis gossypii*. Acetamiprid was later discovered by optimization studies of such cyanoimino compounds (Fig. 2).

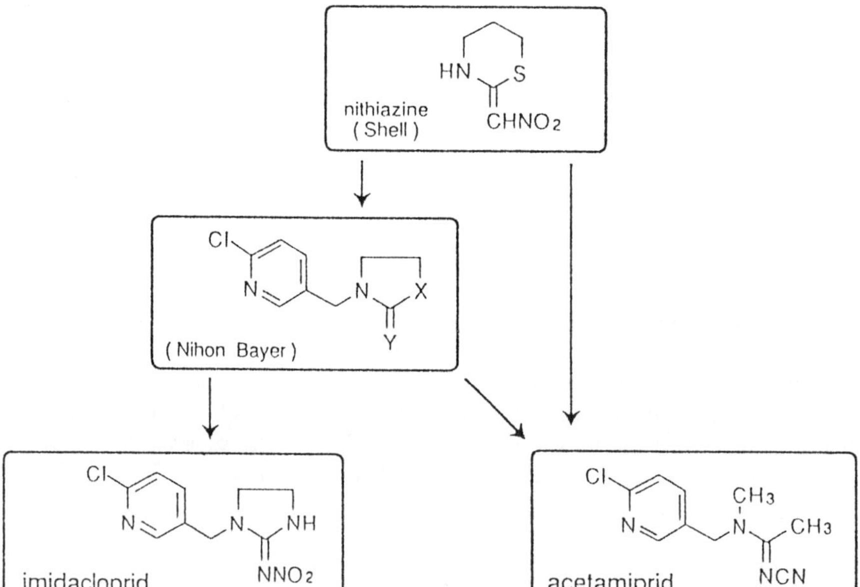

Fig. 2. Chemical structure of acetamiprid and related compounds.

Table 3. Insecticidal activity of imidazolidine derivatives

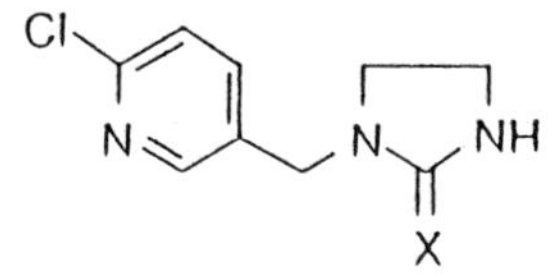

X	≧ LC_{90} (ppm)		Toxic Symptom[c]
	P. separata[a]	*A. gossypii*[b]	
$=CHNO_2$	8	0.12	+
$=NNO_2$	31	0.5	+
=NCN	>125	0.5	++

[a]Leaf-dipping method, second-instar larvae.
[b]Spraying method, first-instar nymphs.
[c]Adult males of the German cockroach, *Blattella germanica*.
+, weak excitation; ++, strong excitation by injection of chemical solution.

Table 4. Insecticidal activity of acetamiprid derivatives

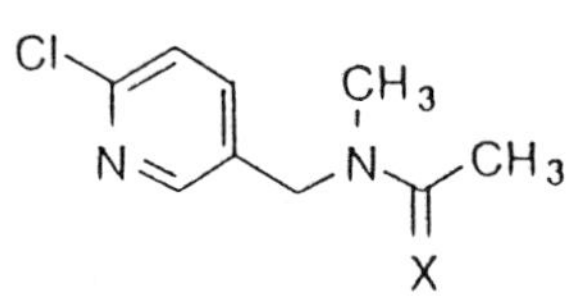

X	≧ LC_{90} (ppm)		Toxic Symptom[c]
	P. separata[a]	*A. gossypii*[b]	
$=CHNO_2$	125	8	+
$=NNO_2$	>125	0.5	−
=NCN	8	0.12	++

[a]Leaf-dipping method, second-instar larvae.
[b]Spraying method, first-instar nymphs.
[c]Adult males of *Blattella germanica*.
-, no symptom; +, weak excitation; ++, strong excitation by injection.

3.2 Chemical Structure and Biological Activities

The effects of the substituent at the 2-position of the imidazolidine ring on the activities were compared for the three compounds shown in Table 3. The compound having cyanoimine structure (X: =NCN) caused more severe excitation in the cockroach than compounds with nitromethylene (X: $=CHNO_2$) and nitroimine (X: $=NNO_2$), but the activities of the cyanoimine derivative against the armyworm and the aphid were insufficient. This suggested that the cyanoimine structure was effective to cause excitation of insects, and that the replacement of the imidazolidine ring by acyclic structures was of interest to develop new generations. Table 4 shows the activities of three compounds without imidazolidine structure. Among them, the nitromethylene and nitroimine derivatives were less active than the corresponding compounds with the imidazolidine structure shown in Table 3. The acyclic cyanoimine derivative, however, exhibited excellent insecticidal activities against the armyworm and the aphid, as well as causing severe excitation of the cockroach.

Table 5. Insecticidal activity of acetamiprid derivatives

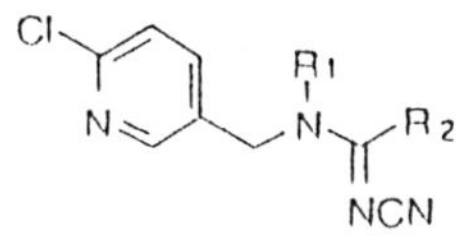

R1	R2	LC_{50} (ppm)	
		P. xylostella[a]	*A. gossypii*[b]
H	CH_3	346	0.05
CH_3	H	> 500	0.2
CH_3	CH_3	19	0.056
CH_3	C_2H_5	100	0.14
CH_3	*i*-C_3H_7	> 500	0.2
CH_3	*n*-C_4H_9	> 500	0.82
C_2H_5	CH_3	437	0.07
n-C_3H_7	CH_3	> 500	14
CH_2F	CH_3	73	0.058
CH_2Cl	CH_3	> 500	1.5
CH_3	CHF_2	> 500	14
phenyl	CH_3	> 500	>125

[a]Leaf-dipping method, second-instar larvae.
[b]Spraying method, first-instar nymphs.

Table 5 shows the substituent effects in the acyclic cyanoimine moiety to the insecticidal activites against the second-instar larvae of the diamondback moth, *Plutella xylostella*, and the first-instar nymphs of the cotton aphid. Regarding the substituent (R_1) on the amino nitrogen, the methyl group exhibited the highest activity against the diamondback moth, and the compounds with hydrogen, methyl, and ethyl groups showed potent activity against the cotton aphid. The larger groups such as *n*-propyl and phenyl at R_1 resulted in decrease of the activities. Among the compounds having various substituents (R_2) at the imino carbon, the compound with methyl group was the most active against the diamondback moth when R_1 was fixed to methyl.

Table 6. Insecticidal activity of acetamiprid derivatives

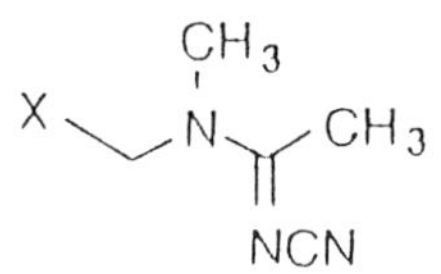

X	LC_{50} (ppm)	
	P. xylostella[a]	*A. gossypii*[b]
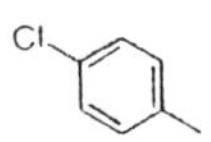	>500	44
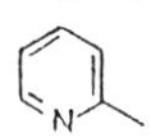	>500	81
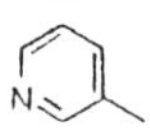	>500	0.62
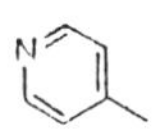	>500	5.5
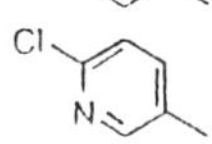	19	0.056
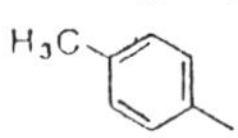	>500	0.76
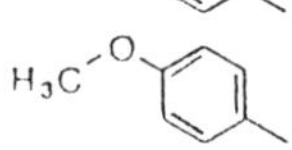	298	2.0
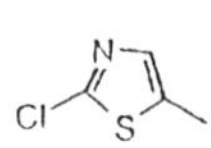	35	0.34

[a] Leaf-dipping method, second-instar larvae.
[b] Spraying method, first-instar nymphs.

The effects of the aromatic ring moiety (X) on activity are shown in Table 6. Among the compounds with nonsubstituted pyridine such as 2-, 3-, and 4-pyridyl groups, the 3-pyridyl derivative was the most active against the cotton aphid, and no compound exhibited activity against the diamondback moth even at 500 ppm. Introduction of a chlorine atom into the 6-position of the 3-pyridyl group significantly enhanced the activities against both insects by more than 10 fold compared with the nonsubstituted 3-pyridyl analogue. The chlorothiazole derivative also exhibited good activities against both insects.

From the results shown, the acyclic cyanoimine that has two methyl groups at the positions of R_1 and R_2 and a 6-chloro-3-pyridyl group at the aromatic ring moiety was considered to be the best compound, exhibiting excellent insecticidal activities. Acetamiprid, which has such substituents, was finally selected from the results of further biological experiments.

3.3 Synthesis of Acetamiprid

The synthetic routes of acetamiprid are shown in Fig. 3. Methyl *N*-cyanoacetimidate (**1**), the key intermediate, is obtained by the reaction of methyl orthoacetate and cyanamide in a good yield (Kishimoto et al. 1993). The reaction of **1** with 3-aminomethyl-6-chloropyridine (**2**) in aqueous methanol gives the condensation product (**3**) quantitatively. Acetamiprid is obtained by the methylation of **3** by dimethylsulfate and inorganic base in an excellent yield (route 1). Alternatively, acetamiprid is synthesized by the reaction of **1** with 6-chloro-3-(methylamino)methylpyridine (**4**) in one step from the intermediate (**1**) (route 2).

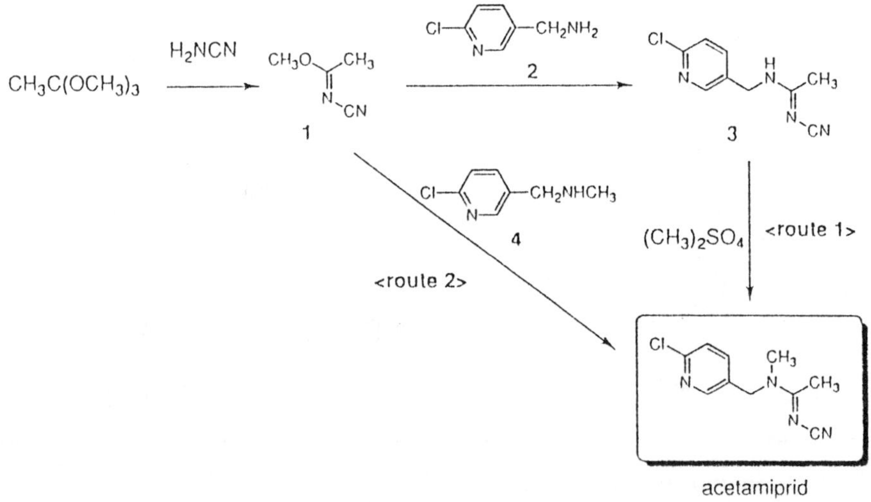

Fig. 3. Synthesis of acetamiprid.

4 Biological Activities of Acetamiprid

4.1 Insecticidal Spectrum

Acetamiprid acts on a wide range of insect pests, e.g., aphids, leafhoppers, whiteflies, and mealybugs in Hemiptera, thrips in Thysanoptera, the diamondback moth, leaf miners and fruit moths in Lepidoptera. LC_{50} values of acetamiprid for 15 pest species and the beneficial honeybee are shown in Table 7 (Takahashi et al. 1992; Iwasa et al. 1993; Mitsui et al. 1993). Acetamiprid has excellent activities against aphids including agriculturally important species, such as the cotton aphid, *Aphis gossypii*, and the green peach aphid, *Myzus persicae*. Other hemipterous

Table 7. Insecticidal spectrum

Species	Stage[a]	LC_{50} (ppm)
Hemiptera		
Aphis craccivola	Mix	0.91
Aphis gossypii	N1	0.056
Aphis spiraecola	N1	0.17
Myzus persicae	N1	0.21
Brevicoryne brassicae	N1	0.039
Ropalosiphum padi	Mix	0.032
Bemisia tabaci	E	4.8
Planococcus citri	N1	1.8
Lepidoptera		
Carposina niponensis	E	2.8
Grapholita molesta	E	3.1
Mamestra brassicae	L2	13
Plutella xylostella	L1	4.4
Spodoptera litura	L1	9.6
Thysanoptera		
Thrips palmi	A	3.4
Isoptera		
Reticulitermes speratus	A	0.16
Hymenoptera		
Apis mellifera	A	>200

[a]Stage: E, egg; N, nymph; L, larva; A, adult; Mix, mixed.

pests, such as the sweet potato whitefly, *Bemisia tabaci*, and the citrus mealybug, *Planococcus citri*, are also very susceptible to the compound. Besides, it is distinctive that the compound possesses strong activity against some important lepidopterous pests, such as the peach fruit moth *Carposina niponensis*, the oriental fruit moth *Grapholita molesta*, and the diamondback moth *Plutella xylostella*. The ovicidal effects on these fruit moths are an especially important feature of the insecticide. Other than these insect pests, acetamiprid has strong activities on melon thrips, *Thrips palmi*, and the termite, *Reticulitermes speratus*.

4.2 Activities in Different Developmental Stages

Activities of acetamiprid on the diamondback moth, *P. xylostella*, at different developmental stages are shown in Table 8 (Mitsui et al. 1993). The LC_{50} value against the 1st-instar larvae of the species is much smaller than the values for the 3rd- and 4th- instar larvae. It was considered that in finding a proper target for this insecticide, application timing was of the most importance.

Table 8. Activity of acetamiprid against the diamondback moth, *Plutella xylostella*, at various developmental stages

Compound	LC_{50} (ppm)						
		Larval instar					
	Egg	1st	2nd	3rd	4th	Pupa	Adult
Acetamiprid	17	4.4	19	35	32	>500	52
Acephate	>500	11	29	28	43	>500	110
Benfuracarb	24	20	23	24	21	>500	-

4.3 Cross-Resistance Between Conventional Insecticides and Acetamiprid

Many insect pests have recently developed resistance to most conventionally used insecticides. In aphids, three resistant strains from the peach-potato aphid, *Myzus persicae*, and one from the cotton aphid, *Aphis gossypii*, were assayed to investigate cross-resistance between conventional insecticides and acetamiprid. As shown in Table 9, one organophosphate-resistant strain of *M. persicae*, the Tokushima strain, showed a 19 fold higher LC_{50} value for acetamiprid than that of the susceptible Odawara strain (Iwasa et al. 1993). The value for the Tokushima strain, 3.9 ppm, is however still lower than the values of acephate or pirimicarb for the susceptible strain. For the diamondback moth, *Plutella xylostella*, two insecticide-resistant strains were assayed (Table 10) (Mitsui et al. 1993). Acetamiprid was effective on both Mizobe and Haibara strains, which were acephate- and cypermethrin resistant, as well as the susceptible Hiratsuka strain. No obvious cross-resistance to acetamiprid in any pest species is known up to this point.

Table 9. Activity of acetamiprid against resistant aphids

	Myzus persicae LC$_{50}$(ppm)			
Compound	Kochi	Wakayama	Tokushima	Odawara (S)
Acetamiprid	0.29	1.1	3.9	0.21
Acephate	140	190	76	16
Pirimicarb	>500	>500	>500	8.7
Cypermethrin	81	>500	3.5	0.47

	Aphis gossypii LC$_{50}$ (ppm)	
Compound	Shizuoka	Odawara (S)
Acetamiprid	0.069	0.056
Malathion	2.2	2.0
Methomyl	2.2	0.64
Cypermethrin	>500	0.21

Table 10. Activity of acetamiprid against resistant strains of the diamondback moth, *Plutella xylostella*

	LC$_{50}$ (ppm)			R/S ratio	
Compound	Mizobe[a]	Haibara[b]	Hiratsuka (S)[c]	Mizobe	Haibara
Acetamiprid	39	40	19	2.1	2.1
Acephate	250	510	29	8.6	18
Cartap	-	210	23	-	9.1
Cypermethrin	76	120	0.22	350	550

[a]Mizobe and [b]Haibara, insecticide-resistant strains, collected at Kagoshima in 1991 and at Shizuoka in 1991, respectively.

[c]Insecticide-susceptible strain.

4.4 Systemic and Translaminar Action

Another noteworthy property of this compound is its systemic or translaminar activity. Because of its high solubility in water, acetamiprid applied to the soil moves into the plant body through the roots and moves upward in the plant. When the roots of plant seedlings were dipped in acetamiprid aqueous solution, LC_{50} values against aphids infesting the leaves were between 0.02 and 0.03 ppm, and the value for the diamondback moth, *P. xylostella,* was 0.31 ppm. Results for aphids (Iwasa et al. 1993) and the diamondback moth (Mitsui et al. 1993) are shown in Tables 11 and 12, respectively. This systemic property generates the possibility that granules are available as one of the acetamiprid formulations.

When acetamiprid was sprayed only on the upper surface of cabbage leaves, mortality of *M. persicae* infesing the undersurface of treated leaves reached 100% on 7 days after treatment. This result (Fig.4), demonstrated the translaminar action of the compound. This character of the compound is effective when pest insects mine plant leaves or live inside the rolled leaves.

Table 11. Systemic activity against aphids by root-dipping method

	LC_{50} (ppm)		
Compound	Cucumber *(A. gossypii)*	Eggplant *(M.persicae)*	Radish *(M.persicae)*
Acetamiprid	0.019	0.031	0.023
Acephate	2.1	0.71	1.2
Benfuracarb	0.5	0.86	1.6

Table 12. Systemic activity against the diamondback moth, *P. xylostella* , by root-dipping method

	LC_{50} (ppm)	
Compound	24 hr	72hr
Acetamiprid	0.73	0.31
Acephate	6.2	1.8
Benfuracarb	4.3	1.8

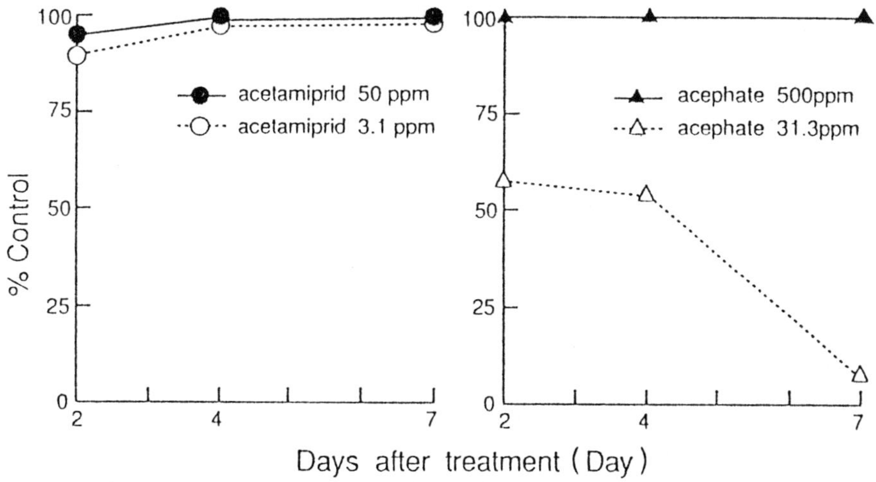

Fig. 4. Translaminar activity of acetamiprid in cabbage leaf against the peach-potato aphid, *Myzus persicae*.

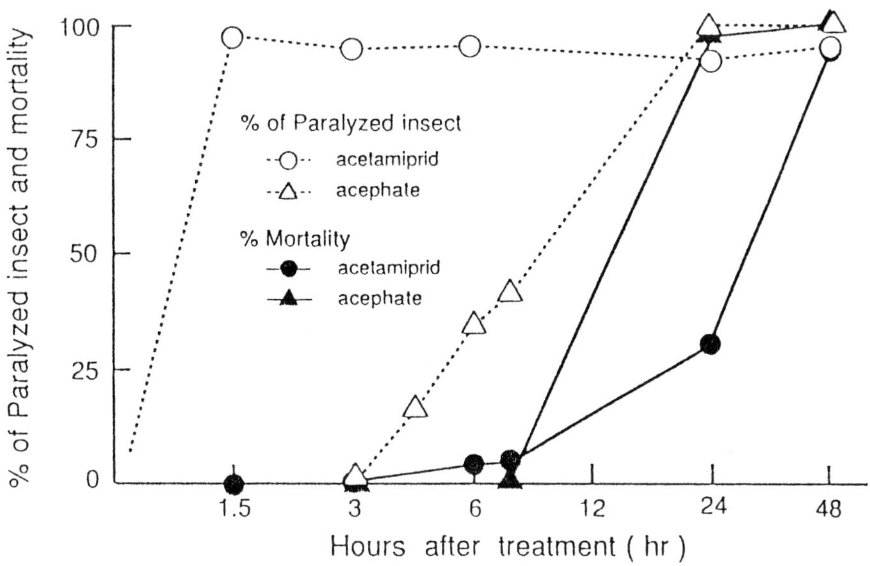

Fig. 5. Speed of acetamiprid action against the common cutworm, *Spodoptera litura*, by cabbage leaf dipping method.

4.5 Speed of Action

To determine the speed of action, larvae of the common cutworm, *Spodoptera litura*, were treated with acetamiprid. All larvae that fed on cabbage leaves dipped in an aqueous solution of acetamiprid exhibited symptoms of intoxication within 90 min after first contact with the leaves. Data are shown in Fig. 5 (Mitsui et al. 1993). Although it took 48 h until mortality of treated larvae reached 100%, they stopped feeding when symptoms appeared.

4.6 Comparison of Activities in Different Administration Methods

Next, we made a comparison of acetamiprid activity in different administration methods (Table 13). When the 5th-instar larvae of the armyworm, *Pseudaletia separata*, were applied topically, the minimum dose required to show 100% mortality was 5 μg/larva. To investigate orally applied activity, a small piece of maize leaf was given a topical application of the chemical, and the starveling *P. separata* larva was allowed to eat all the treated leaf in a short time. The dose required for the oral application method was 1.25 μg/larva. These results (Mitsui et al. 1993) suggested that acetamiprid could act on insect pests by both contact and oral action.

Table 13. Comparison of acetamiprid activities against the armyworm, *Pseudaletia separata* [a], between oral and contact application

	% Mortality	
Dose (μ g/larva)	Oral[b] (24 hr)	Contact[c] (24 hr)
5		100
1.25	100	40
0.31	50	30
0.08	0	0

[a]Fifth-instar larvae.
[b]Feeding method.
[c]Topical application, 1 μ l/larva.

4.7 Persistency of Acetamiprid Activity

To assess the persistency of acetamiprid efficacy on cabbage leaves in a greenhouse condition, potted cabbage was sprayed and kept in a greenhouse until the leaves were cut periodically for bioassay against *P. xylostella* larvae. After the leaves were sprayed with 100 and 200 ppm of acetamiprid, control efficacy lasted 14 and 21 days, respectively. Data are shown in Fig. 6 (Mitsui et al. 1993).

For aphids, the persistent effect of acetamiprid sprayed on cucumber leaves was assayed by periodical inoculation of *A. gossypii*. As shown in Fig. 7, foliar spraying of acetamiprid (50 ppm) suppressed the population increase of the aphid longer than control chemicals (Iwasa et al. 1993).

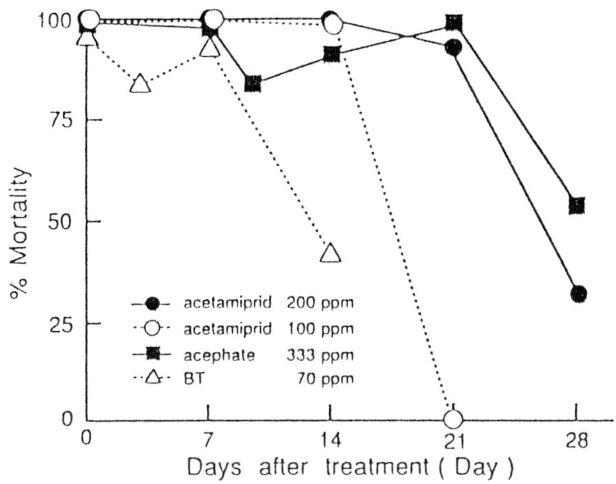

Fig. 6. Persistency of acetamiprid activity on cabbage leaf against the diamondback moth, *Plutella xylostella*, by foliar spraying (greenhouse test).

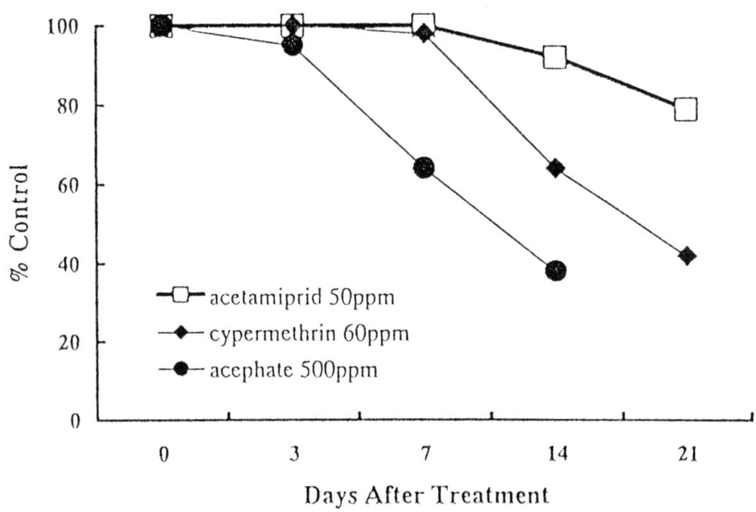

Fig. 7. Persistency of acetamiprid activity on cucumber leaf against the cotton aphid, *Aphis gossypii*, by foliar spraying (greenhouse test).

5 Mode of Action

5.1 Elecrophysiological Study

Acetamiprid-treated insects showed fast knockdown and some symptoms of intoxication, such as excitation, convulsion, and paralysis followed by death, in that order. From these, it was presumed that acetamiprid might act on the central nervous system of insects.

To examine the effect of acetamiprid on the insect central nervous system, a central nerve cord of the American cockroach, *Periplaneta americana*, was used for electrophysiological study. When the nerve preparation was treated by 1×10^{-5}M acetamiprid, bursts, with increases of both frequency and magnitude of spontaneous discharge, appeared immediately. After stopping the bursts, which lasted for a few minutes, electrical stimulation was applied to different positions to determine the blocking site by acetamiprid. When the presynaptic nerve fibers were stimulated, the response of the preparation was normal. When the postsynaptic cercal sensory neurons were stimulated, however, no response appeared. This result means that the axonal conduction was not blocked by acetamiprid treatment but that the synaptic transmission was completely blocked. These results (Yamamoto et al. 1994) (Fig. 8), were identical with the response of nerve preparation treated by nicotine, but were not identical with that by curare because no burst of spontaneous discharge was observed in response after treatment.

For some related compounds, the nitromethylene heterocycle insecticides were reported as possible agonists on postsynaptic acetylcholine receptors of *P. americana* (Shoroeder and Flattum 1984). According to the reports on imidacloprid's mode of action, the compound bound to the cholinergic receptors in the nerve membrane of *P. americana* (Bai et al. 1991) and acted on the synapse of nerve preparation of *P. americana* in a manner similar to nicotine (Sone et al. 1994).

5.2 Receptor-Binding Assay

Another experiment conducted to study the mode of action of the compound used receptor- binding assays. Housefly head membrane preparation was used for nicotinic acetylcholine receptor (nAChR) binding assay using $[^3H]$- α -bungarotoxin (α-BGTX) as a radioligand. In a competition assay, acetamiprid displaced α -BGTX binding at 3.2×10^{-6} M as K_i value (Fig. 9). In the same condition, nicotine had a K_i value of 9.4×10^{-7} M (Kishimoto et al. 1993).

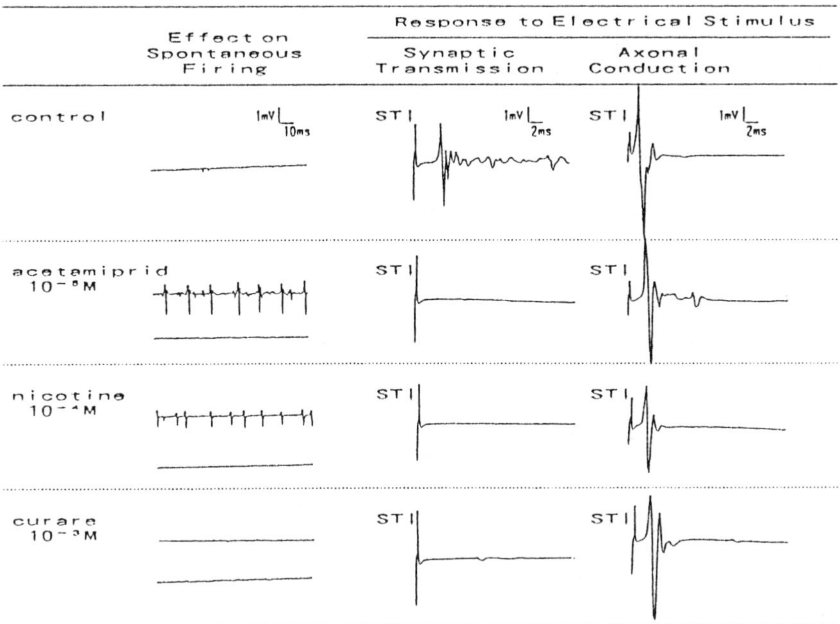

Fig. 8. Effects of acetamiprid on the nerve cord of the American cockroach, *Periplaneta americana*. Effects of chemicals on spontaneous firing, synaptic transmissions and axonal conduction were investigated separately.

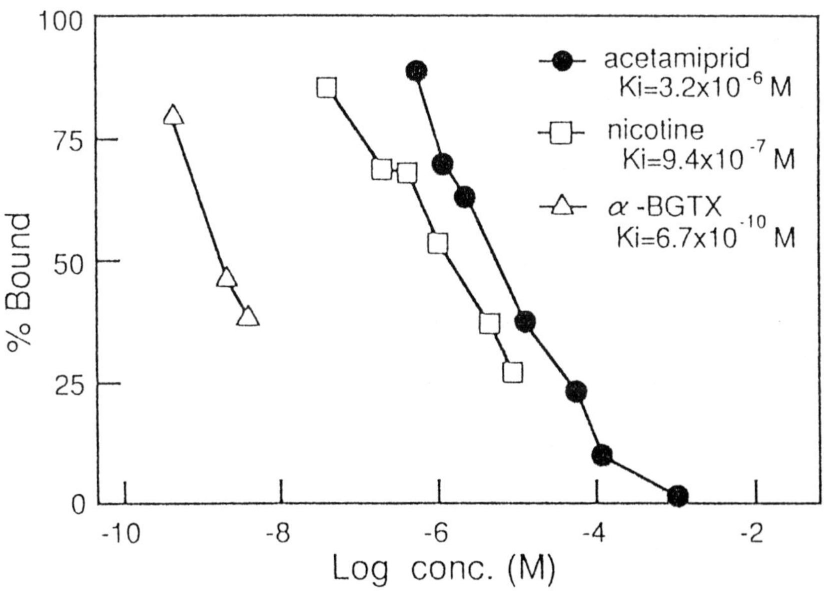

Fig. 9. Inhibition of [^{3}H]α-bungarotoxin (*α-BGTX*) binding to the housefly head membrane preparation by acetamiprid.

Considering the results of these electrophysiological and receptor-binding experiments, acetamiprid seems to act on the nAChR of the insect central nervous system as an agonist as well as the nitromethylene heterocycles and imidacloprid.

6 Efficacy of Acetamiprid in Field Tests

6.1. Efficacy Against Insect Pests on Vegetables

The first field trial was conducted to investigate activities of acetamiprid against the diamondback moth, the important pest on crucifers. The efficacy of acetamiprid against the insect on cabbage by foliar application is shown in Fig. 10. The compound showed about 90% control efficacy against the insect at 100 ppm. Because acetamiprid's action is very characteristic, as mentioned before, and the application timing seems very important to obtain good control efficacy, precise experiments in regard to the spraying timing were carried out. From these experiments, the best efficacy was obtained when the application of the chemical was done in the early timing of the pest insect occurrence (Murahashi et al. 1993).

The efficacy of acetamiprid against the green peach aphid on cabbage by foliar application is shown in Fig. 11. The efficacy of the compound against the insect was excellent at 100 ppm, and it was ascertained that acetamiprid showed sufficient efficacy even at 25 ppm. As there are few insecticides possessing high efficacy against both the diamondback moth and aphids, acetamiprid is very valuable for insect pest control programs in crucifers.

The efficacy of acetamiprid granules against the diamondback moth on cabbage in a planting hole application is shown in Fig. 12. In the beginning of granule formulation research on acetamiprid, an ordinary granule (release-uncontrolled granule) was formulated and it gave good control efficacy against the pests. To obtain much better efficacy, many release-controlled granules were formulated, considering that acetamiprid was relatively soluble in water and easily decomposed in soil, and their efficacies were investigated. The release speed of the active ingredient from the granule is controlled to avoid influence by rainfall and the difference in soil features (Takahashi et al. 1998b). The release-controlled 2% granule of acetamiprid suppressed the insect for 3-4 weeks after application at the dosage of 1 g/plant.

Efficacy against diamondback moth larvae on cabbage by a nursery box application and a planting hole application is shown in Fig. 13. In the planting hole application, granule is applied under the roots, and in the case of the nursery box application, the granule is above the roots. Acetamiprid exhibited excellent efficacy also in the nursery box application at 0.5 g/plant. This application method is labor-conserving because granules are applied on the nursery box before planting by a transplanting machine, so that farmers need not apply granules to

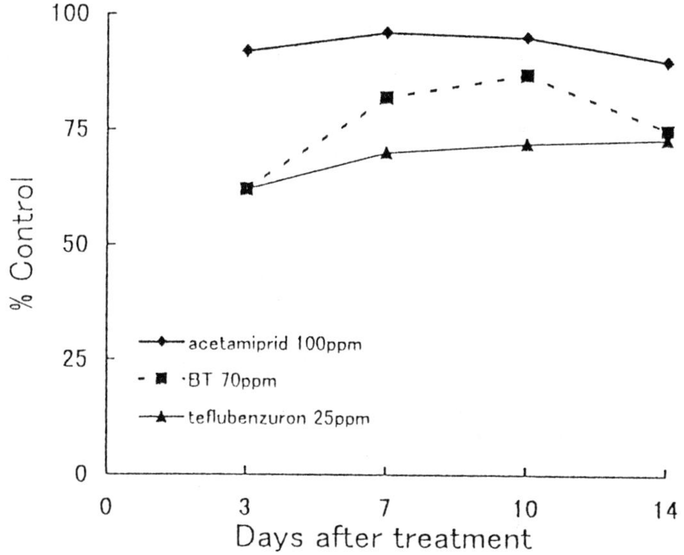

Fig. 10. Control efficacy of acetamiprid against the diamondback moth, *P. xylostella*, on cabbage by foliar application (1992).

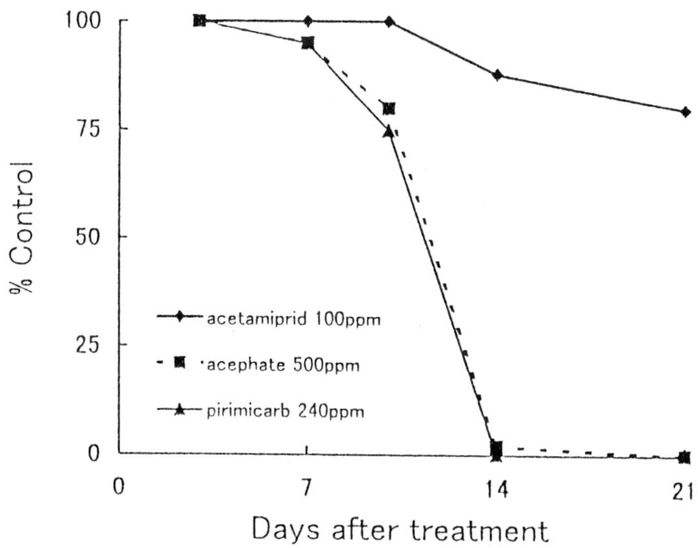

Fig. 11. Control efficacy of acetamiprid against the peach-potato aphid, *M. persicae*, on cabbage by foliar application (1992).

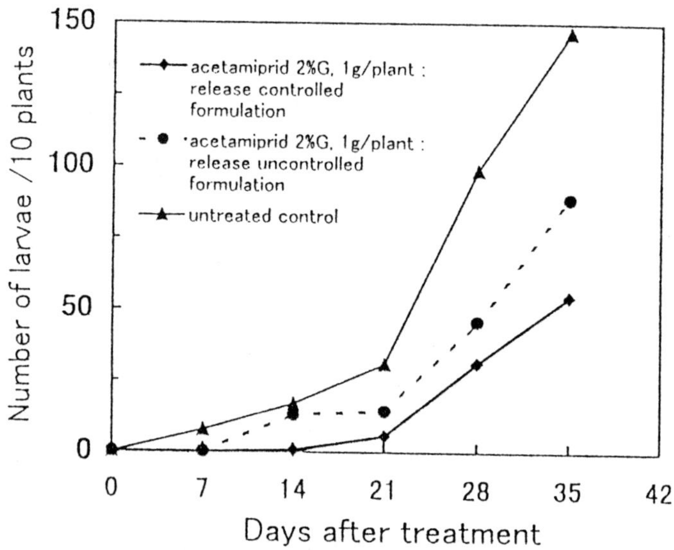

Fig. 12. Comparison of efficacy between two different granular formulations of acetamiprid against larvae of the diamondback moth, *P. xylostella*, on cabbage by planting hole application (1995).

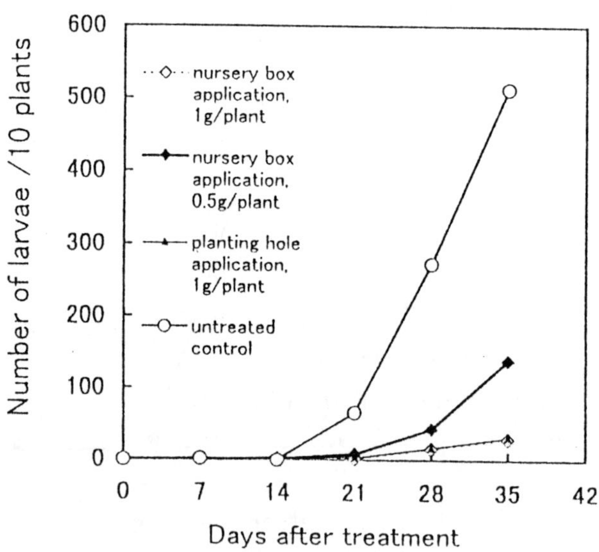

Fig. 13. Efficacy of acetamiprid (2%G) against the diamondback moth, *P. xylostella*, by nursery box and planting hole applications with cabbage seedlings.

individual seedlings. Also, in the case of the planting hole application, manual application of the granule is not so practical, but automatic application of the compound by transplanting machine becomes feasible.

There are many ways of granular application such as plant foot application, row application, overallridge application, or top dressing application in addition to planting hole application and nursery box application. The latter two methods seemed to be the most suitable for acetamiprid granule from the standpoint of both control efficacy and labor-conserving for farmers (Mitsui et al. 1996). It was confirmed that the efficacy of acetamiprid against the aphid was superior to that against the diamondback moth.

Figure 14 shows the efficacy of acetamiprid against the melon thrips, *Thrips palmi,* which has invaded Japan in recent years and become a serious pest of plants such as eggplant, cucumber, and green pepper. Acetamiprid exhibited high efficacy against the pest at 50 ppm by spraying on eggplant. Acetamiprid shows good efficacy also against other thrips, the western flower thrips, *Frankliniella occidentalis,* and the flower thrips, *Frankliniella intonsa*. The compound is able to control aphids and whiteflies in addition to thrips among insect pests on vegetables. As it was considered to be critical in practical use whether acetamiprid affected beneficial insects such as the honeybee and bumblebee, the effect of the compound on these bees was investigated in greenhouse. The effect of acetamiprid on the bumblebee is shown by the number of bite-marks, which is a scar from an insect bite, on the flower of tomato (Fig. 15). In this experiment, bumblebees were released 3 h after the spraying of acetamiprid (100 ppm) solution. The bite-mark in the plot of acetamiprid spraying was almost the same as that of an untreated control. Moreover, it was confirmed by another test that the effect on honeybees was also very slight. These results enhanced the practicability of acetamiprid for insect pest control of vegetables such as tomato and strawberry (Takahashi et al. 1998a).

6.2. Efficacy Against Insect Pests on Fruit Trees

As it was confirmed by the laboratory and greenhouse experiments that acetamiprid possessed high activities against Lepidoptera, excellent ovicidal activity, and translaminar activity, fruit moths and leafminers in Lepidoptera were considered as target insects of the insecticide from the standpoint of their ecology.

Figure 16 shows the efficacy against the oriental fruit moth, *Grapholita molesta*, an important insect pest on pear and apple in Japan, by the rate of damaged fruits. For controlling the insect, ovicidal activity or a fast-acting larvicidal property is required. Acetamiprid at 100 ppm showed superior efficacy to the standard insecticide fenitrothion at 400 ppm (Takahashi et al. 1992). It was verified that the compound was applicable also to the peach fruit moth, *Carposina niponensis*, and the persimmon fruit moth, *Stathmopoda masinissa*, which possess life histories similar to the oriental fruit moth.

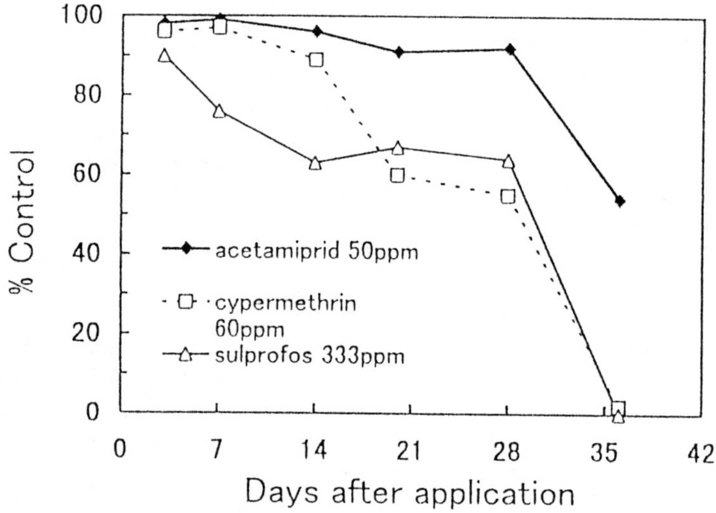

Fig. 14. Control efficacy of acetamiprid against the melon thrips, *Thrips palmi*, on eggplant by foliar application (1991).

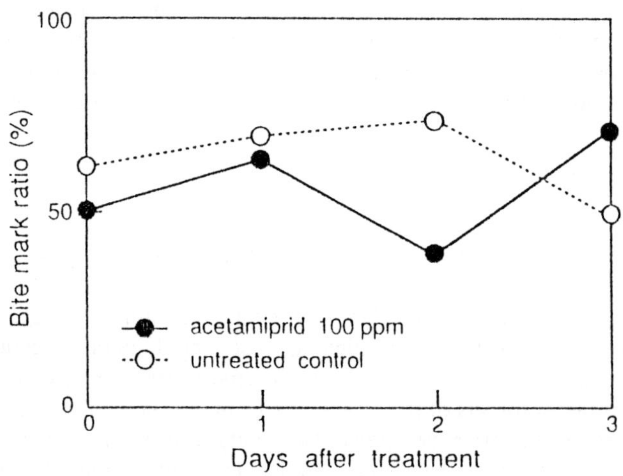

Fig. 15 Influence of acetamiprid on the behavior of the bumblebee, *Bumbus terrestris*, when the insect was released 3 h after spraying. Test was conducted in greenhouse where tomatoes were planted (1994).

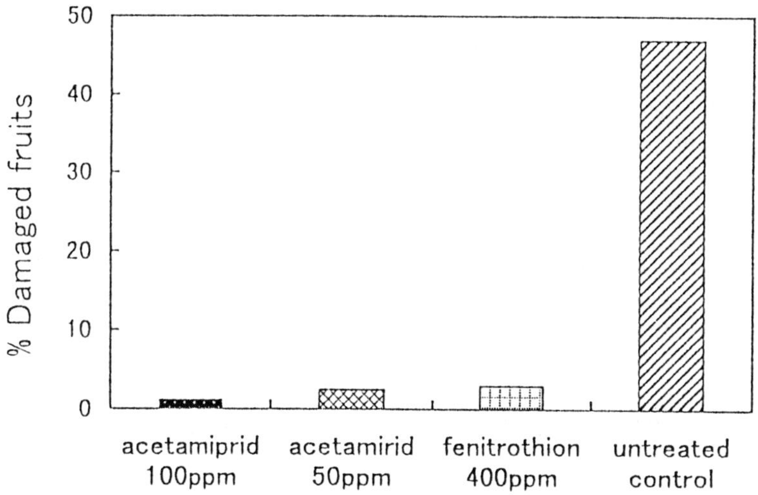

Fig. 16. Control efficacy of acetamiprid against the oriental fruit moth, *Grapholita molesta*, on pear (1991).

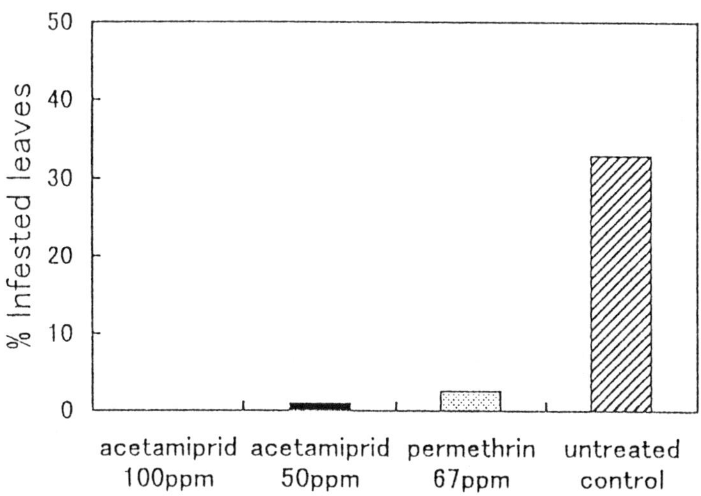

Fig. 17. Control efficacy of acetamiprid against the apple leafminer, *Phyllonorycter ringoniella*, on apple (1991).

The efficacy against the apple leafminer, *Phyllonorycter ringoniella* on apples is shown by the infested-leaves rate in Fig. 17. The insect is a serious pest to apples and mines into leaves; therefore, translaminar activity of insecticide is important to suppress the damage by the insect. Acetamiprid showed almost perfect efficacy against the insect even if spraying was conducted when the developmental stage of the insect was relatively advanced. The compound is also applicable to the citrus leafminer, *Phyllocnistis citrella*, on citrus tree (Hatano et al. 1998) and the peach leafminer, *Lyonetia clerkella*, on peach trees, which possess biological similarities to the apple leafminer.

The compound showed high activities against Coleoptera such as longicorn beetles, leaf beetles and weevils. The efficacy against the white-spotted longicorn beetle, *Anoplophora malasiaca*, on citrus trees is exhibited in Fig. 18. The results are demonstrated by the number of adult-feeding marks on green twigs and leaves. Acetamiprid suppressed damage at 50 ppm, and its efficacy was superior to that of the standard insecticide methidathion (Hatano et al. 1998; Takahashi et al. 1998a). Similar effect of acetamiprid on adults of a kind of forest longicorn beetle (Japanese pine sawyer), *Monochamus alternatus*, was reported as a practical control agent (Abe et al. 1998). It is characteristic of acetamiprid to be able not only to kill the insect, but also to prevent the damage caused by these insects.

Figure 19 shows activity of acetamiprid against the Comstock mealybug, *Pseudococcus comstocki*, on pears. The efficacy was evaluated by counting fruits infested by the insect. The compound showed an excellent efficacy against the insect at both 100 and 50 ppm. It is difficult to control the insects with insecticide because they inhabit a narrow space in the tree. Acetamiprid is effective against the insect even inhabiting narrow spaces of trees because of its translaminar activity (Takahashi et al. 1992). Moreover, it is effective against each developmental stage of the insect.

Acetamiprid also shows superior efficacy against other mealybugs such as the citrus mealybug, *Planococcus citri* and the Japanese mealybug, *Planococcus kraunhiae*. Also, official tests revealed that it was highly effective against scale insects such as the red wax scale, *Ceroplastes rubens*, and the arrowhead scale, *Unaspis yanonensis*.

6.3. Efficacy Against Insect Pests on Tea Trees

In the control of tea tree pests, the tea leafroller, *Caloptilia theivora*, was targeted for acetamiprid by its biological characteristics. The insect is a serious lepidopterous pest on tea, laying eggs on the surface of leaves, and the hatched larvae mine into the leaves. Therefore, an ovicidal activity or a translaminar activity against the larvae is required to suppress the insect. The contamination of infested tea leaves by the pest reduces the quality of tea. The efficacy was evaluated by counting the number of leaves infested by the insect larvae. The compound exhibited good control against this insect at the concentration of 50 ppm (Fig. 20).

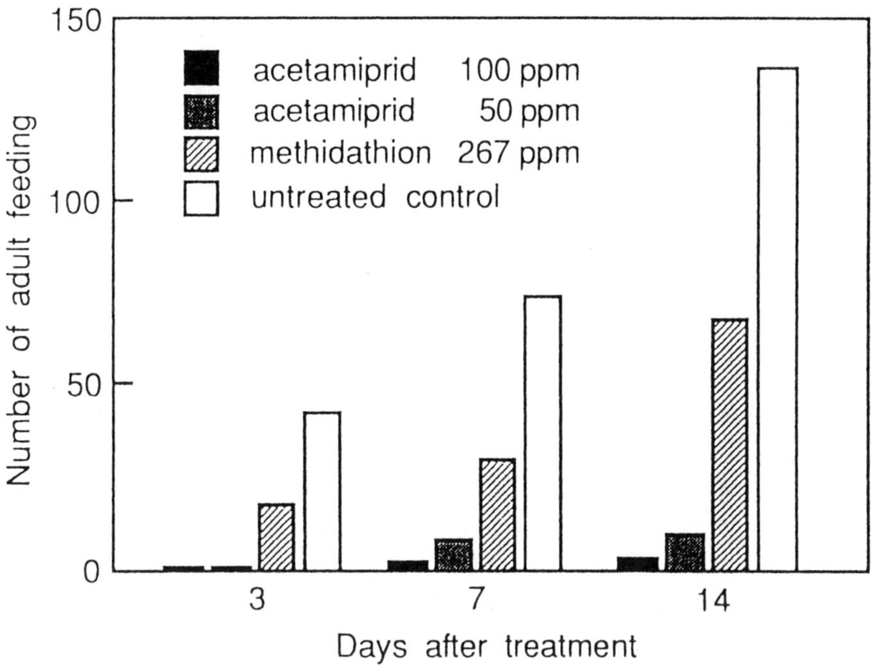

Fig. 18. Efficacy of acetamiprid against the white-spotted longicorn beetle, *Anoplophora malasiaca,* in inhibiting adult feeding on citrus tree (1994).

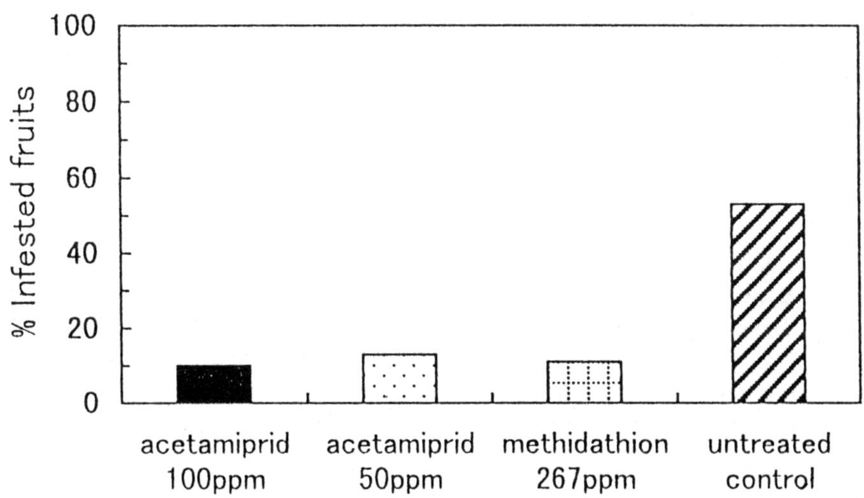

Fig. 19. Control efficacy of acetamiprid the comstock mealybug, *Pseudococcus comstocki*, on pear (1991).

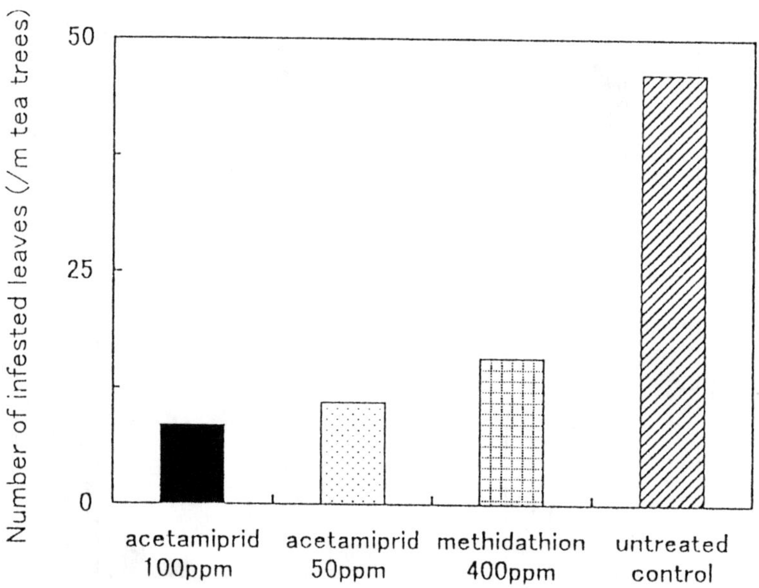

Fig. 20. Control efficacy of acetamiprid against the tea leafroller, *Caloptilia theivora*, on tea tree. (1992)

It was also shown that acetamiprid was applicable to the green leafhopper, *Empoasca onukii*, and the yellow tea thrips, *Scirtothrips dorsalis*. These three kinds of insect pests appear at almost the same time in Japan; therefore, acetamiprid is a practical way of controlling them all simultaneously. Acetamiprid also does not leave stains after spraying because it is formulated as a soluble powder (Takahashi et al. 1992).

7. Conclusion

The aim of acetamiprid research was to develop an environmentally benign insecticide with a strong efficacy against pests normally difficult to control, with no cross-resistance to conventional insecticides. In this research, nithiazine and Nihon Bayer's patented compounds were chosen as the lead compounds, and the synthetic development of their derivatives was focused on the discovery of compounds with both high activities against Lepidoptera and specificity in chemical structure. In the biological evaluations, toxic symptoms of treated insects caused by test compounds as well as their killing effects were observed carefully. As a result of these investigations, a novel insecticide acetamiprid was discovered that turned out to be a powerful tool for the chemical control of insect pests and is expected to be available for integrated pest management.

References

Abe Y, Nakamura K, Takahashi H, Hatano R, Tanaka Y, Matsubara I, Tabata K (1998) Biologacal property and efficacy of acetamiprid, an insecticide, effective control maturationfeeding caused by *Monochamus alternatus*. In: Abstracts of the 109th annual meeting of the Japanese Forestry Society, p 225

Bai D, Lummis SCR, Leicht W, Breer H (1991) Actions of imidacloprid and a related nitromethylene on cholinergic receptors of an identified insect motor neurone. Pestic Sci 33:197-204

Harris M, Price RN, Robinson J, May TE (1986) WL108477 - a novel neurotoxic insecticides. Proceedings of Brighton Crop Protection Conference-Pests and Diseases, pp115-122

Hatano R, Mitsui J, Yano M, Take T, Takahashi H (1998) Control effects and characteristics of acetamiprid (Mospilan®), a novel insecticide, on insect pests of citrus in the field experiments. In: Abstracts of the 23rd annual meeting of the Pesticide Science Society of Japan, p 130

Iwasa T, Takakusa N, Mitsui J, Yano M, Yamamoto A, Hatano R, Matsuda M (1993) Studies on a novel insecticide, NI-25; biological activities and actions against aphids. In: Abstracts of the 18th annual meeting of the Pesticide Science Society of Japan, p 41

Kishimoto T, Suzuki J, Ishimitsu K, Mitsui J, Iwasa T, Yamamoto A, Takakusa N (1993) Studies on a novel insecticide, NI-25; chemical structure and biological activities. In: Abstracts of the 18th annual meeting of the Pesticide Science Society of Japan, p 39

Matsuda M (1995) Discovery of acetamiprid. abstracts of papers, the 20th annual meeting of the Pesticide Science Society of Japan, pp 39-40

Matsuda M, Takahashi H (1996a) Biological activities of acetamiprid. Plant Prot 50:248

Matsuda M, Takahashi H (1996b) Mospilan® (acetamiprid, NI-25) - a new systemic Insecticide,- Agrochem Jpn 68:20-21

Mitsui J, Iwasa T, Yoneda A, Yano M, Yamamoto A, Hatano R, Matsuda M (1993) Studies on a novel insecticide, NI-25; biological activities and actions against Lepidoptera. In: Abstracts of the 18th annual meeting of the Pesticide Science Society of Japan, p 40

Mitsui J, Take T, Yano M, Tanaka H, Nishimura K, Takahashi H (1996) Studies on a novel insecticide, Mospilan®; biological properties and insecticidal activities of Mospilan G evaluated by various applications. In: Abstracts of the 21st annual meeting of the Pesticide Science Society of Japan, p 108

Murahashi K, Take T, Nishimura K, Takahashi H (1993) Studies on a novel insecticide, NI-25; control efficacy of NI-25 against insect pests on cabbage. In: Abstracts of the 18th annual meeting of the Pesticide Science Society of Japan, p 42

Schroeder ME, Flattum RF (1984) The mode of action and neurotoxic properties of the nitromethylene heterocycle insecticides. Pestic Biochem Physiol 22:148-160

Shiokawa K, Tsuboi S, Kagabu S, Moriya K (1985) Japan. Kokai Tokkyo Koho, JP60-172976

Soloway SB, Henry AC, Kollmeyer WD, Padgett WM, Powell JE, Roman SA, Tiemans CA, Corey RA, Horne CA (1978) In: Geissbuehler H (ed) Advances in pesticide science, Vol 2 Pergamon Press, New York, pp 206-217

Sone S, Nagata K, Tsuboi S, Shono T (1994) Toxic symptoms and neural effect of a new class of insecticide, imidacloprid, on the American cockroach, *Periplaneta americana* (L). J Pestic Sci 19:69-72

Takahashi H, Mitsui J, Takakusa N, Matsuda M, Yoneda H, Suzuki J, Ishimitsu K, Kishimoto T (1992) NI-25, a new type of systemic and broad spectrum insecticide. Proceedings of Brighton Crop Protection Conference-Pests and Diseases, pp89-96

Takahashi H, Takakusa N, Suzuki J, Kishimoto T (1998a) Development of a new insecticide, acetamiprid. In: Abstracts of the 23rd annual meeting of the Pesticide Science Society of Japan, p 23

Takahashi H, Mitsui J, Murahashi K, Asai M, Yamada T (1998b) Efficacy of acetamiprid 2% granule against diamondback moth on cabbage by soil treatment. J Pestic Sci 23:275-280

Yamamoto A, Hatano R, Mitsui J, Iwasa T, Matsuda M (1994) Studies on a mode of action of a novel insecticide, NI-25; effect on the insect nervous system. In: Abstracts of the 19th annual meeting of the Pesticide Science Society of Japan, p 92

8 CGA 293'343: A Novel, Broad-Spectrum Neonicotinoid Insecticide

P. Maienfisch*, F. Brandl, W. Kobel, A. Rindlisbacher, and R. Senn
Novartis Crop Protection AG, CH-4002 Basel, Switzerland

1. Introduction

1.1 Historical Background

In the early 1980s, an important discovery was made by Shell researchers when the insecticidal activity of heterocyclic nitromethylene compounds was found (Soloway et al. 1979). After an intensive optimization program, Nithiazin was identified as the most active compound (Figure 1). However, this product was not commercialized. This new chemical class subsequently attracted much attention as a potential tool for insect control, and was taken up by several companies such as Bayer, Takeda, Nippon Soda, Agro Kanesho, Mitsui Toastu and Ciba (since 1996; Novartis).

An important breakthrough in this chemistry was achieved by researchers of Nihon Tokushu Noyaku (now Nippon Bayer) in 1984. By introducing a 3-pyridylmethyl group as a substituent on the heterocyclic nitromethylenes, the insecticidal activity increased dramatically. This finding resulted in the discovery of imidacloprid (Elbert et al. 1990), the first neonicotinoid [1] insecticide introduced to the market by Bayer in 1991. As second and third neonicotinoids, nitenpyram (Minamida et al. 1993) from Takeda and acetamiprid (Takahashi et al. 1992) from Nippon Soda were brought to the market in 1995 and 1996, respectively (Figure 1).

The launch of imidacloprid in 1991 was a tremendous success. Sales increased impressively during the first 5 years after market introduction; in 1995 with sales of $360 million imidacloprid became the second biggest selling insecticide, close behind the organophosphate chlorpyrifos (source, Wood Mackenzie). In 1997 sales of the active ingredient (a.i.), including crop and animal health applications, reached $562 million (source, Agrow) and imidacloprid became the top-selling insecticide. It can be speculated, that the neonicotinoids will become the most or at least one of the most important insecticidal classes of the next decade.

[1] The term neonicotinoid was proposed by Yamamoto (1996).

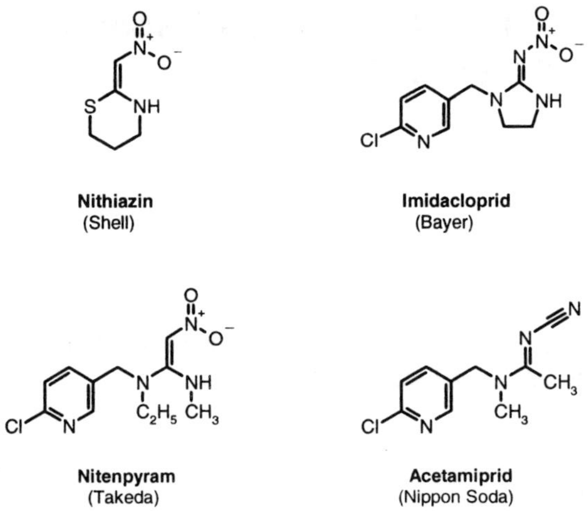

Fig. 1. Chemical structure of nithiazin, imidacloprid, nitenpyram, and acetamiprid

1.2 Properties of Neonicotinoids

The great market success of imidacloprid can be explained by the unique properties of the neonicotinoids. They possess excellent activity especially against homopteran, coleopteran, and lepidopteran pests. As compared to commercial standards, lower application rates are needed. Neonicotinoids have a low partition coefficient log P and good water solubility, which makes these compounds highly systemic. Therefore, such compounds can be used as a seed treatment, and this use is one of the most important strengths of this chemical class. Neonicotinoids act at the nicotinic acetylcholine receptor (Benson 1989; Sattelle et al. 1989; Bai et al. 1991; Tomizawa and Yamamoto 1992, 1993; Liu and Casida 1993; Liu et al. 1993; Tomizawa 1994; Tomizawa et al. 1995). This is a very attractive mode of action, which so far has not been widely used for insecticides; consequently neonicotinoids are important for controlling insects resistant to other commonly used insecticides such as organophosphates, carbamates, and pyrethroids. A real strength is also the favorable safety profile, especially the low mammalian toxicity observed for these compounds.

1.3 Structure-Activity Relationships of Neonicotinoids

The common structure of the neonicotinoids consists of three parts: the pharmacophore, the bridging chain, and the heterocyclic group, as illustrated in Figure 2.

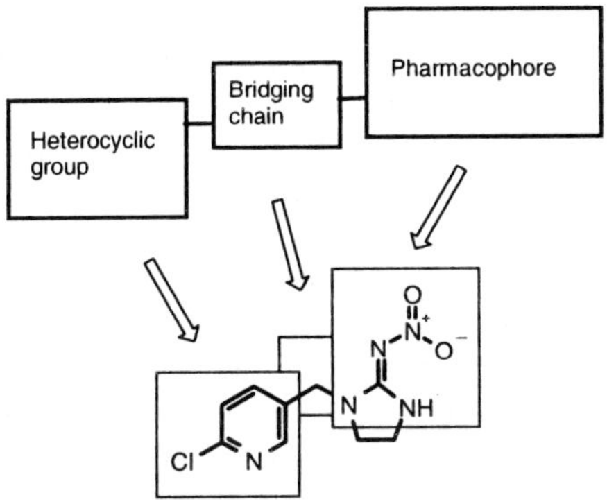

Fig. 2. Structural elements of neonicotinoids

The influence of these different structural elements on the biological activity has already been described (Kagabu et al. 1992; Moriya et al. 1992, 1993; Minamida et al. 1993a,b; Tomizawa and Yamamoto 1993; Tabuchi et al. 1994; Shiokawa et al. 1994, 1995). In general, the pharmacophore can be represented by the group N-C(X)=Y, where Y is an electron-withdrawing group and X is equal to N, C, O, or S. The pharmacophore has a dramatic effect on the insecticidal activity. For example, in a series of imidacloprid analogues insecticidal activity against the green leafhopper decreased in the order Y = CH-NO_2, N-NO_2 > N-CN, > $C(CN)_2$ >> CH-CN, O, NH (Shiokawa et al. 1995). According to the published data, which are in line with our own findings, the best insecticidal activity is observed for compounds possessing the nitroenamine, the nitroamidine, or the cyanoamidine pharmacophore (Table 1). Beside its influence on the biological activity, the pharmacophore is also responsible for some specific properties such as photolytical stability, degradation in soil, metabolism in plants, and toxicity to bees and beneficials. Differences in these properties have a major impact on the practical use and performance in the field.

Table 1. Types of pharmacophores (X = N, C, O or S)

	Nitroenamine Pharmacophore	**Nitroamidine Pharmacophore**	**Cyanoamidine Pharmacophore**
Functional group	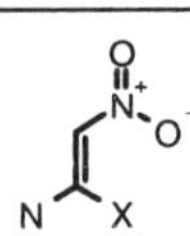	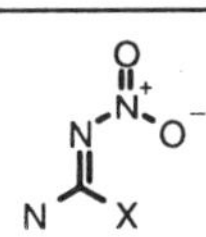	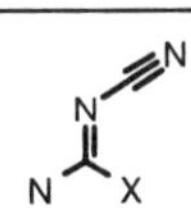
Example	Nitenpyram Nithiazin	Imidacloprid CGA 293'343 TI-435 AKD-1022	Acetamiprid

The methylene group is normally used as the bridging chain. Other groups such as an ethylene or substituted methylene group decrease the biological activity. Successive modifications have demonstrated the major influence of the heterocyclic group on the biological activity. Compounds lacking the heterocyclic group, such as the nitromethylene compounds reported by Shell and compounds where the heterocyclic group is replaced by a phenyl group, show only moderate insecticidal activity compared to the best standards. The best biological activity so far observed was for compounds having a 6-chloro-3-pyridyl or a 2-chloro-5-thiazolyl group. Other N-containing heterocycles and 3-pyridyl- or 5-thiazolyl compounds, where the chloro atom is replaced by a hydrogen, a fluoro, or a bromo atom or by a methyl or trifluomethyl group, have also shown interesting insecticidal activity. Based on their heterocyclic group, the most prominent neonicotinoids can be grouped into three different subclasses as illustrated in Table 2.

Table 2. Neonicotinoid subclasses

Neonicotinoid subclass	Heterocyclic Group	Example	Remark
Nitromethylene compounds	None	Nithiazin	Lead structures
Chloronicotinyl compounds	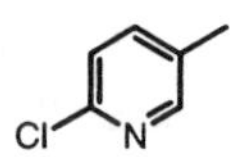	Imidacloprid Nitenpyram Acetamiprid	1st-generation neonicotinoids
Thianicotinyl compounds	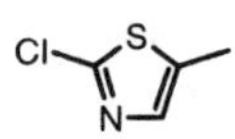	CGA 293'343 TI-435 AKD-1022	2nd-generation neonicotinoids

From the nitromethylene subclass, no compound has been commercialized. However, nithiazin and its analogues have served as lead structures and consequently influenced further developments in this field. The first-generation neonicotinoids imidacloprid, nitenpyram, and acetamiprid, having a chloropyridyl heterocycle, entered the markets between 1991 and 1996 and show a much improved biological performance. For this subclass of the neonicotinoids, the term chloronicotinyl compounds was proposed by Bayer and is now broadly used in the literature. The second-generation neonicotinoids, possessing a chlorothiazole heterocycle, are now in development and beginning to enter the marketplace. CGA 293'343 from Novartis, TI-435 from Takeda, and AKD-1022 from Agro Kanesho are the first representatives of second-generation neonicotinoids (Figure 3). We propose to call this subclass of the neonicotinoids thianicotinyl compounds (Table 2).

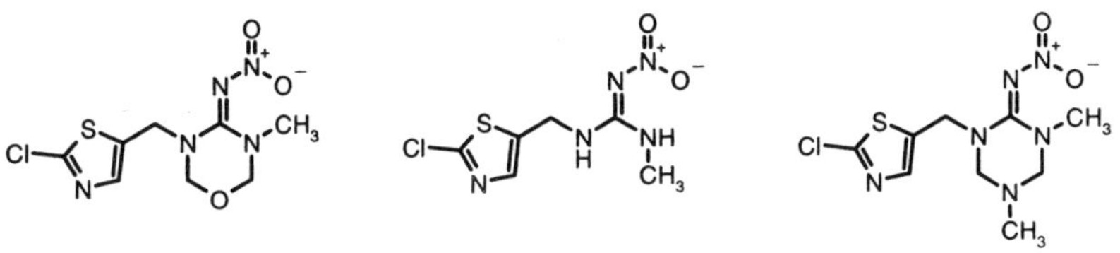

Fig. 3. Second-generation neonicotinoids currently under development

2. Discovery of CGA 293'343

2.1 From the Lead to CGA 293'343

Influenced by the work of Shell and patent applications of Nihon Tokushu Noyaku Seizo, K.K. Ciba (since 1996; Novartis) started in 1985 a derivatization program in this chemistry. In the publication of Soloway et al. (1979) acyclic nitromethylene compounds were described to possess only weak insecticidal activity against corn earworm (*Heliothis zea*) in contrast to the highly active nitromethylene heterocycles such as nithiazin. Nevertheless, these acyclic compounds showed some activity and served as lead structures in our first attempts to discover new neonicotinoids, resulting in the synthesis of acyclic nitroenamine, cyanoamidine, and nitroamidine derivatives **1**, **2,** and **3** respectively (Figure 4). The nitroenamine and nitroamidine compounds **1** and **3,** respectively (Gsell 1987a; Kristiansen 1989), turned out to be highly insecticidal, whereas the cyanoamidines **2** (Gsell 1987b) possess somewhat inferior activity. Independently, parallel work was performed on compounds of type **1** and **3** by Takeda (Minamida et al. 1987; Uneme et al. 1988) and Nihon Tokushu Noyaku (Shiokawa 1988). In this very competitive field, Takeda, Nihon Tokushu Noyaku, and Novartis filed patent applications for compounds of type **1** and **3** with priority dates differing only by days or a few months. For example, Takeda's patent application for nitroenamine derivatives **1** (Minamida et al. 1987), which includes nitenpyram has a priority date of only 6 days before the Novartis application (Gsell 1987a).

After this experience we developed a molecular modeling-based approach for the design of novel structural types of neonicotinoids (Maienfisch et al. 1993, 1997a). This resulted in the synthesis and evaluation of pyrrolidines **4** and isoxazolones **5** (Figure 5). A real breakthrough was achieved in 1991 with the synthesis of nitroimino-oxadiazinane derivatives **6** (Maienfisch and Gsell 1992). After performing an optimization program, CGA 293'343 was identified as the best compound and selected for development.

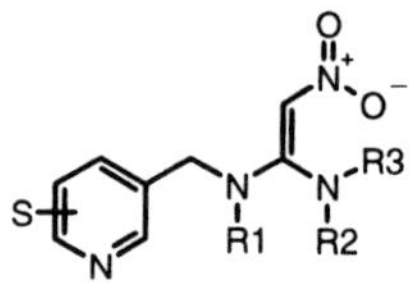

R1, R2, R3 = H, alkyl
S = H, substituent

1

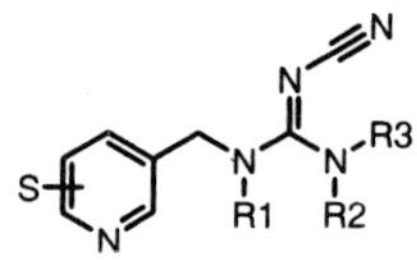

R1, R2, R3 = H, alkyl
S = H, substituent

2

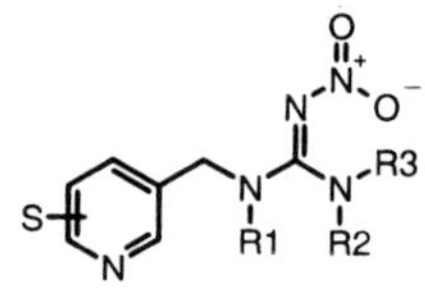

R1, R2, R3 = H, alkyl
S = H, substituent

3

Fig. 4. First acyclic neonicotinoids prepared by Novartis

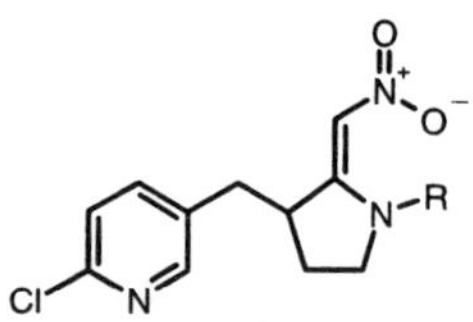

R = H, alkyl

4

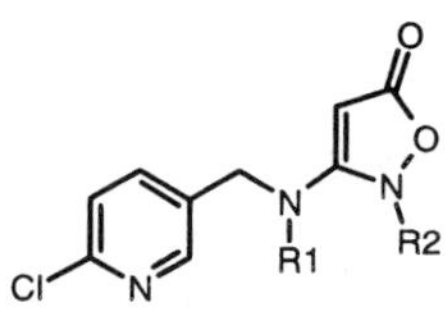

R1, R2 = H, alkyl

5

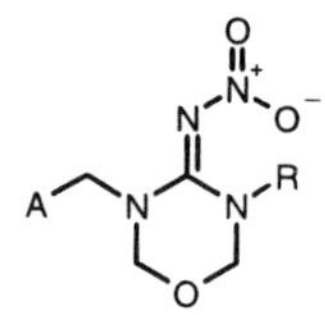

A = heterocycle; R = H, alkyl

6

Fig. 5. Novel structural types of neonicotinoids discovered by Novartis

CGA 293'343 is a novel broad-spectrum insecticide currently under worldwide development by Novartis Crop Protection AG. Thiamethoxam (ISO draft proposal) has been proposed as the common name for this compound. CGA 293'343 provides excellent control of a wide variety of commercially important pests and possesses contact, stomach, and systemic activity. The chemical structure differs remarkably from the three neonicotinoids already introduced to the market: CGA 293'343 is the first commercially available neonicotinoid with a chlorothiazole as heterocyclic group and belongs therefore to the subclass of the thianicotinyl compounds. It possesses a nitroamidine pharmacophore and a novel oxadiazinane ring providing a unique structure.

In 1997, registration was granted in New Zealand and already first sales have been recorded. Registration and sales in further countries will follow soon. CGA 293'343 will be marketed under the trademark Actara™ for foliar and soil treatment and under the trademark Cruiser® for seed treatment. At the 1997 Annual Meeting of the Entomological Society of America in Nashville, TN, the first results on this new development compound were presented (Maienfisch et al. 1997b; Senn et al. 1997; Kobel et al. 1997; Lawson et al. 1997; Brandl et al. 1997; Morton et al. 1997; Steinemann and Lawson 1997; Rindlisbacher et al. 1997).

2.2 Synthesis of CGA 293'343

CGA 293'343 can be synthesized in only a few steps and with high yields from easily accessible starting materials (Maienfisch and Gsell 1992). Starting from nitroguanidine, *N*-methyl-nitroguanidine **8** is prepared according to a published procedure (McKay and Wright 1947) and then converted to the oxadiazinane **9** by treatment with formaldehyde in the presence of formic acid. The subsequent alkylation with the thiazole **10** (prepared according to published procedures: Beck and Heitzer 1986; Uneme et al. 1990), in dimethylformamide and potassium carbonate as a base afforded CGA 293'343 in good yields (Figure 6).

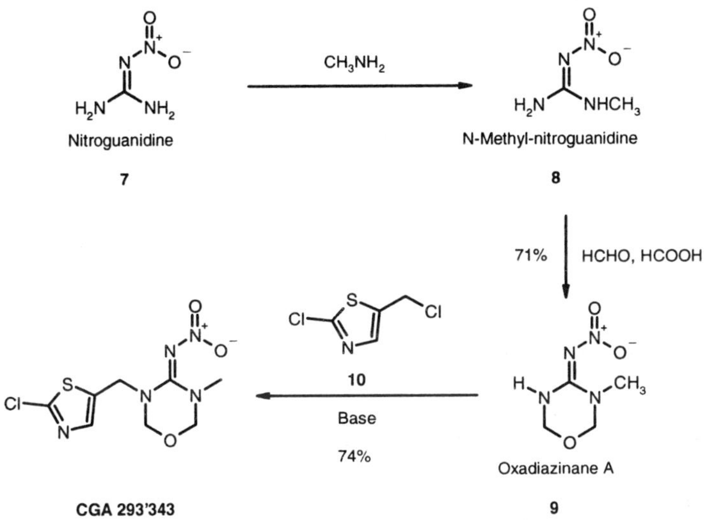

Fig. 6. Synthesis of CGA 293'343

2.3. Chemical and Physical Properties of CGA 293'343

CGA 293'343 is a crystalline, odourless compound with a melting point of 139.1°C. The compound has a relatively high water solubility of 4.1 g/l at 25°C and a low partition coefficient log P_{ow} of -0.13 at pH 6.8. No dissociation is observed within the range of pH 2 to pH 12. CGA 293'343 is hydrolytically very stable at pH 5 (half-life, >1 year at r.t.) and stable at pH 7 (estimated half-life at r.t., approximately 200-300 days). The compound is more labile at pH 9 (half-life, a few days). CGA 293'343 is photolytically rapidly degraded (half-life, ~1 h as a droplet deposit on Teflon). No decomposition was observed after storage of the active ingredient or formulations at 54°C after 2 months. At temperatures above 150°C, exothermic decomposition occurs (Table 3).

Table 3. Chemical and physical properties of CGA 293'343

• Physical state (20°C)	Crystalline powder
• Melting point	139.1°C
• Colour	Slightly cream
• Odour	Odourless
• Vapour pressure at 25°C	6.6×10^{-9} Pa
• Water solubility at 25°C	4'100 mg/l
• Solubility in organic solvents at 25°C	Methanol: 10'200 mg/l Ethanol: 3'210 mg/l Acetonitrile: 78'000 mg/l *n*-Octanol: 630 mg/l Acetone: 42'500 mg/l Ethylacetate: 5'740 mg/l Dichloromethane: 43'000 mg/l Toluene: 630 mg/l Hexane: 0.18 mg/l
• pH value	6.84 (saturated solution in water)
• Partition coefficient *n*-Octanol/water at 25°C	log P_{ow} = -0.13
• Dissociation constant	CGA 293'343 has no dissociation within the range of pH 2 to pH 12
• Hydrolysis	pH 5 / 70°C: ~5% degradation after 261 h pH 7 / 70°C: half-life = 64 h pH 9/ 70°C: half-life = 1.24 h
• Photodecomposition	half-life = ~1h (droplet deposit on Teflon)

3. Biological Laboratory Evaluation of CGA 293'343

CGA 293'343 has undergone extensive testing in the laboratory to characterize its biological activities. Tables 4, 5, and 6 summarize the results obtained for various insect pests after different application methods under laboratory conditions. In Tables 4 and 5, the biological activity is listed as LC_{80} value in ppm,

Table 4. Spectrum of activity (LC_{80} in ppm) of CGA 293'343 and commercial neonicotinoids after foliar spray application

Order / Pest	Stage[a]	Bioassay	LC_{80} in ppm			
			CGA 293'343	IMID[b]	NITE[c]	ACET[d]
Lepidoptera						
Spodoptera littoralis (cotton leafworm)	L3	Soybean	100	50	>100	100
Heliothis virescens (tobacco budworm)	L1	Soybean	12	12	>100	3
Plutella xylostella (diamondback moth)	L3	Soybean	100	100	>100	25
Coleoptera						
Diabrotica balteata (banded cucumber beetle)	L3	Sandy soil	0.2	0.8	>100	12
	L3	Sandy soil / Persistence 30d[e]	0.8	3	n.t.[f]	12
	L3	Field soil	3	3	n.t.	12
	L3	Field soil / Persistence 30d	3	12	n.t.	12
	Adult	Bean	12	12	n.t.	25
Homoptera						
Aphis craccivora (groundnut aphid)	m.p.	Pea	3	3	3	3
Myzus persicae (green peach aphid)	m.p.	Pepper	0.8	0.8	3	3
Bemisia tabaci (cotton whitefly)	N2	Bean	3	3	12	25
Nilaparvata lugens (brown planthopper)	N2	Rice	0.8	0.8	0.8	100
	N2	Rice / Persistence 4d	3	25	12	>100
Nephotettix cincticeps (green leafhopper)	N2	Rice	0.8	0.2	0.2	12

[a] L1, L3: 1st, 3rd instar larvae; N2: 2nd instar nymph; m.p.: mixed population; [b] imidacloprid; [c] nitenpyram; [d] acetamiprid; [e] days after treatment, when infestation with insects was made.

Table 5. Systemic activity (LC_{80} in ppm) of CGA 293'343 and commercial neonicotinoids after into-water or drench application

Order / Pest	Stage[a]	Bioassay	LC_{80} in ppm			
			CGA 293'343	**IMID[b]**	**NITE[c]**	**ACET[d]**
Lepidoptera						
Spodoptera littoralis (cotton leafworm)	L1	Maize / Drench	12	12	n.t.[e]	n.t.
Coleoptera						
Diabrotica balteata (banded cucumber beetle)	Adult	Bean / Drench	0.05	0.8	n.t.	n.t.
Homoptera						
Aphis craccivora (groundnut aphid)	m.p.	Pea / i.w.[f]	12	12	12	12
Myzus persicae (green peach aphid	m.p.	Pepper / Drench	0.8	0.2	3	3
	m.p.	Pepper / Drench / Persistance 28d[g]	0.8	0.8	3	>12
Nilaparvata lugens (brown planthopper)	N2	Rice / i.w.	0.05	0.05	0.05	0.8
	N2	Rice / i.w. / Persistance 21d	0.05	0.2	0.2	3
Thysanoptera						
Frankliniella occidentalis (western flower thrips)	m.p.	Bean / Drench	3	12	n.t.	n.t.

[a] L1: 1st instar larvae; N2: 2nd instar nymph; m.p.: mixed population; [b] imidacloprid; [c] nitenpyram; [d] acetamiprid; [e] not tested; [f] into water; [g] days after treatment, when infestation with insects was made.

Table 6. Activity of CGA 293'343 after seed treatment application

Order / Pest	Stage[a]	Bioassay	Rate (mg a.i.[b] / seed)	% Activity CGA 293'343	% Activity IMID[c]
Lepidoptera					
Spodoptera littoralis	L1	corn / 11d[d]	0.1	28	72
(cotton leafworm)		corn / 32d	0.1	96	93
		corn / 11d	0.03	17	28
		corn / 32d	0.03	77	63
Coleoptera					
Diabrotica balteata	Adult	corn / 11d	0.03	100	100
(banded cucumber beetle)		corn / 47d	0.03	95	95
Diabrotica balteata	L3	corn / 14d	0.1	100	100
(banded cucumber beetle)		corn / 29d	0.1	100	100
		corn / 14d	0.03	100	100
		corn / 29d	0.03	97	100
Homoptera					
Aphis gossypii	m.p.	cotton / 14d	0.1	100	100
(cotton aphid)		cotton / 48d	0.1	100	92
		cotton / 14d	0.03	100	78
		cotton / 48d	0.03	77	20
Myzus persicae	m.p.	cabbage / 14d	0.03	100	100
(green peach aphid)		cabbage / 48d	0.03	100	100
Thysanoptera					
Frankliniella occidentalis	m.p.	cotton /14d	0.3	100	89
(western flower thrips)		cotton /41d	0.3	70	13

[a] L1, L3: 1st, 3rd instar larvae; m.p.: mixed population; [b] active ingredient; [c] imidacloprid; [d] days after sowing, when infestation with insects was made.

and comparative figures are given for the standards imidacloprid (IMID), nitenpyram (NITE) and acetamiprid (ACET). In Table 6 the biological activity is reported as % activity at a given rate in comparison to imidacloprid.

The spectrum of activity seen in the laboratory includes exceptional activity on Lepidoptera, Coleoptera, Thysanoptera, and especially Homoptera. In general the performance is at least as good as for the commercial neonicotinoids imidacloprid, nitenpyram and acetamiprid.

Used as a foliar application (Table 4), CGA 293'343 has proven to be highly effective against a wide variety of agriculturally important pests such as aphids (*Aphis craccivora, Myzus persicae*), planthoppers (*Nilaparvata lugens*), leafhoppers (*Nephotettix cincticeps*), and whiteflies (*Bemisia tabaci*). CGA 293'343 also shows excellent activity against *Diabrotica balteata*. A special benefit of this compound is its outstanding residual effect observed in the laboratory for *Nilaparvata lugens* and *Diabrotica balteata*. The residual activity is clearly superior to that of the commercial neonicotinoids.

CGA 293'343 exhibits high systemic activity after into-water or drench application (Table 5). At very low concentrations, it is highly effective against homopteran, thysanopteran, and coleopteran pests and possesses an excellent residual activity. LC_{80} values of 3 to 12 ppm have been observed for *Spodoptera littoralis, Aphis craccivora,* and *Franklinella occidentalis*. For *Diabrotica balteata, Myzus persicae,* and *Nilaparvata lugens,* very low LC_{80} values of 0.05 to 0.8 ppm were found. Overall, the systemic efficacy clearly surpasses that of the commercial neonicotinoids.

After use as a seed treatment, CGA 293'343 has performed well against all insect pests tested in our laboratory trials (Table 6). Against *Diabrotica balteata, Myzus persicae,* and *Aphis gossypii* CGA 293'343 gives full initial protection even at the lower rate of 0.03 mg a.i./seed and shows also excellent residual activity. The performance against *Spodoptera littoralis* is somewhat weaker. In general, the activity against *Diabrotica balteata, Myzus persicae,* and *Spodoptera littoralis* is comparable to imidacloprid, whereas CGA 293'343 is clearly superior against *Aphis gossypii* and *Frankliniella occidentalis.*

4. Field Performance of CGA 293'343

Under field conditions, CGA 293'343 controls sucking and some chewing coleopteran and lepidopteran pests in vegetables, field crops, and fruit trees. Based on its high systemic activity, the compound has been profiled since 1994 not only for foliar use but also for soil and seed treatments.

4.1. Foliar Use

CGA 293'343 is active at very low rates of 10-50 g a.i./ha against aphids (*Aphis gossypii, Myzus persicae, Macrosiphum euphorbiae*), jassids, hoppers in rice, and Colorado potato beetle, as well as rice water weevil. Higher rates of about 50 g a.i./ha are needed for the control of whiteflies (*Trialeurodes vaporariorum, Bemisia* spp.) as well as against lepidopteran leafminers (Table 7).

Table 7. Use rates for foliar application

Crops	Target pest	Recommended use rates
Vegetables	Aphids, jassids	2.5 - 5 g a.i./hl (25 - 50 g a.i./ha)
Vegetables	Whiteflies	5 g a.i./hl (50 g a.i./ha)
Cotton	Aphids	30 - 50 g a.i./ha
Cotton	Whiteflies	100 g a.i./ha
Potato	Aphids, Colorado potato beetle	10 - 20 g a.i./ha
Tobacco	Aphids	50 g a.i./ha
Pomes fruits	Aphids, jassids	2.5 - 5 g a.i./hl (25 - 50 g a.i./ha)
Pomes fruits	Leafminers	25 - 50 g a.i./ha
Citrus	Aphids	1.5 - 3 g a.i./hl (45 - 90 g a.i./ha)
Rice	Hoppers	12.5 - 25 g a.i./hl (foliar use) 1 g a.i./seedling box
Rice	Rice water weevil	0.25 - 0.5 g a.i./ seedling box

4.1.1 Pome and Stone Fruits, Citrus

CGA 293'343 is active at low dosage rates (2.5-5 g a.i./hl) against aphids (*Aphis pomi, Myzus persicae, Schizaphis piricola,* and *Aphis gossypii*), including aphid species such as *Dysaphis devecta* and *Dysaphis plantaginea* that are normally difficult to control (Figure 7). At higher rates (5-10 g a.i./hl), the control of mining moths (*Phyllocnistis citrella* and *Phyllonorycter* sp.) was observed. Application of CGA 293'343 to the trunk in citrus gave very good control of scale insects and leafminers in South Africa, Thailand, and Brazil (Figure 8).

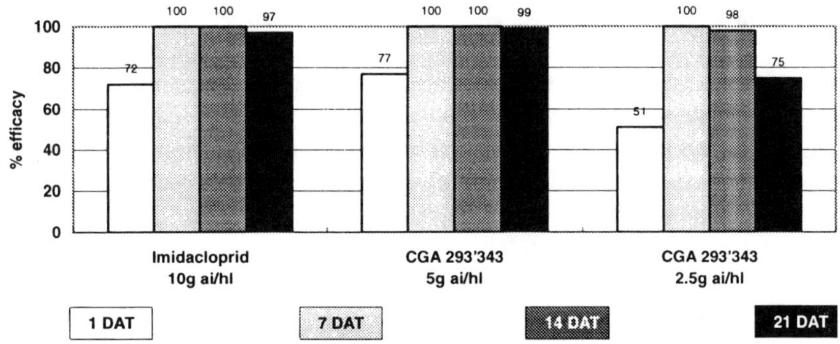

Fig. 7: Efficacy of CGA 293'343 against *Dysaphis plantaginea* on apples, France, 1996 (DAT, days after treatment)

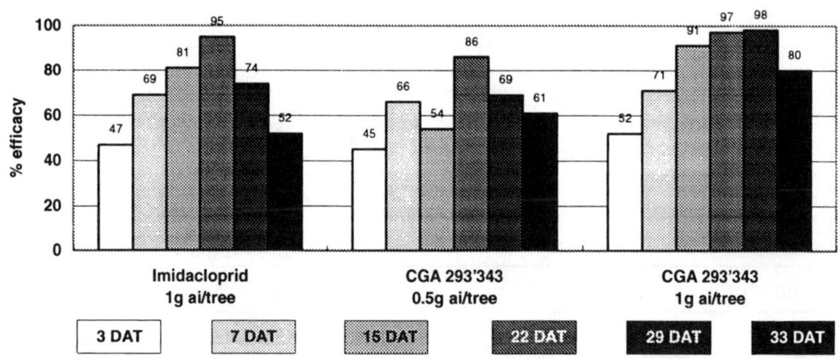

Fig. 8. Efficacy of CGA 293'343 applied as trunk application against *Phyllocnistis citrella* on citrus, Thailand, 1996

4.1.2 Vegetables

Trials were carried out in many vegetable crops all over the world. CGA 293'343 is very active against aphids (*Aphis fabae, Aphis gossypii, Aulacorthum solani, Myzus persicae, Brevicoryne brassicae*) (Figure 9) and jassids (*Empoasca devastans, Empoasca lybica*) on tomato, cucumber, cabbage, eggplant, okra, squash, and beans at dosage rates of about 2.5 g a.i./hl. For control of adult as well as larval stages of whiteflies (*Bemisia tabaci, Bemisia argentifolii, Trialeurodes vaporariorum*), higher dose rates of about 5 g a.i./hl are needed (Figures 10a, 10b). CGA 293'343 at half the dose rate of imidacloprid usually reached the same efficacy for the control of aphids and jassids as well as whiteflies. For thrips, good results were obtained against *Thrips palmi* in Japan on eggplant and cucumbers, where a granular application at transplanting was applied.

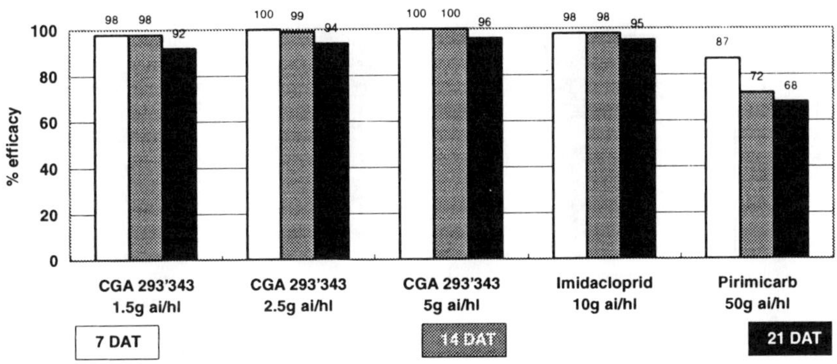

Fig. 9. Efficacy of CGA 293'343 against *Aphis gossypii* on melon, Spain, 1996 (Application date: June 17, 1996; Spray volume: 1250 l/ha; Untreated check: 66, 75, and 22 aphids per leaf (7, 14, 21 DAT))

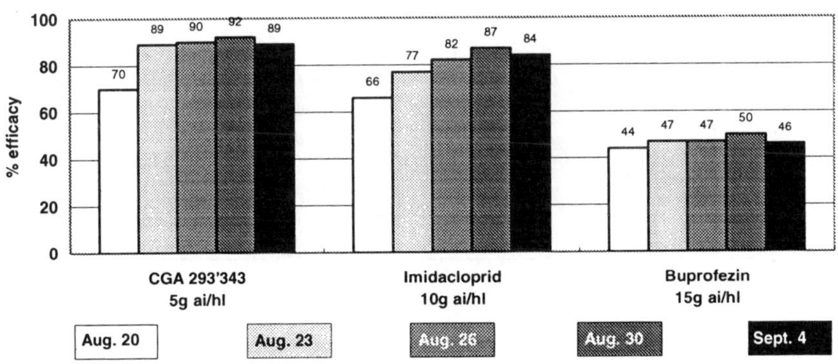

Fig. 10a. Efficacy of CGA 293'343 against *Bemisia tabaci* (adults) on sweet pepper, Spain, 1996 (Application dates: August 13 and 20, 1996; Spray volume: 1200 l/ha; Untreated check: 36, 51, 78, 68, and 52 adult whiteflies per leaf)

4.1.3 Potatoes

CGA 293'343 was tested in the United States, Italy, and Switzerland for the control of Colorado potato beetle (*Leptinotarsa decemlineata*). The compound showed excellent activity at dose rates of about 25 g a.i./ha and was superior to the standards, imidacloprid and fenvalerate (Figure 11). At these rates, aphids (*Myzus persicae, Macrosiphum euphorbiae*) and jassids (*Empoasca fabae*) were also controlled. Against the same pest spectrum, CGA 293'343 may also be applied into the furrow at planting time (see also Section 4.2).

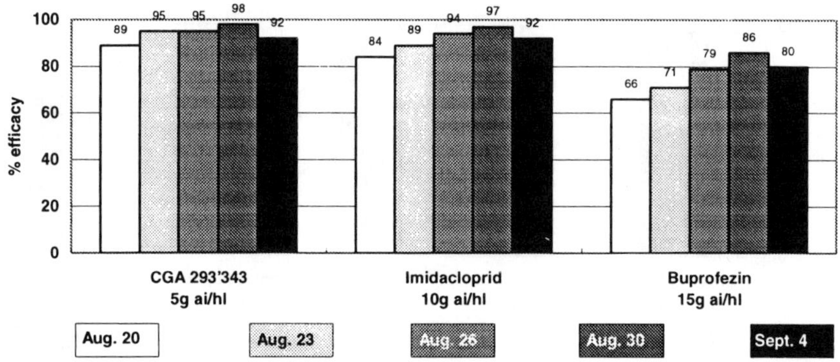

Fig. 10b. Efficacy of CGA 293'343 against *Bemisia tabaci* (nymphs) on sweet pepper, Spain, 1996 (Application dates: August 13 and 20, 1996; Spray volume: 1200 l/ha; Untreated check: 73, 70, 66, 93, and 65 nymphs per leaf)

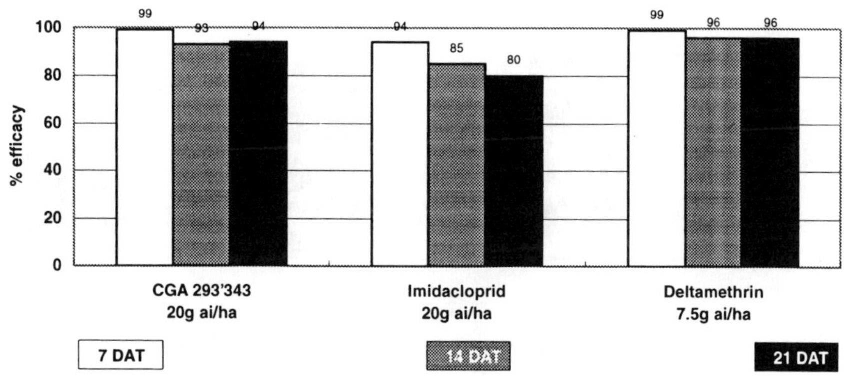

Fig. 11. Efficacy of CGA 293'343 against *Leptinotarsa decemlineata* on potato Switzerland 1997 (Application date: June 9, 1997; Spray volume: 500 l/ha; Untreated check: 1437, 1674 larvae per plot of 16 square meters (7 and 14 DAT) and % leaf surface dstroyed (21 DAT)

4.1.4 Rice

CGA 293'343 was tested on natural pest populations in Japan and Taiwan in 1996 for the control of rice hoppers and rice water weevil in different application systems (foliar, seedling box, into-water application). The spectrum of activity and the applied dose rates are shown in Table 8.

Table 8. Spectrum of activity and dose rates of CGA 293'343 in rice

Method	Target pests	Dose rate
Seedling box application	*Lissorhoptrus oryzophilus* (rice water weevil) *Nephotettix cincticep* (green rice leafhopper) *Laodelphax striatellus* (smaller brown planthopper) *Sogatella furcifera* (white-back planthopper)	0.5 - 1 g a.i./box
	Nilaparvata lugens (brown planthopper)	1.0 g a.i./box
Foliar application	*Nephotettix cincticeps* (green rice leafhopper) *Laodelphax striatellus* (smaller brown planthopper) *Sogatella furcifera* (white-back planthopper) *Nilaparvata lugens* (brown planthopper)	12.5-25 g a.i./ha

The performance of CGA 293'343 against the rice water weevil by seedling box application is very promising (Figure 12). Hoppers, which occur relatively early in the rice-growing season, are also controlled by seedling box application at the low dose rates. For season-long control of the brown planthopper, 1.0 g a.i./box is required. Foliar application of CGA 293'343 at rates of 12.5 to 25 g a.i./ha provided planthopper control equal to the best standard (Figure 13).

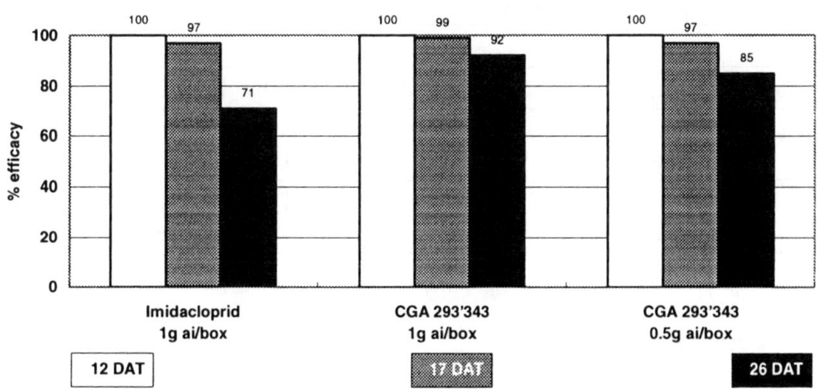

Fig. 12. Efficacy of CGA 293'343 (GR3) applied to seedling box against *Lissorhoptrus oryzophilus* in rice, Japan, 1996

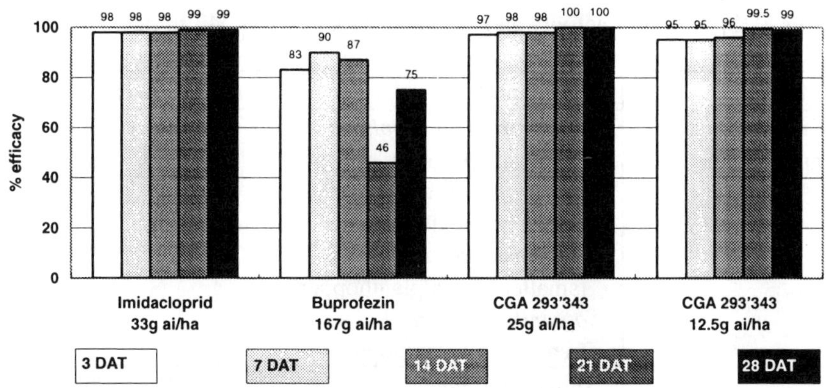

Fig. 13. Efficacy of CGA 293'343 applied as foliar spray against *Nilaparvata lugens* on rice (average of 4 trials), Taiwan, 1996

4.2 Soil Application Use

CGA 293'343 shows very good efficacy when applied to the soil. The activity spectrum is broader than that of foliar applications. In addition to aphids, whiteflies, and Coleoptera, thrips and some Lepidoptera are also controlled (Table 9).

Table 9. Use rates for soil application

Crops	Target pest	Recommended Use Rates
Vegetables	Aphids, Jassids	5 - 15 mg a.i./linear meter
Vegetables	Whiteflies, Thrips	10 - 20 mg a.i./linear meter
Lettuce	Aphids	2.5 - 5 mg a.i./linear meter
Cotton	Aphids, Jassids, Whiteflies, Thrips	10 - 14 mg a.i./linear meter
Potato	Aphids, Jassids, Colorado Potato Beetle	5 - 10 mg a.i./linear meter
Tobacco	Aphids, *Epitrix fasciate*, *Faustinus cubae*	70 - 140 g a.i./ha (drench after transplanting)

4.2.1. Vegetables

For the control of early- to midseason sucking and chewing pests in direct-sown and in transplanted vegetables, CGA 293'343 can be used as a granular planting hole application (Figure 14), as a soil drench, or as an in-furrow spray application

(Figure 15). The compound can also be used curatively as a side-dress application after reaching a threshold level. Main characteristics of these application methods are the long persistence of its activity as well as the convenience of its application for the farmer.

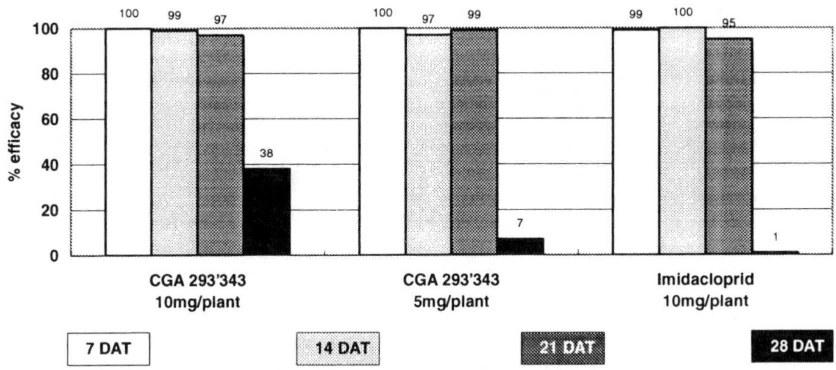

Fig. 14. Efficacy of CGA 293'343 applied as granules to planting hole at transplanting against *Thrips palmi* on eggplant, Japan 1995 (Application date: July 31, 1995; Untreated check: 36, 73, and 37 thrips per 20 leaves (7, 14, 21, and 28 DAT))

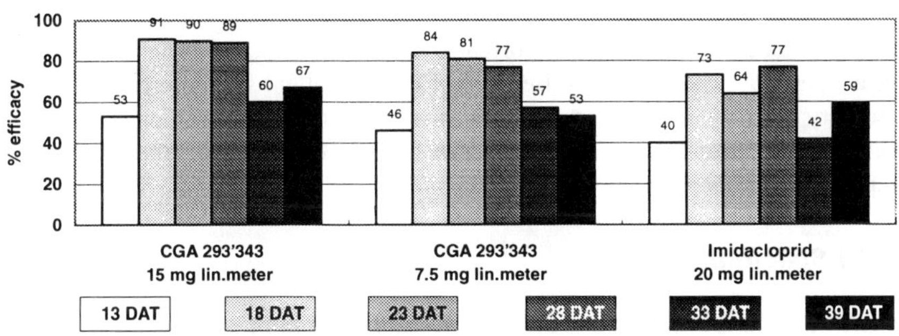

Fig. 15. Efficacy of CGA 293'343 applied in-furrow against *Bemisia tabaci* on squash, Egypt 1996 (Application date: April 17, 1996; Untreated check: 1230, 5230, 4520, 6050, 997, and 914 adult whiteflies per 40 plants (13, 18, 23, 28, 33, and 39 DAT)

4.2.2 Other Crops

Trials in potatoes, applied in-furrow at planting or about 4 weeks after planting, showed good control of *Leptinotarsa decemlineata* as well as aphids over 60 to 80 days. In tobacco, the compound was applied as a posttransplanting drench application in Brazil. In addition to *Myzus persicae* (Figure 16), leaf-feeding Coleopteran pests such *as Epitrix fasciata* and *Fausinus cubae* also were controlled for up to 90 days after application. Early- to midseason cotton pests, such as whiteflies, jassids, and aphids, can be controlled by in-furrow application at sowing or later as a side-dress application.

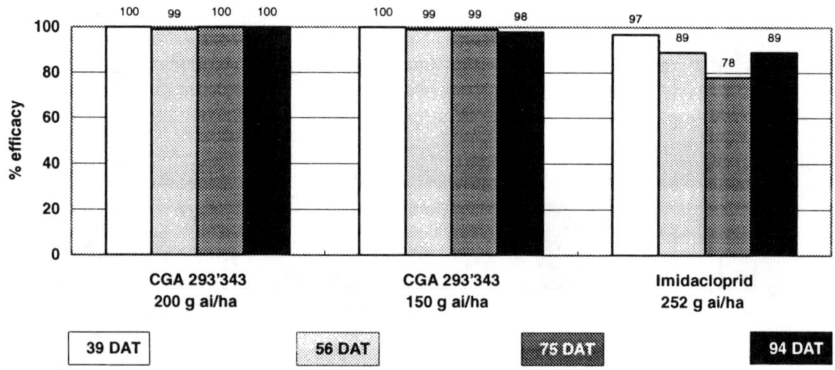

Fig. 16. Efficacy of CGA 293'343 applied as drench against *Myzus persicae* on tobacco, Brazil 1995/96 (Application date: October 17, 1995; Untreated check: 962, 1583, 420, and 93 aphids per plant (39, 56, 75, and 94 DAT))

4.3 Seed Treatment Use

The neonicotinoids offer the opportunity to set a new standard in the control of insect pests by seed treatment. Because of their excellent systemic properties, they are very well suited for seed treatment use to control soil-dwelling, early-season sucking and leaf-feeding insects. CGA 293'343 is one of the most active representatives of this new chemical class.

4.3.1 Spectrum of Activity

CGA 293'343 is highly active against a broad spectrum of soil-dwelling insects. It also offers effective control of a wide range of early-season, leaf-feeding (Coleoptera and Lepidoptera), and sucking insects (Homoptera and Thysanoptera). Due to its fast action on sucking insects, it limits the transmission of plant pathogenic viruses (e.g., in cereals and sugar beets). CGA 293'343 is under worldwide testing and registration in all major crops such as corn, sorghum, cotton, canola, cereals, rice, sugar beets, sunflowers, peas, beans, and potatoes as a seed treatment. The spectrum of activity for seed treatment use is shown in Table 10.

Table 10. Spectrum of activity of CGA 293'343 as a seed treatment

Crop	Targeted soil-dwelling insects	Targeted early leaf-feeding and sucking insects
Corn / maize	*Agriotes* spp. (wireworms) *Heteronychus arator* (black maize beetle) *Tanymecus* spp. (ground beetle) *Elasmopalpus* spp. (stalk borer) *Delia platura* (seed corn maggot)	*Listronotus spp.* (Argentine stem weevil) *Oscinella frit* (fruit fly) *Rhopalosiphum maidis* (corn leaf aphid) *Zyginidia scutellaris* (jassid) *Cicadulina mbila* (jassid) *Spodoptera spp.* (armyworm) *Sesamia spp.*
Sorghum	*Agriotes* spp. (wireworms) *Pterohelaeus spp.* (false wireworms) *Solenopsis germinata* (fire ants)	*Rhopalosiphum maydis* (corn leaf aphid) *Schizaphis graminum* (green bug) *Blissus leucopterus* (chinch bug)
Cotton	*Eutinobothrus* spp.	*Aphis gossypii* (cotton aphid) *Thrips tabaci* *Frankliniella spp.* (thrips) *Empoasca devastans* (jassid) *Alabama argillacea* (cotton leaf worm)
Canola (oilseed rape)	*Delia anthomyiidae* (cabbage root fly)	*Psylliodes chrysocephalus* (cabbage stem flea beetle) *Phyllotreta spp.* (turnip flea beetle) *Brevicoryne brassicae* (cabbage aphid) *Myzus persicae* (green peach aphid) *Athalia rosae* (turnip sawfly)

Table 10 ff: Spectrum of activity of CGA 293'343 as a seed treatment

Crop	Targeted soil dwelling insects	Targeted early leaf-feeding and sucking insects
Cereals	*Agriotes* spp. (wireworms) *Zabrus tenbrioides* (ground beetle)	*Rhopalosiphum padi* (bird cherry aphid) *Sitobion avenae* (grain aphid) *Diuraphis noxia* (Russian wheat aphid) *Mayetiola destructor* (Hessian fly)
Rice	*Reticulitermes* spp. (termites) *Elasmopalpus* spp.	*Stechaenothrips spp.* *Deois spp.*
Sugar beet	*Atomaria linearis* (pygmy mangold beetle) *Agriotes* spp. (wireworms) *Blaniulus* spp. (millipedes) *Onychiurus* (springtails) *Bothynoderes* spp. (beet root weevil)	*Myzus persicae* (green peach aphid) *Aphis fabae* (black bean aphid) *Pegomya spp.* (beet leafminer) *Chaetocnema tibialis* (flea beetle) *Tanymecus spp.* (ground beetle)
Sunflower	*Agriotes* spp. (wireworms) *Tanymecus* spp. (ground beetle)	*Brachycaudus spp.* (aphid) *Zygogramma exclamationis*
Peas		*Sitona lineatus* (pea and bean weevil) *Acyrthosiphon spp.* (pea aphid)
Beans		*Aphis fabae* (black bean aphid) *Sitona lineatus* (pea and bean weevil) *Bemisia spp.* (white fly)
Potato		*Leptinotarsa decemlineata* (Colorado potato beetle) *Myzus persicae* (green peach aphid) *Empoasca fabae* (jassid)
Peanuts		*Frankliniella fusca* (thrips)

4.3.2 Lasting Activity and Systemicity

CGA 293'343 is highly systemic in plant roots, which was demonstrated by using radioactive-labeled material applied to corn seeds. After germination, the compound rapidly penetrates into the seed (Figure 17). In combination with the retention of the compound in the drilling zone (Figure 18) and the formation of a treatment halo around the seed, CGA 293'343 provides an efficient protection shield for the germinating seed against soil-dwelling insects.

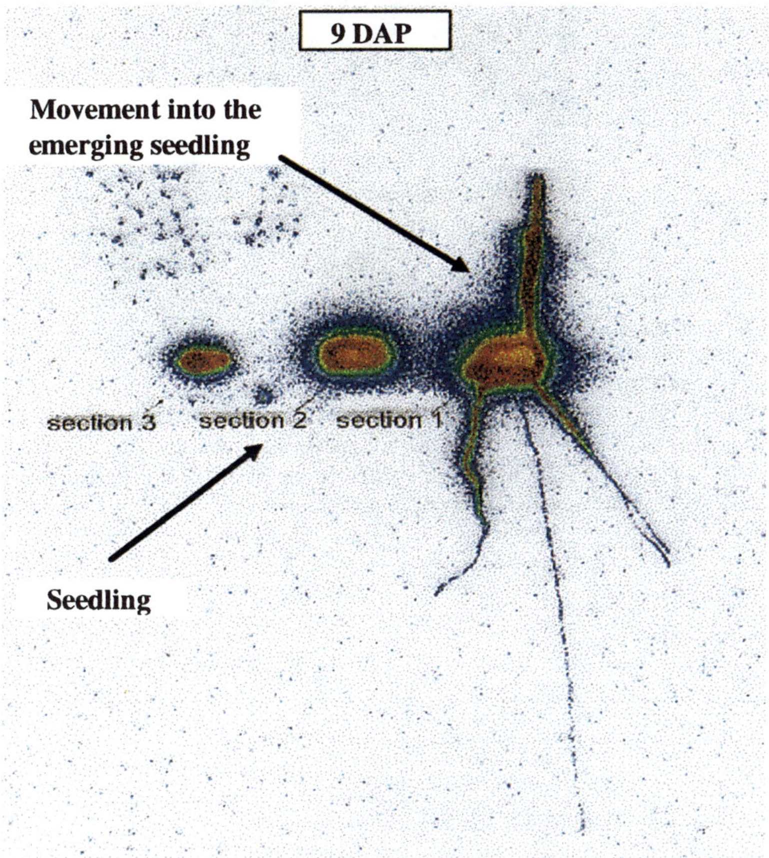

Fig. 17. Penetration of CGA 293'343 into the germinating (^{14}C labeled material; red: high concentration of labeled material; blue: low concentration of labeled material; DAP: days after planting)

The same studies have shown how the compound is taken up by the roots and translocated by the xylem into the coleoptile, cotyledons, and further into the young leaves (Figure 18). This systemic property facilitates excellent control

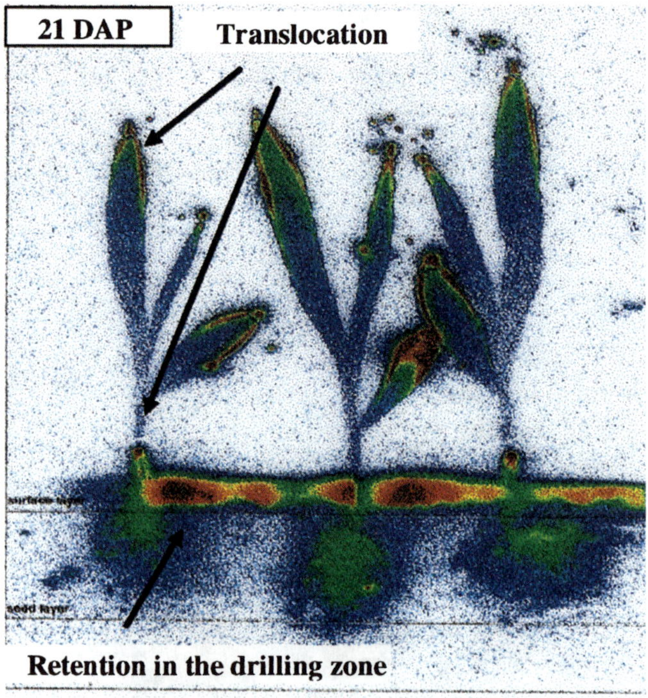

Fig. 18. Retention of CGA 293'343 in the drilling zone and translocation into the young leaves (^{14}C labeled material; red: high concentration of labeled material; blue: low concentration of labeled material; DAP: days after planting)

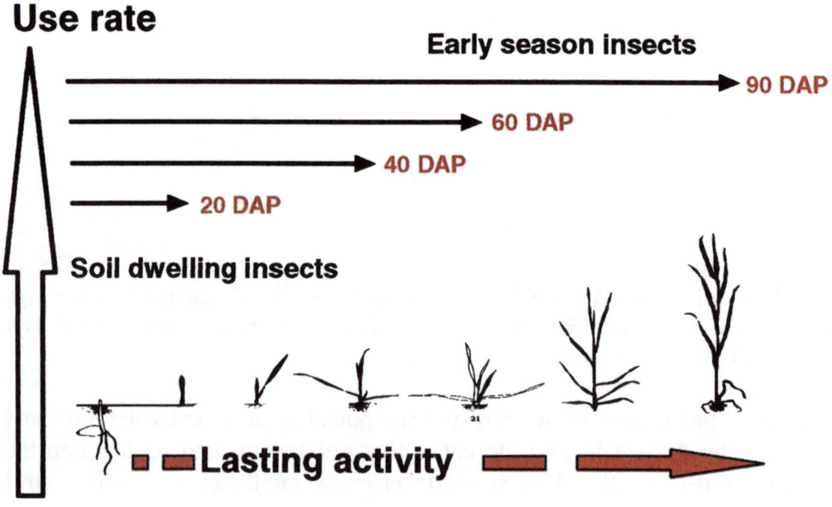

Fig. 19. Lasting activity of CGA 293'343 (DAP: days after planting)

against early leaf-feeding and sucking insects. CGA 293'343 can replace the need for soil-applied granular insecticides at planting and early foliar insecticide treatments after the emergence of the crop. Depending on the use rate, the crop species, and the target insect, CGA 293'343 offers the possibility of controlling early-season insects up to 90 days after crop emergence (Figure 19).

4.3.3 Crop Tolerance

CGA 293'343 has been tested on all major seed varieties of all important crops. Based on a large database, CGA 293'343 is very safe for the germinating seed and young seedling at the recommended use rates and even at 1.5 times use rates for registration requirements. Crop safety is essential for the establishment of a good crop stand, the foundation for high yield and high quality. The characteristics of CGA 293'343 regarding crop tolerance include no delay in germination of the treated seed, uniform emergence, and strong vigor of the seedling and young plant.

4.3.4 Use Rates

CGA 293'343 has been tested worldwide at different use rates according to crop-specific insect pest problems and needed control periods. Generally, CGA 293'343 is a low-rate technology. For most crop/pest combinations, CGA 293'343 provides equal or even superior activity at a lower rate than currently available compounds from the same chemical class (Figure 20).

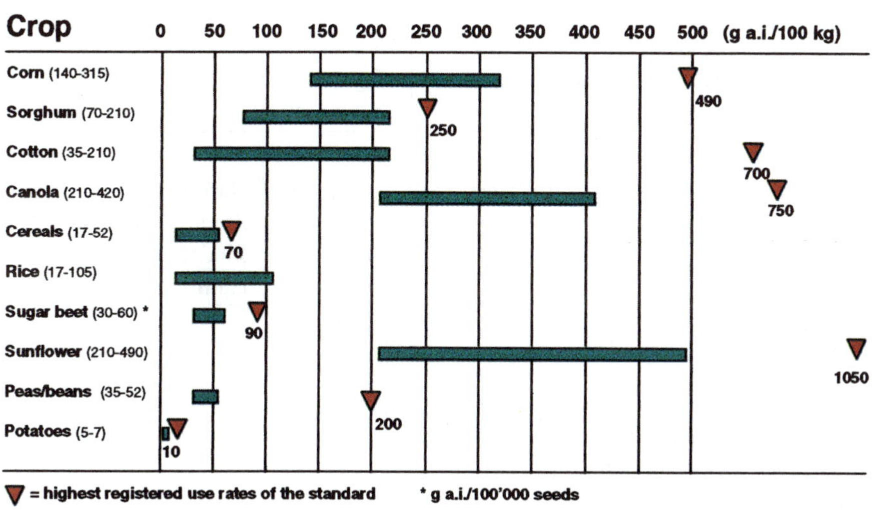

Fig. 20. Range of use rates of CGA 293'343 in different crops

5. Effect of CGA 293'343 on Beneficial Arthropods

The effects of CGA 293'343 after foliar applications on beneficials have been tested in laboratory and field trials. Based on the available data, it can be classified as slightly to moderately harmful to most beneficial insects, but safe to predatory mites in the field (Table 11). This rating is quite similar to other neonicotinoid compounds. CGA 293'343 is more selective than most of the established insecticides (e.g., pyrethroids). These properties allow it to be recommended for use in many crops where integrated pest management (IPM) is important. In pome and stone fruit, a limited number of preferably postbloom applications will be recommended. In glasshouses the inherent toxicity of CGA 293'343 on bumle bees can be overcome when applied as a drench or via the irrigation water. For soil application (e.g., into-irrigation water application) and seed treatment, CGA 293'343 has little or no impact on most beneficial insects.

Table 11. Selectivity data with CGA 293'343 against beneficial arthropods (IOBC rating)

Pest	**Dose rate**	**Effect (IOBC[a] rating)**
Laboratory		
Orius laevigatus (flower bug)	10 g a.i./hl	Moderately harmful
Coccinella septempunctata (ladybeetle)	5 g a.i./hl	Slightly harmful
Poecilus cupreus (groundbeetle)	5 g a.i./hl	Slightly harmful
Chrysoperla carnea (lacewing)	5 g a.i./hl	Slightly harmful
Typhlodromus pyri (predatory mite)	5 g a.i./hl	Harmless
Amblyseius fallacis (predatory mite)	5 g a.i./hl	Slightly harmful
Bombus terrestris (bumble bee) contact	5 g a.i./hl	Harmful
Field		
Cyrtorhinus lividipennis (mirid bug)[b]	6 g a.i./ha	Harmful
Spiders[b]	15 g a.i./ha	Slightly harmful
Ladybeetles[c]	10 g a.i./hl	Moderately harmful
Syrphids[d]	5 g a.i./hl	Harmless

[a] International organisation for biological and integrated control of noxious animals and plants
[b] Field trials performed at Cikampek, Indonesia, 1994
[c] Field trials performed at Saillon, Switzerland, 1994
[d] Field trials performed at Etoy, Switzerland, 1994

6. Safety Aspects of CGA 293'343

6.1 Animal Metabolism

When CGA 293'343 is administered orally to rats, there is rapid and almost complete absorption. Excretion is very rapid and almost complete, occurring predominantly via the urine. Tissue residues are, therefore, very low. Because there is no retention potential, no accumulation is assumed. There is no sex, dose, or label differences in toxicokinetics. Metabolism in goats and hens is similar to that in rats. While the majority of CGA 293'343 is excreted as an unchanged molecule, one major metabolite and a number of smaller fractions were found and identified.

6.2 Mammalian Toxicology

The toxicological properties have been evaluated in extensive studies, and the following conclusions can be drawn:

- *Acute toxicity:* CGA 293'343 is only slightly toxic following exposure to a single dose by the oral, dermal, or inhalatory route. It is not irritant to skin and eyes and is devoid of a sensitizing potential (Table 12).
- *Short-term toxicity:* In repeat dose studies in rats and dogs, effects were noted in food consumption and body weight development at toxic dose levels. Liver and kidney were determined to be the target organs. The lowest NOAEL (no-observed-adverse-effect level) relevant to humans was found in dogs at 8.2 mg/kg. Dermal administrations over 28 days were well tolerated by rats; the NOAEL was at least 60 mg/kg.
- *Mutagenicity:* No mutagenic potential was detected in five tests covering specified genes, chromosomes, and DNA primary structure in bacteria or eukaryotic cells or in vivo.
- *Reproduction toxicity:* Developmental toxicity studies in rats and rabbits indicated no evidence of teratogenicity. A two-generation study in rats indicated no influence on reproduction.
- *Chronic toxicity/oncogenicity:* Liver masses appear to be increased at high dose levels in mice, but not in rats.
- *A preliminary acceptable daily intake (ADI)* can be derived: using an uncertainty factor of 2000 the ADI will be at least 0.004 mg/kg/day, which translates into a maximum permissibly intake (MPI) of at least 0.24 mg/person/day.

Table 12: Acute toxicity of CGA 293'343

Acute Toxicity Test	Species	Results	EPA toxicity category
Oral LD_{50}	Rat	1563 mg/kg	III
Dermal LD_{50}	Rat	> 2000 mg/kg	III
Inhalation LC_{50} (4 h)	Rat	>3720 mg/m^3	III
Skin irritation	Rabbit	Nonirritant	IV
Eye irritation	Rabbit	Nonirritant	IV
Skin sensitization	Guinea pig	Nonsensitizing	IV

6.3 Plant Metabolism, Residues and Consumer Safety

Metabolism is being investigated in maize, paddy rice, cucumbers, and pears using the highest recommended use rates. The parent was found to be the major constituent with one metabolite found, in most investigations, below or around the threshold values of 0.05 ppm or 10% of TRR (total residual radioactivity). The residue definition covers the parent and the major metabolite. Maximum residue levels (MRLs) have been established for numerous fruits, vegetables, and dairy and meat products (Table 13).

When calculating consumer exposure with the TMDI (theoretical maximum daily intake) method on the basis of the W.H.O. consumption figures (global population), a TMDI of 0.0438 mg/person/day is found. This value represents only 18% of the already conservative MPI of 0.24 mg/person/day. Therefore, consumers are not at risk.

Table 13. Maximum residue levels (MRL)

Crop	MRL (mg/kg)	Crop / product	MRL (mg/kg)
Apple	0.05	Maize	0.05
Pear	0.02	Rice	0.05
Tomato	0.1	Meat	0.02
Peas	0.05	Milk	0.02
Sugar beet	0.02	Eggs	0.02
Potato	0.1		

6.4 Ecotoxicity

CGA 293'343 has favorable ecological toxicology characteristics for the majority of test species (Table 14). According to the EPA categorization for birds, it is slightly toxic by gavage and practically nontoxic by the dietary route. It is practically nontoxic to fish, *Daphnia*, and mollusks. Algae and earthworms are insensitive to CGA 293'343. It is only moderately toxic to mysid shrimp. CGA 293'343 is, however, highly toxic to honeybees, and adequate risk management will be required.

6.5 Environmental Fate

In laboratory soils, CGA 293'343 degrades at moderate to slow rates. The half-life ranges from 34 to 75 days under favorable conditions, but may increase by a factor of three under unfavorable conditions. Under field conditions, degradation is generally faster, because field soils usually have higher microbial activity and exposure to light is another important degradation pathway. CGA 293'343 and its main metabolite have a low adsorption capacity to particles, and they are considered to be moderately mobile in soils. In a lysimeter study, the concentrations of CGA 293'343 in the leachate were below the E.U. (European Union) trigger value of 0.1 µg/l. In water, hydrolytic degradation occurs at slightly alkaline conditions. The compound is photolyzed rapidly in water. In natural aquatic systems (e.g., paddy rice), degradation also occurs in the absence of light by microbial degradation. Based on the low vapor pressure and results of soil volatility studies, significant volatilization is not expected.

7. Conclusions

In 1985 Ciba (since 1996; Novartis) started a derivatization program on neonicotinoids. A real breakthrough was achieved in 1991 with the synthesis of CGA 293'343. CGA 293'343, a new, broad-spectrum neonicotinoid insecticide under development by Novartis Crop Protection, offers excellent control of a wide variety of commercially important pests in many crops. Control of most insect pests with CGA 293'343 is superior or equivalent to currently registered neonicotinoid insecticides. CGA 293'343 exhibits contact, stomach, and systemic activity. Its good stability and excellent systemic characteristics make this compound appropriate for foliar, granular, and seed treatment application. The long lasting residual effect is a special benefit of this compound. Low use rates, flexible application methods, excellent efficacy, and the favorable safety profile make this new insecticide wellsuited for modern integrated pest management programs in many cropping systems. CGA 293'343 is the first commercially available second-generation neonicotinoid and will be marketed under the trademarks Actara™ for foliar and soil treatment and under the trademarks Cruiser® for seed treatment.

Table 14. Ecological toxicology characteristics of CGA 293'343

Acute toxicity test	**Species**	**Results**	**EPA Toxicity Categy**
Avian oral LD_{50}	Bobwhite quail Mallard duck	1552 mg/kg 576 mg/kg	Slightly toxic Slightly toxic
Avian dietary LC_{50}	Bobwhite quail Mallard duck	> 5200 ppm > 5200 ppm	Practically nontoxic Practically nontoxic
Freshwater fish LC_{50} (96 h)	Rainbow trout Bluegill	>125 mg/l >114 mg/l	Practically nontoxic Practically nontoxic
Marine fish LC_{50} (96 h)	Sheepshead minnow	>111 mg/l	Practically nontoxic
Freshwater invertebrate EC_{50} (48 h)	*Daphnia magna*	>100 mg/l	Practically nontoxic
Marine invertebrate EC_{50} (96h)	Mysid shrimp Eastern oyster	6.9 mg/l >119 mg/l	Moderately toxic Practically nontoxic
Algae EC_{50} (72 h)	Green algae	>81.8 mg/l	None
Earthworm EC_{50} (14 d)	*Eisenia foetida*	>1000 mg/kg soil	None
Beneficial insect contact LD_{50}	Honeybee	0.024 µg /bee	Highly toxic

Acknowledgment

We would like to thank all colleagues from chemistry, biology, development, product management, human safety, and environmental safety at Novartis Crop Protection who contributed to this article, in particular D. Allemann, M. Angst, M. Bachmann, M. Bolsinger, T. Bridges, H. Buholzer, F. Buholzer, H-R. Dettwiler, B. Duverger, J. Ehrenfreund, H. Elmsheuser, S. Ferguson, W. Fischer, C. Flückiger, A. Fougeroux, L. Gsell, R.G. Hall, H. Hammann, Y. Hashino, J. Hättenschwiler, D. Hofer, W. Hofherr, T. Hoppe, J. Hosang, H-P. Hürlimann, J.P. Koenig, O. Kristinansen, D.S. Lawson, P. Mair, B. Minton, S. Moore, H.V. Morton, M.C. Neale, N. Ngo, R. Phaff, T. Pitterna, T. Rapold, W. Reiner, G. Röhrl, P. Sandmeier, H. Scheffler, B. Sechser, E. Sieger, A. Steinemann, P. Stöcklin, H-P. Streibert, H. Szczepanski, A. Tally, B. Thede, L. Thedford, P. Thanei, M. Walti, S. Uk, M. Urban, H-J. Widmer, and P. Wyss.

References

Bai D, Lummis SCR, Leicht W, et al (1991) Actions of imidacloprid and related nitromethylene on cholinergic receptors of an identified insect motor neurone. Pestic Sci 33:197-204

Beck G, Heitzer H (1986) Manufacture of 2-chloro-5-(chloromethyl)thiazole. Ger Offen, DE 3631538 A1 880324

Benson JA (1989) Insect nicotinic acetylcholine receptors as targets for insecticides. BCPC Monogr. 43 (Prog. Prospects Insect Control), pp 59-70

Brandl F, Moore S, Fougeroux A (1997) CGA 293'343 - a novel insecticide for the growing needs of the global seed treatment market. 1997 Annual Meeting of the Entomological Society of America, paper no 499

Elbert A, Overbeck H, Iwaya K, Tsuboi S (1990) Imidacloprid, a novel systemic nitromethylene analogue insecticide for crop protection. Brighton Crop Protection Conferences - Pest and Diseases 1, pp 21-28

Gsell L (1987a) Preparation and testing of [(pyridylmethyl)amino]nitroethylenes as insecticides, acaricides, and ectoparasiticides. Eur Pat Appl EP 302833 A2 890208

Gsell L (1987b) Preparation of pyridylmethylcyanoguanidines as insecticides and acaricides. Eur Pat Appl EP 306696 A1 890315

Kagabu S, Moriya K, Shibuya K, et al (1992) 1-(6-Halonicotinyl)-2-nitromethylene-imidazolidines as potential new insecticides. Biosci Biotechnol Biochem 56:362-363

Kobel W, Tally A, Thede B (1997) CGA 293'343 - a novel broad spectrum insecticide: safety aspects. 1997 Annual Meeting of the Entomological Society of America, paper no 497

Kristiansen O, Maienfisch P, Gsell L (1989) Guanidine derivatives as insecticides. Eur Pat Appl EP 418199 A2 910320

Lawson DS, Bridges T, Ferguson S, et al (1997) CGA 293'343: a new insecticide from Novartis Crop Protection, for use from the ground up. 1997 Annual Meeting of the Entomological Society of America, paper no 498

Liu M-YJ, Cassida JE (1993) High affinity binding of ^{3}H-imidacloprid in the insect acethylcholine receptor. Pestic Biochem Physiol 46:40-46

Liu M-YJ, Lanford J, Cassida JE (1993) Relevance of ^{3}H-imidacloprid binding site in house fly head acethylcholine receptor to insecticidal activity of 2-nitromethylene- and 2-nitroimino-imidazolidines. Pestic Biochem Physiol 46:200-206

Maienfisch P, Gsell L (1992) Preparation of 3-(heterocyclylmethyl)-4-iminoperhydro-1,3,5-oxadiazine derivatives as pesticides. Eur Pat Appl EP 580553 A2 940126

Maienfisch P, Gonda J, Jacob O, Gsell L (1993) Novel 2-nitromethylidene/2-cyanimino/2-nitroimino-pyrrolidines and -piperidines, intermediates, and their use as pesticides. PCT Int Appl WO 9424124 A1 941027

Maienfisch P, Gonda J, Jacob O, et al (1997a) Synthesis and insecticidal activity of novel nicotinoids. Abstracts, 214th ACS National Meeting, Las Vegas, NV, September 7-11, 1997, AGRO-018. American Chemical Society, Washington, DC

Maienfisch P, Gsell L, Rindlisbacher A, Allemann D (1997b) CGA 293'343 - a novel broad spectrum insecticide: chemistry and biological laboratory evaluation. 1997 Annual Meeting of the Entomological Society of America, paper no 495

McKay AF, Wright GF (1947) Preparation and properties of *N*-methyl-*N*-nitroso-*N'*-nitroguanidine. J Am Chem Soc 69:3028-3030

Minamida I, Iwanaga K, Okauchi T (1987) Alpha-unsaturated amines, their production and use. Eur Pat Appl EP 302389 A2 88112210.5

Minamida I, Iwanaga K, Tabuchi T et al, (1993a) Synthesis and insecticidal activity of acyclic nitroethene compounds containing a heteroarymethylamino group. J Pesticide Sci 18: 41-48

Minamida I, Iwanaga K, Tabuchi T, et al (1993b) Synthesis and insecticidal activity of acyclic nitroethene compounds containing a 3-pyridylmethylamino group. J Pestic Sci 18:31-40

Moriya K, Shibuya K, Hattori Y, et al (1992) 1-(6-Halonicotinyl)-2-nitroimino-imidazolidines and related compounds as potential new insecticides. Biosci Biotechnol Biochem 56:364-365

Moriya K, Shibuya K, Hattori Y, et al (1993) 1-Diazinylmethyl-2-nitromethylene- and 2-nitroimino-imidazolidines as potential new insecticides. J Pestic Sci 18:119-123

Morton HV, Moore S, Ngo N (1997) CGA 293'343: a new molecule for seed-delivered insect control in U.S. field crops. Annual Meeting of the Entomological Society of America, paper no 500

Rindlisbacher A, Buholzer H, Stöcklin P, Reiner W (1997) CGA 293'343 seed treatment activity in lab and greenhouse trials. 1997 Annual Meeting of the Entomological Society of America, poster no D184

Sattelle DB, Buckingham SD, Wafford KA, et al (1989) Actions of the insecticide 2-(nitromethylene)tetrahydro-1,3-thiazine on insect and vertebrate nicotinic acetylcholine receptors. Proc R Soc London (Biol) 237:501-514

Senn R, Tally A, Flückiger C, Hashino Y (1997) CGA 293'343 - a novel broad spectrum insecticide: global field performance. 1997 Annual Meeting of the Entomological Society of America, paper no 496

Shiokawa K, Tsuboi S, Moriya K, et al (1988) Insecticidally active nitro compounds. Eur Pat Appl EP 375907 A1 891211180.7

Shiokawa K, Tsuboi S, Moriya K, Kagabu S (1995) Chloronicotinyl insecticides: development of imidacloprid. In: Ragsdale NN, Kearney PC, Plimmer JR (Eds) Eight International Congress of Pesticide Chemistry - Option 2000, ACS Conference Proceeding Series, pp 49-59

Shiokawa K, Tsuboi S, Iwaya K, Moriya K (1994) Development of a chloronicotinyl insecticide, imidacloprid. Nippon Noyaku Gakkaishi 19:209-217; 329-332

Soloway SB, Henry AC, Kollmeyer WD, et al (1979) Nitromethylene insecticides. In: Geissbühler H, Brooks GT, Kearney PC (Eds) Advances in pesticide science, part 2. Pergamon Press, New York, pp 206-227

Steinemann A, Lawson S (1997) Uptake and translocation of CGA 293'343. 1997 Annual Meeting of the Entomological Society of America, poster no D185

Tabuchi T, Fusaka T, Iwanaga K, et al (1994) Synthesis and insecticidal activity of acyclic nitroethene compounds containing (6-substituted)-3-pyridylamino group. J Pestic Sci 19:119-125

Takahashi H, Mitsui J, Takakusa N, et al (1992) NI-25, a new type of systemic and broad spectrum insecticide. Brighton Crop Protection Conferences - Pest and Diseases 1, pp 89-96

Tomizawa M (1994) Structure-activity relationships relationships of nicotinoids and the related compounds. J Pestic Sci 19:229-240

Tomizawa M, Yamamoto I (1992) Binding of nicotinoids and related compounds to the insect nicotinic acethylcholine receptor. J Pestic Sci 17:231-236

Tomizawa M, Yamamoto I (1993) Structure-activity relationships of nicotinoids and imidacloprid analogs. J Pestic Sci 18:91-98

Tomizawa M, Otsuka H, Miyamoto I, et al (1995) Pharmacological characteristics of insect nicotinic acethylcholine receptor with its ion channel and comparison of the effect of nicotinoids and neonicotinoids. J Pestic Sci 20:57-64

Uneme H, Higuchi N, Minamida I (1990) Preparation of chlorothiazole derivatives. Eur Pat Appl EP 446913 A1 910918

Uneme H, Higuchi N, Minamida I, Okauchi T (1988) Guanidine derivatives, their production and insecticides. Eur Pat Appl EP 376279 A2 89124012.9

Yamamoto I (1996) Neonicotinoids-mode of action and selectivity. Agrochem Jpn 68:14-15

Part 3

Nicotinoid Insecticide Toxicology

9 Imidacloprid: Toxicology and Metabolism

J. Thyssen * and L. Machemer **
*Toxicology, Bayer Corporation, Agriculture Division, Stillwell, Kansas USA
**Institute of Toxicology, Bayer AG, BG Health, Wuppertal, Germany

1 Introduction

Imidacloprid (Figure 1) is the first active ingredient of the chemical class of nicotinoid insecticides to reach the market. Toxicological information on this class of substances is nearly nonexistent in open literature. Only limited toxicological information on the structurally related compound nitroguanidine is available, and it indicates no critical toxicological properties (GRA & I 1986-1990).

The mode of action of the nicotinoid insecticides was described by Schroeder and Flattum (1984) as an effect on the cholinergic synapses that resulted in a consequent postsynaptic blockage in the nervous system of the American cockroach. We now understand that Imidacloprid binds itself to the nicotinergic acetylcholine receptors . This binding is highly specific, involving only nicotinic and not muscarinic receptors, which trigger a defined toxicity picture typical for nicotinic reactions.

There are strong differences in the receptor binding potency between insects and mammals. Methfessel (1992) reported for Imidacloprid a 100 fold difference in binding potency between insect and mammalian acetylcholine receptors, insects being more sensitive than mammals.

We discuss the next three areas relevant to understanding the safety of Imidacloprid: metabolism from a toxicological point of view, hazard assessment (mammalian toxicity), and the safety of applicators and consumers.

2 Metabolism in Rats

Two major routes of metabolism exist in mammalian systems (see Figure 2). The first is the oxidative cleavage to imidazolidine, which does not appear to be metabolized further, and 6-Cl nicotinic acid. The imidazolidine moiety is excreted directly via the urine, and the nicotinic moiety is degraded via GSH conjugation to a mercapturic acid derivative and further to methyl

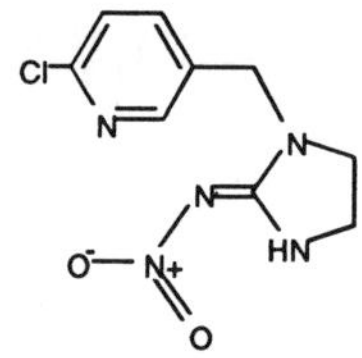

1-[(6-Chloro-3-pyridinyl)methyl] - *N*-nitro-2-imidazolidinimine

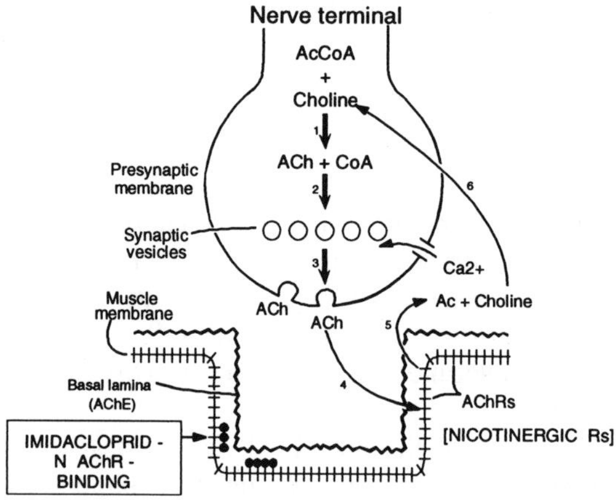

Fig. 1. Imidacloprid, its site of action.

mercaptonicotinic acid. It is also conjugated with glycine to form a hippuric acid compound. Since the two different conjugations represent only 7% each of the total biotransformation, a possible gluthion or glycine depletion is not of toxicological importance.

The second important biotransformation is the hydroxylation of the intact molecule in the imidazolidine ring followed by the elimination of water under the formation of an unsaturated metabolite. There were no qualitative differences between male and female rats after oral administration of a low dose of 1 mg/kg body weight. At a high dose of 20 mg/kg body weight, orally administered female rats showed a slightly higher renal elimination than males, who also metabolized Imidacloprid, only at a higher rate. There were no qualitative differences because all metabolites were found in both sexes and doses. More than 90% of a given dose was eliminated within 24 hours with a total excretion after 48 hours. Eighty percent was excreted via the urine, the rest via the feces, with no qualitative differences in the metabolite picture. Imidacloprid was readily absorbed and distributed to all organs within 1 hour. Whole-body autoradiography revealed no distribution to the fatty tissue, to the central nervous system, nor to the mineral parts of the bones. This would indicate that at a minimum dose of 20 mg Imidacloprid/kg body weight no accumulation potential exists and the blood-brain barrier is intact. These same studies also showed, after prolonged exposure, that higher concentrations existed in the kidneys due to the elimination tasks of this organ system.

In conclusion, Imidacloprid is quickly absorbed from the intestinal tract and quickly distributed into the mammalian system, followed by a rapid and total excretion.

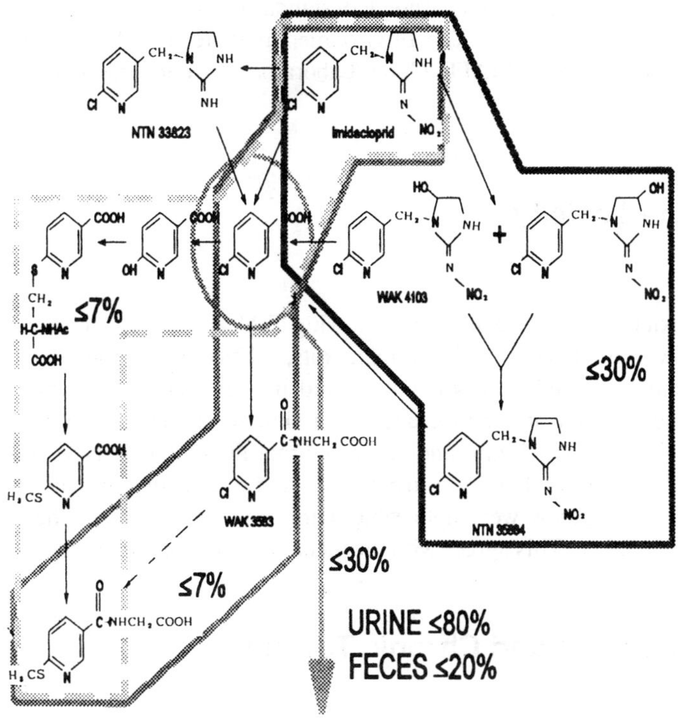

Fig. 2. Imidacloprid, its biotransformation pathway.

3 Mammalian Toxicology - General Study Conduct

Technical grade Imidacloprid, with a purity of 94-96%, was used in all toxicology studies. The material accountability was always defined. All studies were performed according to the Federal Insecticide, Fungicide, and Rodenticide Act (FIFRA), Toxic Substance Control Act (TSCA), Organization for Economic Co-operation and Development (OECD), European Economic Community (ECC), and the Japanese Ministry of Agriculture, Forestry and Fisheries (MAFF) testing guidelines. Specifically bred animals were used for the studies, which were performed according to Good Laboratory Practice (GLP) standards: rats, mice, guinea pigs, rabbits, and dogs. Standardized holding conditions were available and complied with animal health laws. Temperature, humidity, airflow, and light/dark cycles were defined. The animals were fed standardized diets and all health standards were followed. The Environmental Protection Agency (EPA), OECD and Japanese MAFF Good Laboratory Practice standards were followed.

4 Acute Toxicity

Acute toxicity data are seen in the following Table 1: Overall, Imidacloprid had a low to moderate toxicity in mice, rats, rabbits, and guinea pigs after oral, dermal, and inhalatory exposure and in skin and eye testing. The compound was not an irritant and not a skin sensitizer. In high doses, tremors, incoordinated gait, and decreased motility and activity were seen. There was a red nasal stain, a urine stain, and overall the animals looked apathetic. Symptoms were noticeable within minutes; however they were quickly reversible. This agrees with the metabolism picture described earlier.

5 Subchronic and Chronic Toxicity

As mentioned earlier, all studies were performed according to regulatory guidelines. The objective of these study types was to develop a toxicological profile after medium- and long-term dietary exposure. A dose- response relationship, the determination of a no-observed-effect level (NOEL), and the reversibility of the effects were investigated. In these studies rats and dogs

Table 1. Imidacloprid: Hazard assessment, acute

Animal species	Route of exposure	LD_{50} / LC_{50} (mg/kg BW / mg/m^3 air)
Mouse (M/F)	Oral	131-168
Rat (M/F)	Oral	424-475
Rat (M/F)	Dermal	> 5000
Rat (M/F)	Inhalation AE 4h	> 69 *
Rat (M/F)	Inhalation Dust 4h	> 5323 **
Rabbit [1]	Skin	Not an irritant
Rabbit	Eye	Not an irritant
Guinea Pig (M)	Skin	Not a skin sensitizer ***

* Aerodynamic droplet size < 5 μM; 100%; max conc.

** Aerodynamic particle size < 5 μM; 4-11%.

*** Magmusson - Kligman Test.

[1] Sex not identified on irritancy tests

Symptoms: tremors, incoordinated gait, decreased motility/activity, red nasal stain, urine stain, apathy onset of symptoms in minutes, fast reversibility

were used; rats were exposed for 3 months or 24 months while dogs were exposed for 3 months and 1 year. In all studies, a NOEL was evaluated and a maximum tolerated dose (MTD) was always achieved. In all studies the main target was the liver with a slight increase in weight and minor activity increases in cytochrome P-450 and its dependent enzymes (microsomal hepatic enzymes and mixed-function oxidases). Cell hypertrophy was seen and in male animals, sporadic cellular necrosis. These symptoms were high-dose phenomena and combined with up to 10% body weight gain reduction.

Secondary targets were the testes. Only rats showed a slight degeneration of the testicular tubuli in the highest dose, which was accompanied by a slightly longer blood clotting time and mineralization of the thyroid follicles. This is a spontaneous finding in aged rats, and thyroid functions were undisturbed. All effects observed were strictly dose dependent, the severity of effects decreasing with decreasing dose. The effects were also reversible. The following NOELs could be established: rat, 5.7 mg/kg body weight per day, and dog, 15 mg/kg body weight per day.

6 Oncogenicity

Rats and mice were exposed for 24 months or 18 months via their diet. All studies were performed under MTD guidelines. The chronic results seen in the rat studies supported the MTD, and in mice up to 29% body weight gain reduction, reduced food and water intake, and liver effects were seen. No evidence for oncogenic potential was apparent. Therefore, EPA classified Imidacloprid "E", which means evidence of noncarcinogenicity for humans.

7 Mutagenicity

The tests and results are shown in Table 2. *In vitro* point mutation tests were negative. Chromosomal aberration tests *in vitro* were negative at noncytotoxic concentrations and showed slightly positive effects at cytotoxic concentrations only. *In vivo* chromosomal aberration tests were all negative. Mitotic recombination in yeast, rec assay with *Bacillus subtilis*, and the Unscheduled DNA Synthesis (UDS) test were negative. Overall, the compound does not have any mutagenic potential.

8 Developmental Toxicity

Embryotoxicity testing was performed in rabbits and rats. Animals were orally treated during days 6 to 18 (rabbits) and 6 to 15 (rats) with doses up to 72 mg/kg body weight per day and 100 mg/kg body weight per day, respectively. Cesarean section and examination of embryo and fetal development, including fetal skeletal alterations, were evaluated at days 28 and 21 postcoitum in the rabbit and rat, respectively. At the high dose, rats showed symptoms of maternal toxicity, which was combined with delayed embryo development. Wavy ribs could be seen, which was, however, a reversible finding. The maternal NOEL was 10 mg/kg body weight per day.

The fetal NOEL was 30 mg/kg body weight per day. In rabbits the high dose produced strong maternal toxicity and deaths. Abortions or complete resorptions, delayed ossification, and reduced fetal weights were the results. The maternal NOEL was 8 mg/kg body weight per day, and the fetal NOEL was 24 mg/kg body weight per day, meaning that embryotoxicity was seen only at maternally toxic doses. No fetal malformations were observed in either the rabbit or rat at any dose level.

Table 2. Imidacloprid: Hazard assessment, mutagenicity

● Point mutation	
Salmonella microsome (AMES) test	Negative
Reverse mutation test *E. COLI*	Negative
HPRT Chinese hamster ovary (CHO)	Negative
● Chromosomal aberration	*In vitro*
Cytogenetics human lymphocytes	Slightly positive at cytotoxic concentrations only
Sister chromatid exchange Chinese hamster ovary cells	Slightly positive at cytotoxic concentrations only
● Both tests negative at non-cytotoxic concentrations	
● Chromosomal aberration	*In vivo*
Micronucleus mouse bone marrow	Negative
Sister chromatid exchange Chinese hamster bone marrow	Negative
Cytogenetics Chinese hamster bone marrow	Negative
Cytogenetics mouse spermatogonia	Negative
● Other genotoxicity tests	
Mitotic recombination yeast	Negative
Rec assay *B. subtilis*	Negative
Unscheduled DNA synthesis rat hepatocytes	Negative

Table 3. Imidacloprid: Hazard assessment

Reproductive toxicity

Two-generation / two-litter study in rats

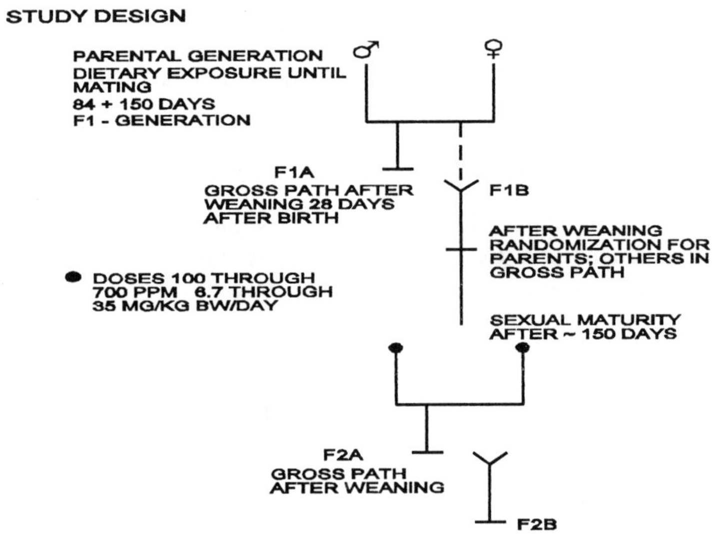

9 Reproductive Toxicity

The study design of the two-generation, two-litter study in rats is given in Table 3. Maternal toxicity at the high dose was defined as body weight gain reduction, lower food intake, and liver enzyme induction. The first two symptoms were pronounced during lactation. The NOEL for dams was 6.7 mg/kg body weight per day. At the high dose, reproductive toxicity in pups was pronounced as a body weight gain reduction only, with a pup reproductive NOEL of 12.5 mg/kg body weight per day. Specifically, no NOEL on mating behavior, fertility, gestation, conception, litter size, or mortality was observed. Malformation did not occur, and gross pathology and histopathology were free of observed defects.

10 Neurotoxicity

Neurotoxicity studies were performed in rats with the objective of evaluating the potential to produce neurotoxic effects in a screening system. Functional observational battery grip strength, foot splay, motor and locomotive activity in a figure eight maze, habituation, and a full pathology with special interest in neuronal tissue were investigated. Animals were exposed acutely, once orally or over 3 months in their diet. At highly toxic or lethal doses, symptoms of reversible cholinergic effects and nonspecific reduced motor activity were seen. Specifically, tremors, which were expected due to the mode of action, could be noted. There was no primary neurotoxicity, and the NOELs for neurotoxicity were 307 mg/kg body weight per day after acute exposures and 196 mg/kg body weight per day in the 3-month study. This result agrees with the highly protective blood-brain barrier in vertebrate animals (Yamamoto et al. 1995).

11 Conclusions

- Imidacloprid is a specific nicotinergic receptor binder with very low mammalian binding potential.
- Imidacloprid is rapidly absorbed, metabolized in the liver, and excreted mostly via urine.
- Metabolism of Imidacloprid is straightforward; no open ring structures with potential for other toxicological properties.
- Imidacloprid does not penetrate the blood-brain barrier.
- This lack of penetration leads to a low toxicity after acute oral, dermal, inhalatory, and long-term dietary exposure.
- Symptoms of Imidacloprid toxicity are nonspecific.
- Imidacloprid shows specific neurotoxicological symptoms only at lethal doses (tremors).
- Primary target is the liver (weight is increased, microsomal enzymes are increased).
- Secondary targets (seen only at MTD doses) are testes (slight tubuli degeneration) and thyroid (mineralization of the thyroid follicles).
- Imidacloprid is not oncogenic, not mutagenic, not a primary embryotoxin, not a reproductive toxin, and not a neurotoxin.
- Imidacloprid has no worker exposure-related toxic potential.
- Imidacloprid has no consumer-related toxic potential: RfD = 0.057 mg/kg body weight per day based on chronic rat study.
- Imidacloprid is a highly effective and safe representative of a new class of insecticides.

References

GRA & I (1986-1990) Government reports announcements & index (U.S.), Issues 5,11,12,15,16,17,19,23

MAFF, Japan (1984) Ministry of Agriculture, Forestry and Fisheries. On good laboratory practice standards for toxicological studies on agriculture chemicals. 59 NohSan No. 3850, August 10

MAFF, Japan (1985) Ministry of Agriculture, Forestry and Fisheries. Guidance on toxicology study data for application of agriculture chemical registration. 59 NohSan No. 4200, January 28

Methfessel C (1992) Action of Imidacloprid on the nicotinergic acetylcholine receptors in rat muscle. Pflanz Nachr Bayer 45:369-380

OECD (1981) Organization for Economic Co-operation and Development. Guidelines for testing of chemicals. Paris

OECD (1981) Principles of good laboratory practice. C(81)30 (Final) Annex 2, Paris

Schroeder ME, Flattum RF (1984) The mode of action of neurotoxic properties of the nitromethylene heterocyclic insecticides. Pestic Biochem Physiol 22:148-160

US-EPA-FIFRA (1991) Pesticide assessment guidelines, subdivision F, hazard evaluation: human and domestic animals

US-EPA-FIFRA (1991) Good laboratory practice standards, 40 CFR part 160

Yamamoto I, Yabuta G, Tomizawa M, Saito T, Miyamota T, Kagabu S (1984) Molecular mechanisms for selective toxicity of nicotinoids and neonicotinoids. J Pestic Sci 20:33-40

10 The Action of Nicotine in the Mammalian Brain

Satoshi Fujii, Elisabeth C. Walcott, and Katumi Sumikawa

Department of Psychobiology, University of California, Irvine, CA 92697-4550, USA

1. The Cholinergic System in the Brain

Acetylcholine (ACh) is one of the predominant neurotransmitters in the brain. The majority of cholinergic cells are found in the medial septal nucleus, the basal forebrain, the striatum, and the brainstem (Struble et al. 1986). The major projection areas include cortex, hippocampus, striatum, substantia nigra, and medial habenula (Butcher 1995; Woolf et al. 1984). ACh is involved in the regulation of cortical arousal (Semba 1991), attention (Murphy and Sillito 1991), and sleep-wake cycles (Hobson 1990). Thus, the actions of ACh are manifold. The actions of ACh are mediated by two different types of receptors: the ionotropic nicotinic type and the metabotropic muscarinic type. Each of these classes of ACh receptor (AChR) has multiple subtypes with unique structural and functional characteristics, and thus ACh released from a nerve terminal may contribute to a wide variety of brain functions by activating different intracellular pathways depending on the distribution of the receptor types.

Nicotinic sites in the brain were detected long before the receptors were cloned (Clarke et al. 1985); however, the widespread distribution of the different subtypes of nicotinic acetylcholine receptor (nAChR) was not determined until the cloning of a total of 11 different subunits (reviewed in Chavez et al. 1997; Clarke 1993; Delbono et al. 1997; Role and Berg 1996; Sargent 1993). Moreover, the functional significance of these receptors was a mystery for many years largely because of the inability to record nicotinic currents in neurons in the brain. Recently, it has been shown that nicotine facilitates the release of a number of different neurotransmitters by activating presynaptic or preterminal receptors (Gray et al. 1996; McGehee et al. 1995; Summers and Giacobini 1995; Wilkie et al. 1996). This suggests that nAChRs are acting primarily as modulators rather than as direct mediators of fast excitatory transmission (Clarke and Reuben 1996; Fu and Liu 1997; Wilkie et al. 1996). This modulatory role may be one of the most important functions of these receptors in the brain. These findings also help to explain the difficulties in detecting nicotinic currents in the brain, despite the evidence for widespread distribution of receptors. Perhaps the most compelling reason to believe that the nAChR is an important component of central nervous system (CNS) synapses is the evidence for pharmacological manipulation of behavior with compounds that specifically activate or block nAChR (Dani and Heinemann 1996; McGehee and Role 1996). The behavioral and physiological endpoints known to be manipulated by nicotine include enhanced learning and memory, arousal, concentration, and attention, as well as decreased anxiety and pain perception (Benowitz et al. 1989; Clarke 1993; Rosecrans and Karan 1993).

2. The Action of Nicotine in the Hippocampus

Tobacco smoking in humans as well as acute and chronic administration of nicotine in animals can enhance cognitive function (Levin 1992, 1993; Abdulla et al. 1993; Poincheval-Fuhrman and Sara 1993). Although we now know that the nAChR is important for the addictive properties of nicotine (Picciotto et al. 1998), the mechanism underlying these effects of nicotine, however, is unknown. One possibility is that nicotine-induced cognitive enhancement is mediated by nAChRs in the hippocampus (Gray et al. 1996), a brain area known to be important for learning and memory.

There are at least eight different ligand binding subunits (α2-α9) and three structural subunits (β2-β4) with a common structural design in the brain (Heinemann et al. 1991; Lindstrom et al. 1991; McGehee and Role 1995; Role and Berg 1996). Functionally diverse nAChRs are generated by the assembly of different combinations of α and β subunits, each of which may have a different role in mediating the action of nicotine. One nAChR subtype consisting of α4 and β2 forms high-affinity nicotine-binding sites and appears to account for most of the nicotine-binding sites in rat brain (Whiting et al. 1991). α7-bearing nAChRs are capable of binding to α-bungarotoxin (α-BuTX) (Couturier et al. 1990; Seguela et al. 1993). Hippocampal neurons appear to express at least three different nAChR subtypes, including the two major subtypes, α4β2- and α7-bearing nAChRs (Albuquerque et al. 1995). Although recent studies suggest that these receptors are pre- and postsynaptically located (Role and Berg 1996; Gray et al. 1996; Alkondon et al. 1996a,b), their functions in the hippocampus largely remain unknown. Previous studies suggest that nicotine has differential effects on different neuronal cell types in the hippocampal CA1 region (Freund et al. 1988; Reece and Schwartzkroin 1991). Thus, the net effect of nicotine on pyramidal cells (major hippocampal output neurons) appears to be the sum of complex local circuit interactions.

The strength of synaptic connections can be increased by application of a brief volley of high-frequency stimulation to afferent (incoming) nerve fibers (Bliss and Lomo 1973; Bliss and Gardner-Medwin 1973). The stimulation-induced change in synaptic strength can persist for relatively long periods of time and is thus termed long-term potentiation (LTP). Because LTP is long lasting, it is considered a leading candidate for a cellular mechanism underlying learning and memory. Drugs that block the formation of LTP also block some forms of memory, providing additional support for this conclusion (Morris et al. 1986). Evidence also indicates that drugs that facilitate the induction of LTP enhance some forms of memory (Arai et al. 1996). Therefore, we examined if nicotine facilitates the induction of LTP.

Transverse slices (500 μm) were cut from the hippocampi of 19 to 38 day old rats, submerged, and continuously perfused at 2-3 ml/min with oxygenated artificial cerebrospinal fluid (ACSF) at 30°C. A bipolar stimulating electrode was placed in the stratum radiatum of the CA1 region to stimulate the Schaffer collateral/commissural pathway (Fig. 1A). The evoked field excitatory postsynaptic potentials (f-EPSP) and population spike (PS) were recorded simultaneously in the dendritic region with electrode R1 and in the cell body layer with electrode R2, respectively (Fig. 1A). At the beginning of each experiment, a stimulus/response curve was established by measuring the amplitude of the PS. The strength of the stimulus was then adjusted to elicit an amplitude of the PS that was 40-60% of the maximum. The intensity and duration (0.2 ms) of each stimulus pulse remained invariant thereafter for each experiment. Responses were amplified using a preamplifier and stored sequentially using a microcomputer. Baseline responses were recorded following delivery of test stimuli every 20 s via the stimulating electrode, checking the stability for more than 15 min. A tetanus consisting of either 15 or 100 pulses at 100 Hz was then given. After the tetanus, the delivery of the test stimulus was resumed once every 20 s for more than 60 min. To evaluate the effects of different numbers of tetanus pulses on the induction of LTP, the slope of the f-EPSP, and the amplitude of the PS were measured before and after the tetanus (Fig. 1B). As shown in Fig. 1C, maximum LTP can be induced by a tetanus consisting of 100 pulses at 100 Hz. In contrast, a weak tetanus consisting of 15 pulses failed to produce stable LTP, although a short-term potentiation that declined gradually within 20-30 min was produced (Fig. 1D). We then tested whether the short tetanus stimulation (subthreshold for LTP) induced LTP when in the presence of nicotine. The weak tetanic stimulus, that is, the subthreshold for LTP in hippocampal CA1 cells in the absence of nicotine, induced LTP in the presence of 1 μM nicotine (Fig. 2). An application of 1 μM nicotine alone did not induce the LTP (data not shown). Furthermore, when the weak tetanic stimulus was given in the presence of 1 μM nicotine together with 3 μM mecamylamine, a

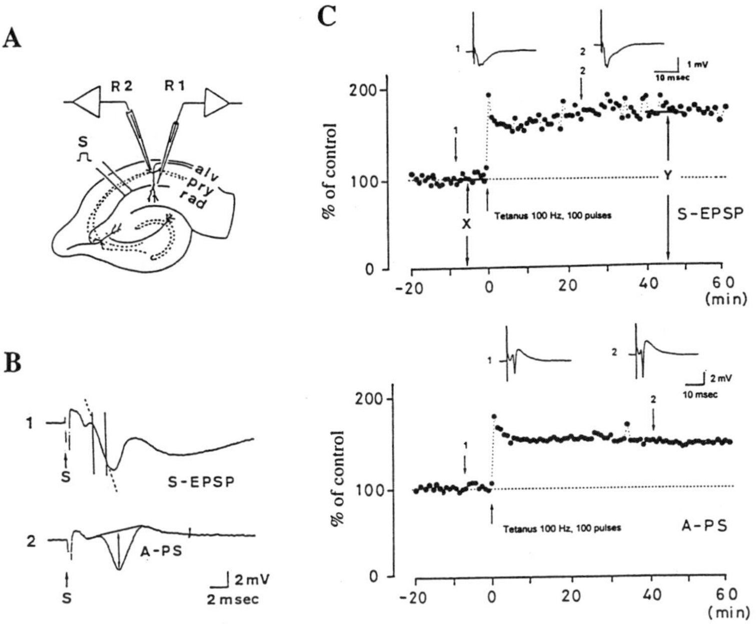

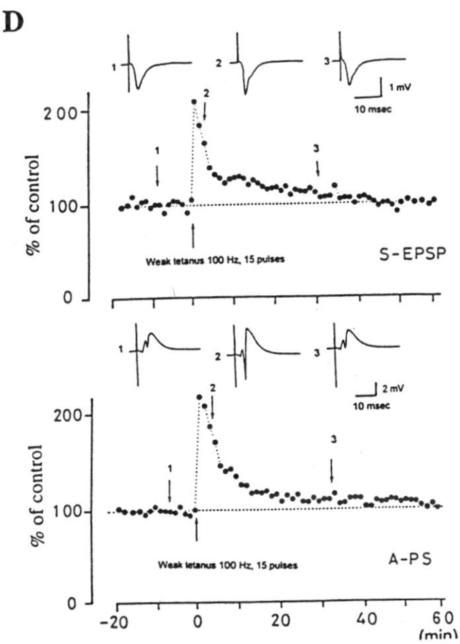

Fig. 1A-D. Effects of tetanic stimuli with different numbers of pulses on the induction of long-term potentiation (LTP) in the hippocampal CA1 region. **A** A schematic depiction of the hippocampal slice and the placement of a stimulating (*S*) and two recording electrodes (*R1*, *R2*). *alv*, alveus; *pyr*, stratum pyramidale; *rad*, stratum radiatum. **B** Evoked responses recorded from the dendritic layer with electrode R1 *(1)* and from the pyramidal cell body layer with R2 (*2*). The parameters measured were the *S-EPSP*, slope of the initial phase of the field EPSP; *A-PS*, amplitude of the population spike; and *S*, stimulus artifact. **C, D** Long-term changes in S-EPSP and in A-PS induced by different tetanic stimulations

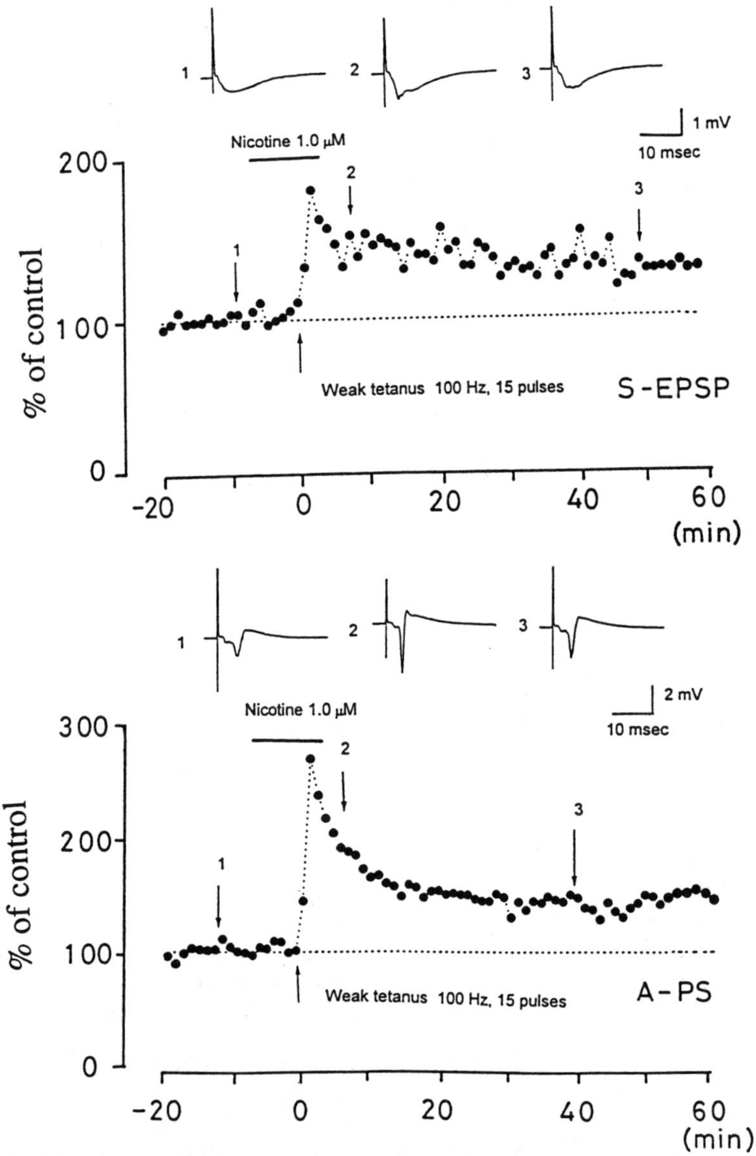

Fig. 2. Nicotine facilitates induction of LTP in the hippocampal CA1 region. Nicotine (1 μM) was added to the perfusion medium; after 5 to 7 min exposure, a tetanus of 15 pulses at 100 Hz was applied. Nicotine facilitated LTP induction. The administration of drugs is indicated by the ***horizontal bars***

nicotine antagonist, induction of LTP was completely blocked (data not shown). This result strengthens the conclusion that nicotine facilitates the induction of LTP by interacting with nAChRs specifically. The finding may provide an explanation of the cellular basis of nicotine-induced cognitive enhancement.

3. The Action of Nicotine in the Septohippocampal Pathway

It has been known for some time that the central cholinergic system is compromised in the brains of aged patients and patients with Alzheimer's disease (AD) (Bartus et al. 1982; Nordberg et al. 1992; Sparks et al. 1992). Damage is observed in the basal forebrain cholinergic neurons, which undergo degeneration and participate in the formation of neurofibrillary tangles, and in the distal axons and terminals, which become part of the developing senile plaque (reviewed in Price et al. 1993). Eventually, as a result of the neuronal degeneration, the chemical markers for the cholinergic system such as choline acetyltransferase and acetylcholinesterase (AChE) activity decline in the target areas. Because AD is a degenerative disorder in which cognitive function becomes severely impaired, and the cholinergic system has been implicated in arousal, attention, and memory (Struble et al. 1986; Bartus et al. 1982), it appears that the dysfunction in the cholinergic system is at least partially responsible for the memory losses seen in patients with AD. Interestingly, a negative correlation between chronic tobacco smoking and the incidence of AD has been found (Brenner et al. 1993; Lee 1994; Court and Perry 1994). The mechanism underlying this effect of nicotine, however, is unknown.

It is well established that the cholinergic component of the septohippocampal pathway originates in large neurons in the medial septal nucleus and the nucleus of the vertical limb of the diagonal band of Broca (Amaral and Kurz

1985; Benardo and Prince 1982; Cole and Nicoll 1984; Freund and Antal 1988; Leranth and Frotscher 1989; Madison et al. 1987). The axonal pathway traverses through the fimbria-fornix and the terminals are located in all layers of the hippocampus (Lewis and Shute 1967; Woolf et al. 1984). Lesions of the medial septum/diagonal band complex or the fimbria-fornix pathway produce significant deficits in spatial and working memory (Beatty and Carbone 1980; Decker et al. 1995; Hepler et al. 1985; Leanza et al. 1996). Many of these deficits can be temporarily reversed by implanting cholinergic-rich fetal grafts (Dunnett et al. 1982; Ridley et al. 1991).

Two pieces of evidence propose a role for α7-containing AChRs in the survival or the growth of fibers. First, nicotine and nicotinic agonists are reported to be neuroprotective in cortical cultures, an effect that is blocked by methyllycaconitine and α-BuTX (Donnelly et al. 1996). Second, blocking α7-containing AChRs, with intraventricular injections of α-BuTX in rats, upregulates nerve growth factor and brain-derived neurotrophic factor mRNA in the hippocampus (Freedman et al. 1993). Thus, if chronic nicotine administration leads to functional blockade of the α7-containing AChRs via desensitization, the same mechanism might be activated whereby induction of growth factors would serve for maintenance of cholinergic synapses. This could be a potential mechanism that prevents smokers from functional decline in AD.

Therefore, we examined the effects of chronic nicotine on the cholinergic innervation pattern in septohippocampal co-cultures that have been used as a model system to study the formation of functional synapses in vitro (Gähwiler 1988; Gähwiler and Brown 1985; Muller et al. 1993). Pieces of septum and hippocampus were cultured for at least 3 days and then were treated with 20 μM nicotine for 8 days. AChE histochemistry was performed to localize cholinergic fibers on the co-cultures after fixation. The sparsely labeled co-cultures were omitted from the analysis, because there were no clear AChE-positive neurons present in the septum, and therefore there was no observable innervation present. This effect, which results

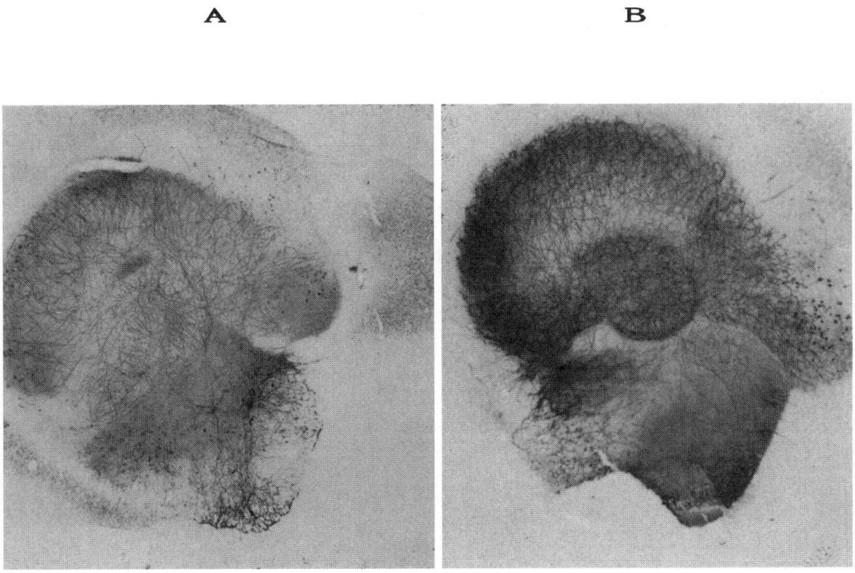

Fig. 3A, B. AChE staining of co-cultures after 11 to 14 days in culture shows two categories of AChE labeling: moderate (A), dense (B)

Table 1. Effects of nicotine treatment on AChE staining[a]

	Moderate staining	Dense staining
Control	35%	41%
Nicotine[b]	22%	55%

[a] Results were based on analysis of 36 co-cultures, 18 control and 18 nicotine-treated co-cultures.
[b] Co-cultures were treated with 20 μM nicotine for 8 days.

from natural variability within the co-culture system, is observed in roughly 20% of the cultures. For data analysis, the remaining co-cultures were divided into two groups: moderately labeled and densely labeled. These distinctions were not difficult to make. Figure 3 shows examples of the two labeling groups. Chronic nicotine treatment caused a small effect on the prevalence of densely stained co-cultures treated with nicotine (Table 1). Results are based on analysis of 36 co-cultures, 18 treated with nicotine and 18 control co-cultures. Although further studies are required, the observed effect of chronic nicotine indicates that nicotine exerts a neurotrophic action.

4. Conclusion

Because nicotine improves the cognitive function of AD patients (Jones et al. 1992; Newhouse et al. 1993; Sahakian and Coull 1994), the mechanism underlying the acute effect of nicotine observed here may be preserved in AD patients. The mechanism that underlies the increased numbers of cholinergic fibers in septohippocampal co-cultures through chronic nicotine exposure may contribute to maintenance of cholinergic synapses in the smoker's brain. In addition, a recent report demonstrated that nicotine modulates the neurotoxic effect of β-amyloid protein in hippocampal cultures (Zamani et al. 1997). Each of these effects may contribute to the negative correlation between tobacco smoking and AD. Understanding the mechanisms underlying nicotine-mediated effects in the hippocampus may aid in the development of more effective therapeutic agents.

Acknowledgment

Research considered in this review was supported by UC-Tobacco-Related Disease Research Program and Smokeless Tobacco Research Council, Inc..

References

Abdulla FA, Calaminici M-R, Stephenson JD, et al (1993) Chronic treatment with cholinoceptor drugs influences spatial learning in rats. Psychopharmacology 111:508-511

Albuquerque EX, Pereira EFR, Castro NG, et al (1995) Neuronal nicotinic receptors: function, modulation and structure. Sem Neurosci 7:91-101

Alkondon M, Pereira EFR, Albuquerque EX (1996a) Mapping the location of functional nicotinic and γ-aminobutyric acid A receptors on hippocampal neurons. J Pharmacol Exp Ther 279:1491-1506

Alkondon M, Pereira EFR, Albuquerque EX (1996b) Characterization of nicotinic acetylcholine receptors in CA1 neurons of rat hippocampal slices. Soc Neurosci Abstr 22:1267

Amaral DG, Kurz J (1985) An analysis of the origins of the cholinergic and noncholinergic septal projections to the hippocampal formation of the rat. J Comp Neurol 240:37-59

Arai A, Guidotti A, Costa E, et al (1996) Effect of the AMPA receptor modulator IDRA 21 on LTP in hippocampal slices. Neuroreport 7:2211-2215

Bartus RT, Dean RD, Beer B, et al (1982) The cholinergic hypothesis of geriatric memory dysfunction. Science 217:408-414

Beatty WW, Carbone CP (1980) Septal lesions, intramaze cues and spatial behavior in rats. Physiol Behav 24:675-678

Benardo LS, Prince DA (1982) Cholinergic excitation of mammalian hippocampal pyramidal cells. Brain Res 249:315-331

Benowitz NL, Porchet H, Jacob P (1989) Nicotine dependence and tolerance in man: pharmacokinetic and pharmacodynamic investigations. Prog Brain Res 79:279-287

Bliss TVP, Lomo T (1973) Long-lasting potentiation of synaptic transmission in dentate area of the anesthetized rabbit following stimulation of perforant path. J Physiol 232:331-356

Bliss TVP, Gardner-Medwin AR (1973) Long-lasting potentiation of synaptic transmission in the dentate area of the unanesthetized rabbit following stimulation of perforant path. J Physiol 232:357-374

Brenner DE, Kukull WA, Vanbelle G, et al (1993) Relationship between cigarette smoking and Alzheimer's disease in a population-based case-control study. Neurology 43:293-300

Butcher LL (1995) Cholinergic neurons and networks. The rat nervous system. Academic Press, New York, pp 1003-1015

Chavez NL, Crona JH, Washburn MS, et al (1997) Pharmacological characterization of recombinant human neuronal nicotinic acetylcholine receptors α2β2, α2β4, α3β2, α3β4, α4β2, α4β4 and α7 expressed in *Xenopus* oocytes. J Pharmacol Exp Ther 280:346-356

Clarke PB (1993) Nicotinic receptors in mammalian brain: localization and relation to cholinergic innervation. Prog Brain Res 98:77-83

Clarke PB, Reuben M (1996) Release of [^{3}H]-noradrenaline from rat hippocampal synaptosomes by nicotine: mediation by different nicotinic receptor subtypes from striatal [^{3}H]-dopamine release. Br J Pharmacol 117:595-606

Clarke PB, Schwartz RD, Paul SM, et al (1985) Nicotinic binding in rat brain: autoradiographic comparison of [^{3}H]acetylcholine, [^{3}H]nicotine, and [^{125}I]-

α-bungarotoxin. J Neurosci 5:1307-1315
Cole AE, Nicoll RA (1984) The pharmacology of cholinergic excitatory responses in hippocampal pyramidal cells. Brain Res 305:283-290
Court JA, Perry EK (1994) CNS nicotinic receptors. Possible therapeutic targets in neurodegenerative disorders. CNS Drugs 2:216-233
Couturier S, Bertrand D, Matter J-M, et al (1990) A neuronal nicotinic acetylcholine receptor subunit (α7) is developmentally regulated and forms a homo-oligomeric channel blocked by α-BTX. Neuron 5:847-856
Dani JA, Heinemann S (1996) Molecular and cellular aspects of nicotine abuse. Neuron 16:905-908
Decker MW, Curzon P, Brioni JD (1995) Influence of separate and combined septal and amygdala lesions on memory, acoustic startle, anxiety, and locomotor activity in rats. Neurobiol Learn Mem 64:156-168
Delbono O, Gopalakrishnan M, Renganathan M, et al (1997) Activation of the recombinant humanα7 nicotinic acetylcholine receptor significantly raises intracellular free calcium. J Pharmacol Exp Ther 280:428-438
Donnelly RD, Xue IC, Arneric SP, et al (1996) In vitro neuroprotective properties of the novel cholinergic channel activator (ChCA), ABT-418. Brain Res 719:36-44
Dunnett SB, Low WC, Iverson SD, et al (1982) Septal transplants restore maze learning in rats with fornix-fimbria lesions. Brain Res 251:348-355
Freedman R, Wetmore C, Stromberg I, et al (1993) α-Bungarotoxin binding to hippocampal interneurons: immunocytochemical characterization and effects on growth factor expression. J Neurosci 13:1965-1975
Freund TF, Antal M (1988) GABA-containing neurons in the septum control inhibitory interneurons in the hippocampus. Nature (Lond) 336:170-173
Freund RK, Jungschaffer DA, Collins AC, et al (1988) Evidence for modulation of GABAergic neurotransmission by nicotine. Brain Res 453:215-220
Fu WM, Liu JJ (1997) Regulation of acetylcholine release by presynaptic nicotinic receptors at developing neuromuscular synapses. Mol Pharmacol 51:390-398
Gähwiler BH (1988) Organotypic cultures of neural tissue. Trends Neurosci 11:484-489
Gähwiler BH, Brown DA (1985) Functional innervation of cultured hippocampal neurons by cholinergic afferents from co-cultured septal explants. Nature (Lond) 313:577-579
Gray R, Rajan AS, Radcliffe KA, et al (1996) Hippocampal synaptic transmission enhanced by low concentrations of nicotine. Nature (Lond) 383:713-716
Heinemann S, Boulter J, Connolly J, et al (1991) The nicotinic receptor genes. Clin Neuropharmacol 14:S45-S61
Hepler DJ, Wenk GL, Cribbs BL, et al (1985) Memory impairments following basal forebrain lesions. Brain Res 346:8-14
Hobson JA (1990) Sleeping and dreaming. J Neurosci 10:371-382
Jones GMM, Sahakian BJ, Levy R, et al (1992) Effects of acute subcutaneous nicotine on attention, information processing and short-term memory in Alzheimer's disease. Psychopharmacology 108:485-494
Leanza G, Muir J, Nilsson OG, et al (1996) Selective immunolesioning of the basal forebrain cholinergic system disrupts short-term memory in rats. Eur J Neurosci 8:1535-1544
Lee PN (1994) Smoking and Alzheimer's disease: a review of the epidemiological evidence. Neuroepidemiology 13:131-144
Leranth C, Frotscher M (1989) Organization of the septal region in the rat brain: cholinergic-GABAergic interconnections and the termination of hippocampo-

septal fibers. J Comp Neurol 289:304-314
Levin ED (1992) Nicotine systems and cognitive function. Psychopharmacology 108:417-431
Levin ED (1993) Nicotinic involvement in cognitive function: possible therapeutic applications. Med Chem Res 2:612-627
Lewis PR, Shute CCD (1967) The cholinergic limbic system: projections to hippocampal formation, medial cortex, nuclei of the ascending cholinergic reticular system and subfornical organ and supraoptic crest. Brain 90:521-540
Lindstrom J, Schoepfer R, Conroy W, et al (1991) The nicotinic acetylcholine receptor gene family. Adv Exp Med Biol 287:255-278
Madison DV, Lancaster B, Nicoll RA (1987) Voltage clamp analysis of cholinergic action in the hippocampus. J Neurosci 7:733-741
McGehee DS, Heath MJ, Gelber S, et al (1995) Nicotine enhancement of fast excitatory synaptic transmission in CNS by presynaptic receptors. Science 269:1692-1696
McGehee DS, Role LW (1995) Physiological diversity of nicotinic acetylcholine receptors expressed by vertebrate neurons. Annu Rev Physiol 57:521-546
McGehee DS, Role LW (1996) Presynaptic ionotropic receptors. Curr Opin Neurobiol 6:342-349
Morris RGM, Anderson E, Lynch GS, et al (1986) Selective impairment of learning and blockade of long-term potentiation by an N-methyl-D-aspartate receptor antagonist, AP5. Nature (Lond) 319:774-776
Muller D, Buchs PA, Stoppini L (1993) Time course of synaptic development in hippocampal organotypic cultures. Dev Brain Res 71:93-100
Murphy PC, Sillito AM (1991) Cholinergic enhancement of direction selectivity in the visual cortex of the cat. Neuroscience 40:13-20
Newhouse PA, Potter A, Lenox RH (1993) The effects of nicotinic agents on human cognition: possible therapeutic applications in Alzheimer's and Parkinson's diseases. Med Chem Res 2:628-642
Nordberg A, Romanelli L, Sundwall A, et al (1989) Effect of acute and subchronic nicotine treatment on cortical acetylcholine release and on nicotinic receptors in rats and guinea-pigs. Br J Pharmacol 98:71-78
Picciotto MR, Zoli M, Rimondini R, et al (1998) Acetylcholine receptors containing the beta 2 subunit are involved in the reinforcing properties of nicotine. Nature (Lond) 391:173-177
Poincheval-Fuhrman S, Sara SJ (1993) Chronic nicotine ingestion improves radial arm mazeperformance in rats. Behav Pharmacol 4:535-539
Price DL, Koliatsos VE, Clatterbuck RC (1993) Cholinergic systems: human diseases, animal models, and prospects for therapy. Prog Brain Res 98:51-60
Reece LJ, Schwartzkroin PA (1991) Nicotine exerts the differential effects on different CA1 hippocampal cell types. Brain Res 540:287-290
Ridley RM, Thornley HD, Baker HF, et al (1991) Cholinergic neural transplants into hippocampus restore learning ability in monkeys with fornix lesions. Exp Brain Res 83:533-538
Role LW, Berg DK (1996) Nicotinic receptors in the development and modulation of CNS synapses. Neuron 16:1077-1085
Rosecrans JA, Karan LD (1993) Neurobehavioral mechanisms of nicotine action: role in the initiation and maintenance of tobacco dependence. J Subst Abuse Treat 10:161-170
Sahakian BJ, Coull JT (1994) Nicotine and tetrahydroaminoacridine: evidence for improved attention in patients with dementia of the Alzheimer type. Drug Dev Res 31:80-88
Sargent PB (1993) The diversity of neuronal nicotinic acetylcholine receptors. Annu Rev

Neurosci 16:403-443
Seguela P, Wadiche J, Dineley MK, et al (1993) Molecular cloning, functional properties, and distribution of rat brain α7: a nicotinic cation channel highly permeable to calcium. J. Neurosci 13:596-604
Semba K (1991) The cholinergic basal forebrain: a critical role in cortical arousal. In: Napier TC (ed) The basal forebrain. Plenum Press, New York, pp 197-218
Sparks DL, Hunsaker JD, Slevin JT, et al (1992) Monoaminergic and cholinergic synaptic markers in the nucleus basalis of Meynert (nbM): normal age-related changes and the effect of heart disease and Alzheimer's disease. Ann Neurol 31:611-620
Struble RG, Lehmann J, Mitchell SJ, et al (1986) Basal forebrain neurons provide major cholinergic innervation of primate neocortex. Neurosci Lett 66:215-220
Summers KL, Giacobini E (1995) Effects of local and repeated systemic administration of (-)-nicotine on extracellular levels of acetylcholine, norepinephrine, dopamine, and serotonin in rat cortex. Neurochem Res 20:753-759
Whiting PJ, Schoepfer R, Conroy, WG, et al (1991) Expression of nicotinic acetylcholine receptor subtypes in brain and retina. Mol Brain Res 10:61-70
Wilkie GI, Hutson P, Sullivan JP, et al (1996) Pharmacological characterization of a nicotinic autoreceptor in rat hippocampal synaptosomes. Neurochem Res 21:1141-1148
Woolf NJ, Eckenstein F, Butcher LL (1984) Cholinergic systems in the rat brain. I. Projections to the limbic telencephalon. Brain Res Bull 13:751-784
Zamani MR, Allen YS, Owen GP, et al (1997) Nicotine modulates the neurotoxic effect of β-amyloid protein $_{(25-35)}$ in hippocampal cultures. Neuroreport 8:513-517

11 Nicotine Analogs: Structure-Affinity Relationships For Central Nicotinic Acetylcholinergic Receptor Binding

Richard A. Glennon and Małgorzata Dukat
Department of Medicinal Chemistry, Virginia Commonwealth University, Richmond, VA 23298-0540, USA

1 Introduction

Cholinergic receptors are one of the oldest known populations of neurotransmitter receptors, and acetylcholine is their natural ligand. Shortly after the discovery of acetylcholinergic receptors it was realized that the actions of cholinergic agents could best be described by invoking two distinct populations of receptors: muscarinic acetylcholinergic receptors and nicotinic acetylcholinergic receptors (nAChRs). Investigations during the 1950s to the 1980s identified a number of interesting nAChR agonists and antagonists using, for the most part, isolated peripheral tissue or organ preparations. Indeed, nicotinic pharmacophores were identified and are still the subject of continuing investigations. Relatively less has been done with nAChRs at the level of the central nervous system. However, the last decade has witnessed a significant resurgence of interest in nAChRs, particularly in central nAChRs, for several reasons. Nicotinic receptors were identified in brain homogenates, and techniques and radioligands were developed for their investigation. Furthermore, there exists today a better understanding of the structure of nAChRs. These receptors are pentameric units that are directly associated with an ion channel. The receptors are composed of various α (α_2-α_9), β (β_1-β_4), γ, and δ subunit combinations (see Shacka and Robinson 1996; Holladay et al. 1997) and their exact composition likely dictates their pharmacology and binding characteristics. There is also a difference between peripheral and central nAChRs, and the major population of brain nAChRs appears to be of the $\alpha_4\beta_2$ type. Further fueling the interest in central nAChRs is evidence that such receptors may be involved in appetite, memory, analgesia, and various other physiological processes as well as in anxiety, memory, and certain mental and neurological disorders

(Arneric and Brioni 1998). Although nicotine (**1a**), a naturally occurring nAChR ligand, is associated with a variety of toxic side effects, there is no reason to believe that these side effects are inextricably linked to the beneficial effects of nicotinic ligands. Thus, we and others have begun investigations to identify the structure-activity relationships (SAR) for nicotinic agonist and antagonist activity, and structure-affinity relationships (SAFIR) for central nAChR binding, to ultimately develop novel nicotinic agents with greater selectivity and reduced toxicity that may be useful for disorders involving nAChRs.

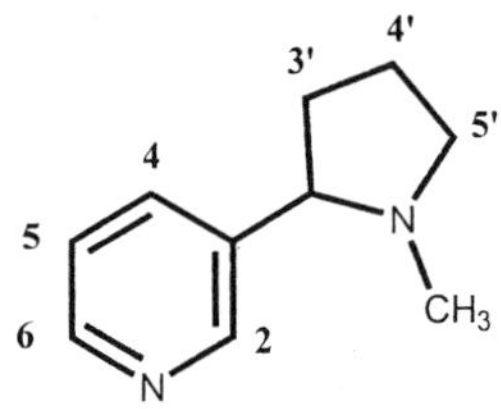

1a

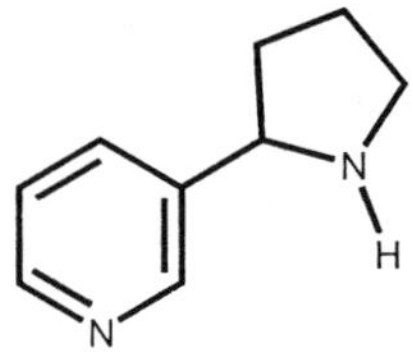

2

At the same time, others have found that certain nicotinic agents represent a novel class of insecticides and are investigating the insecticidal actions of these agents (see other chapters in this book). Obviously, useful nicotinic insecticides would be those with low affinity for mammalian nAChRs and/or with low toxicity in mammals. Also, there do appear to be differences between human and insect nicotinic receptors (e.g., see Tomizawa et al. 1996). The two fields of research are of a convergent and complementary nature, then, in that an understanding of what is good or bad for the binding of nicotinic ligands at mammalian nAChRs is applicable to the design of nAChRs intended for use as insecticides. From this perspective, knowledge of mammalian nAChR SAR and SAFIR would be useful. Specifically, what is it about nicotine that makes it a good ligand for nAChRs?

2 Structure-Activity Studies on Nicotine Analogs

At the time our investigations began, essentially nothing was known about the binding requirements of nicotine at central nAChRs. We initiated our investigation shortly after the commercial introduction of [^{3}H]nicotine. We have used rat brain homogenate nAChRs for the bulk of our studies and, although these homogenates are known to possesses more than one type of nAChRs, the predominant receptor is of the $\alpha_4\beta_2$ type. A second type of receptor has also been detected, α_7 receptors, but nicotine displays low affinity for these receptors.

The first questions to be addressed included: (a) what is the role of nicotine's stereochemistry on binding? (b) what is (are) the preferred conformation(s) for the binding/activity of nicotine? (c) what is the role of the pyrrolidine *N*-methyl group? (d) is the intact pyrrolidine ring necessary for binding? and (e) is the intact pyridine ring necessary for binding? Simplistic as these questions are, answers were unavailable in the literature at the time our studies began. Any attempt to design novel nAChR agents would require an understanding of these concepts. We were also interested in reexamining proposed nicotinic pharmacophores, which were developed using peripheral data, in order to determine if they were applicable to central nAChRs.

2.1 Role of Stereochemistry

Nicotine possesses an asymmetric center and exists as a pair of optical isomers. (-)-Nicotine (K_i = 1-5 nM) is the naturally occurring optical isomer. We and others have determined that (-)nicotine binds with about 5 to 30 times the affinity of (+)- or unnatural nicotine. So, although binding is not stereospecific, it is stereoselective. *N*-Demethylation of (-)nicotine affords (-)nornicotine. (-)Nornicotine binds with about 10 to 30 times lower affinity than (-)nicotine. Interestingly, nornicotine (**2**) displays relatively little stereoselectivity in that both optical isomers bind with nearly equal affinity. It would appear then that the *N*-methyl group somehow contributes to the stereoselective binding displayed by nicotine. For further discussion see Glennon et al. (1994).

2.2 Minimal Structural Requirements for the Binding of Nicotine

2.2.1 The Pyrrolidine Ring

Is the intact pyrrolidine ring of nicotine necessary for binding? The pyrrolidine ring of nicotine was excised and then rebuilt in a stepwise process. Table 1 shows that the simple aminomethylpyridine (i.e., AMP) **3a** lacks affinity for nAChRs (i.e., K_i > 10,000 nM). Simple secondary amines in which R is a small alkyl group bind with low affinity ($K_i \geq$ 1,000 nM), whereas tertiary amines bind with somewhat higher affinity (K_i < 1,000 nM). Optimal affinity seems to be associated with tertiary amines where one of the alkyl groups is a methyl group; the *N*-ethyl-*N*-methyl derivative **3e** (K_i = 28 nM) was identified as the highest affinity member of the series (Dukat et al. 1996). Thus, although the intact pyrrolidine ring does not appear to be an essential structural feature for binding, ring opening results in reduced

Table 1. Binding affinities of selected aminomethylpyridine (AMP) derivatives at central nAChRs[a].

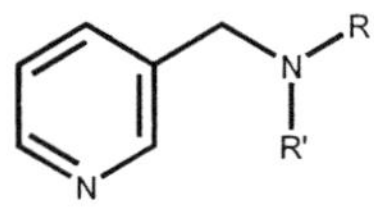

	Agent	R'	R	K_i, nM
3a	AMP	-H	-H	> 10,000
3b	*N*-Me AMP	-H	-Me	> 10,000
3c	*N*-Et AMP	-H	-Et	970
3d	*N,N*-Dimethyl AMP	-Me	-Me	540
3e	*N*-Ethyl-*N*-methyl AMP	-Me	-Et	28
3f	*N*-Methyl-*N*-*n*-propyl AMP	-Me	-*n*Pr	1,140
3g	*N*-Methyl-*N*-*i*-propyl AMP	-Me	-*i*Pr	159
3h	*N*-Methyl-*N*-*c*-propyl AMP	-Me	-*cyclo*Pr	840
3i	*N*-*n*-Butyl-*N*-methyl AMP	-Me	-*n*Bu	> 10,000
3j	*N*-*t*-Butyl-*N*-methyl AMP	-Me	-*t*Bu	195
3k	*N*-*c*-Butyl-*N*-methyl AMP	-Me	-*cyclo*Bu	2,170
3l	*N,N*-Diethyl AMP	-Et	-Et	122

[a] For purpose of comparison, (-)nicotine binds with $K_i = 2.3$ nM. Data taken from Dukat et al. (1996).

affinity by a factor of at least 10. Ring expansion of nornicotine to the piperidine derivative anabasine also results in decreased affinity. However, Abood et al. (1987) demonstrated that ring contraction of nicotine to the four-membered ring azetidine derivative **4** has essentially no effect on affinity. There is evidence, then, for a lack of bulk tolerance in this vicinity of the nAChRs. Supportive of this concept are the recent findings of Lin et al. (1994) who demonstrated that incorporation of relatively small substituents at the 3'-, 4'- or 5'-positions of nicotine results in decreased affinity. Glassco et al. (1993a) have further shown that homologation of the *N*-methyl group of nicotine also results in compounds with low affinity for nicotine receptors.

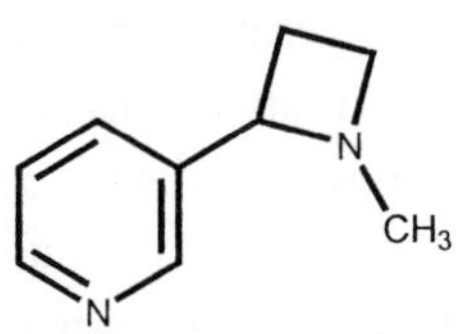

4

2.2.2 The Pyridine Ring

Relatively less has been done with the pyridine ring portion of nicotine. It has been suggested that the pyridine nitrogen atom of nicotine interacts with nAChRs via formation of a hydrogen bond (reviewed by Glennon and Dukat 1996). If this ring nitrogen atom is required for binding, its replacement with an sp^2-hybridized carbon atom should result in reduced affinity. This has been found to be the case; compound **5**, for example, binds with low affinity (K_i = 860 nM) (Glennon et al. 1993). The Abbott group has been instrumental in replacing the pyridine ring of nicotine with other heterocycles. ABT-418 (**6**), for example, retains many of the actions of nicotine (Garvey et al. 1994). Although pyridine ring substitution has not been extensively explored, in general such substitution either reduces nicotinic receptor affinity or results in agents with selectivity for certain populations of nAChRs (reviewed by Glennon and Dukat 1996). This is discussed further here.

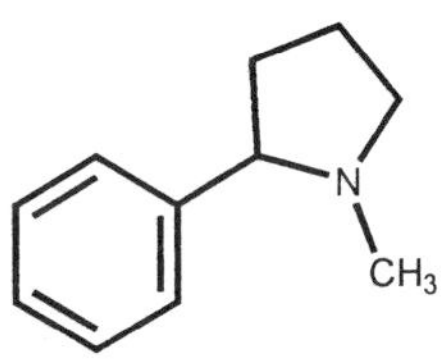

5

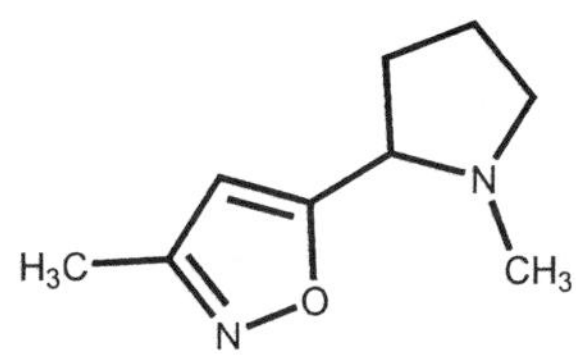

6

2.3 Conformational Investigations

The specific conformation(s) required of nicotine for nicotinic activity/binding remain to be identified. However, several conformationally restricted analogs have been prepared and consideration of their binding may be instructive. Bridged nicotine, **7**, binds with low affinity (K_i > 1,000 nM) at nicotine receptors (Glassco et al. 1993b). This may be a consequence of the structure being restricted in an undesirable conformation for binding, or may simply reflect a lack of bulk tolerance associated with the presence of substitution at the pyridine 4-position. This could be further explored by comparing the affinity of, for example, 4-methylnicotine with that of nicotine. However, introduction of a 4-methyl group could result in an unnatural (i.e., nonnicotine-like) conformation for the resulting derivative. That is, the presence of a 4-position substituent would force the pyrrolidine ring away from this region because of an unfavorable steric interaction. Nevertheless, to further explore this issue, we examined and compared the affinities of 2-methyl *N*-ethyl-*N*-methyl AMP (**3m**), 4-methyl *N*-ethyl-*N*-methyl AMP (**3n**), and *N*-ethyl-*N*-methyl

AMP (**3e**). Both 2- and 4-methyl substitution resulted in substantially (≈ 100-fold) decreased affinity (Dukat et al. 1996). Although the question of substituent intolerance versus conformational alteration could not be addressed directly by these compounds, it was hoped that some light might have been shed on the problem. In another approach, we prepared two conformationally restricted analogs **8** and **9**. Both compounds retained the simple *N,N*-dimethyl AMP structure but differ with respect to the location of the more basic nitrogen atom relative to the pyridine centroid. Although low affinity for both agents would fail to resolve the steric bulk versus conformation problem, the finding that **8** (K_i = 18 nM) binds with 30-fold higher affinity than *N,N*-dimethyl AMP (**3e**; K_i = 540 nM) suggests that some bulk at the pyridine 4-position is indeed tolerated (Dukat et al. 1996). Compound **9** (K_i = 165 nM) also displayed higher affinity than **3e**, suggesting that some bulk is also tolerated at the pyridine 2-position. However, the higher affinity of **8** relative to **9** indicates that the former may more closely represent the bioactive conformer of nicotine. Ring expansion of **8** to **10** (K_i = 780 nM), however, suggests that bulk tolerance in the vicinity of the pyridine 4-position region may be limited.

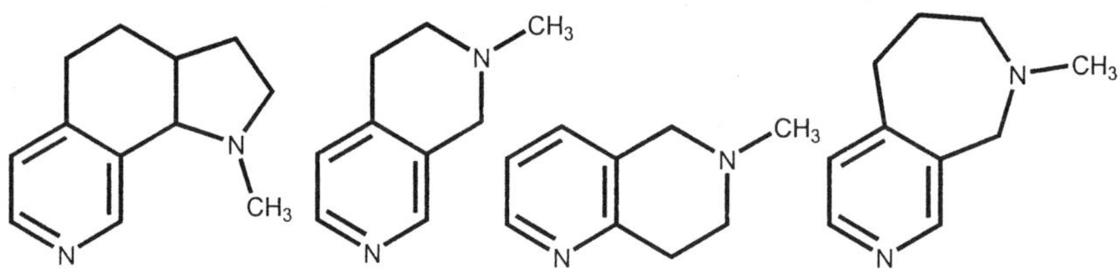

7 **8** **9** **10**

2.4 Internitrogen Distance

Various investigations using peripheral assays have prompted attempts to formulate a nicotinic pharmacophore (reviewed by Glennon and Dukat 1998). Although the concept has been challenged (Barlow and Johnson 1989), it is commonly accepted that an internitrogen distance (i.e., distance between the pyridine and pyrrolidine nitrogen atoms) of 4.8 Å may be an important feature for nicotinic activity (Beers and Reich 1970; Sheridan et al. 1986). However, this distance was derived from studies on a limited number of agents that were available at the time and did not consider the affinity of the agents for central nAChRs. Is this distance important for binding of nicotinic ligands at central nAChRs? We explored a series of agents and found that those with high affinity did indeed possess an internitrogen distance approximating 4.8 Å.

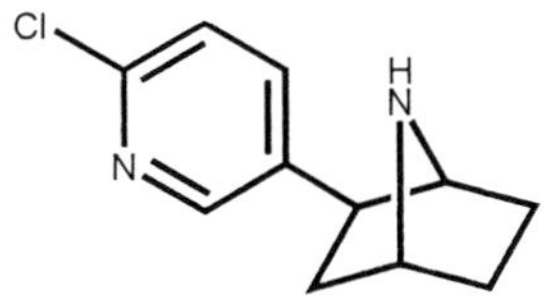

11

While our studies were in progress, Daly and co-workers reported the isolation of epibatidine (**11**), a naturally occurring compound with potent analgesic activity (Spande et al. 1992; reviewed by Dukat, 1994). We measured the affinity of epibatidine for nAChRs (Ki ≈ 0.05 nM) and found it to bind with significantly higher affinity than nicotine itself (Dukat et al. 1994). We also performed molecular modeling studies and found epibatidine to possess an internitrogen distance of 5.5 Å (Dukat et al. 1994). Compounds with an even longer internitrogen distance were prepared, and although they did not bind with as high an affinity as epibatidine, they did display low nanomolar-range affinity for nAChRs (Glennon et al. 1994). These findings seriously questioned the importance of the 4.8 Å internitrogen distance as being a key feature for the binding of nicotinic ligands at nAChRs. Alternatively, it may be the exclusivity of the 4.8 Å distance that should be questioned; that is, more than one mode of (overlapping) binding at nAChRs may be possible (Glennon et al. 1994; Glennon and Dukat 1998). This problem requires investigation.

The seven-membered conformationally-restricted AMP **10** was found to bind with low affinity, and this reduced affinity was initially ascribed to limited bulk tolerance associated with the pyridine 4-position binding region at nAChRs (Cheng et al. 1995). Now, another explanation was possible. The calculated internitrogen distance in **10** is only 4.6 Å relative to that (4.8 Å) for nicotine and **3e**. Thus, reduced affinity might (alternatively or in addition) be related to this shorter distance. Moving the nitrogen atom of **10** by one ring position, to afford **12**, results in retention of the general shape and size of the ring, but also results in an increase in internitrogen distance to 5.5 Å. Compound **12** (R = H; K_i = 46 nM) was prepared and found to bind with enhanced affinity (Cheng et al. 1995). It must be assumed that nAChRs can accommodate somewhat more bulk at the pyridine 4-position region than initially thought.

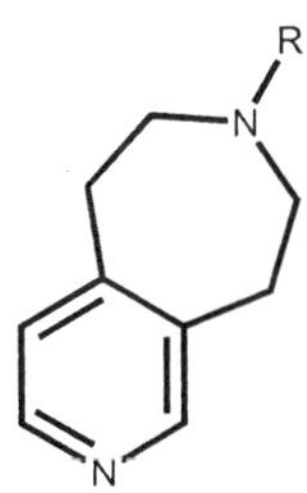

12

2.5 Chain-Extended Analogs

Compound **12**, although displaying higher affinity for nAChRs than **10**, is no longer an aminomethylpyridine or nicotine analog. That is, two methylene groups, not one methylene group, separate the terminal amine of **12** from the pyridine ring. Is it possible that chain-extended analogs of the AMPs **3** might display enhanced affinity? We prepared a series of analogs in which the terminal amine was separated from the pyridine ring by two and three methylene groups; these are the aminoethylpyridine (AEP; **13**) and aminopropylpyridine (APP; **14**) analogs, respectively. Analogs **13** may be viewed as ring-opened derivatives of **12**. Table 2 shows some selected data for the AMP, AEP, and APP analogs. The AEP analogs **13** were found to bind with higher affinity than their corresponding AMP derivatives, and the APP derivatives **14** bind with low affinity (K_i > 10,000 nM). Interestingly, comparing the AMPs and the AEPs, there appears to be a lack of parallel effect upon parallel substituent modification, suggesting that the two series may not be binding in the same manner.

Table 2. Binding affinities of selected chain-extended aminomethylpyridines[a].

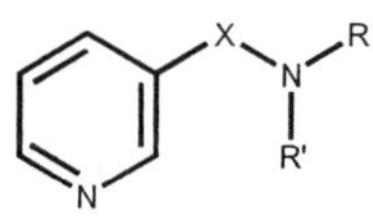

Agent	X	nAChR Affinity (K_i, nM) -R/-R' -H / -Me	-Me / -Me	-Me / -Et
3	-CH_2-	>10,000	540	28
13	-CH_2-CH_2-	290	47	18
14	-CH_2-CH_2-CH_2-	>10,000	>10,000	>10,000
15	-O-CH_2-CH_2-	35	21	22

[a] Data for analogs of **3** are from Table 1; remaining data are unpublished from Cheng, Dukat, Dowd, Fiedler, Martin, Damaj, and Glennon (manuscript submitted).

In an attempt to introduce some conformational restriction, unsaturation was introduced into the propyl side chain of the APP analogs (i.e., alkenyl and alkynyl derivatives were prepared and examined). Although these agents failed to bind with the affinity of the AEP analogs they, nevertheless, displayed higher affinity than their corresponding saturated APP derivatives. More interesting, however, was the observation of similar affinity between some *cis* and *trans* alkenyl derivatives, suggesting that unsaturation may enhance affinity via its electronic rather than steric character. Accordingly, we prepared and evaluated a series of aminoethoxypyridines (AXPs, **15**) (Table 2). The AXPs displayed an affinity at least equivalent to their corresponding AEP analogs, and less variation in affinity than their corresponding AMP analogs. Again, parallel structural changes failed to result in parallel shifts in affinity, suggesting differences in modes of binding for the different series or for certain members within the series.

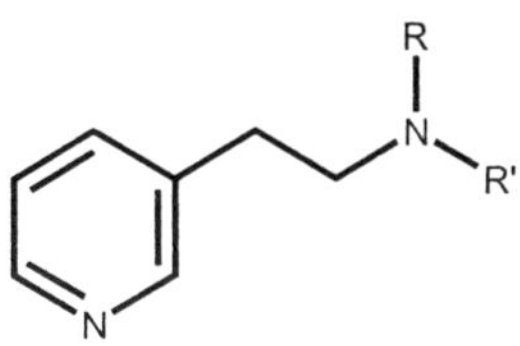

13

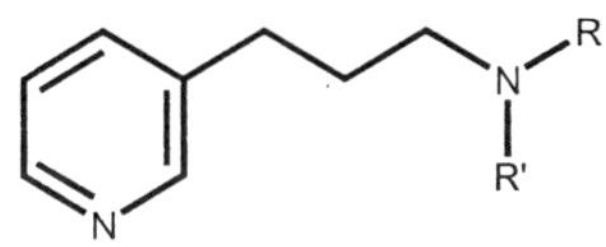

14

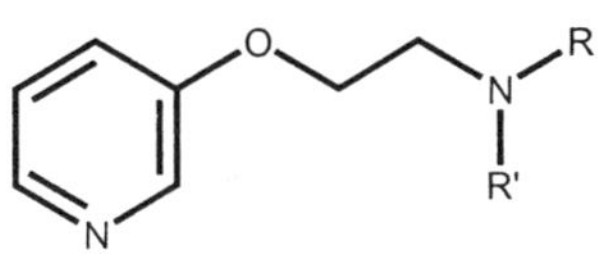

15

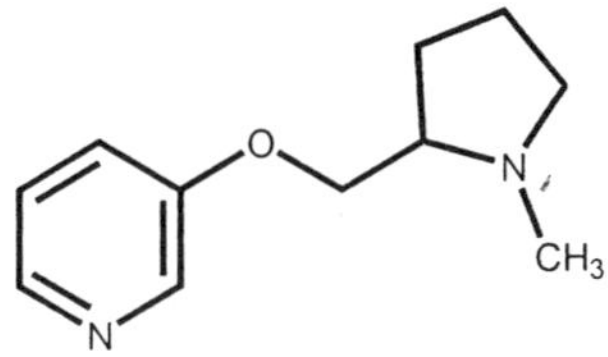

16

While our studies were in progress, Abreo et al. (1996) reported a series of pyridyl ethers similar in nature to the AXP derivatives. For example, A-84543 (**16**) binds with subnanomolar affinity (S(-)isomer $K_i = 0.15$ nM). Ring contraction of the N-desmethyl derivative of A-84543 to an azetidine results in A-85380, an agent with yet higher affinity (S(-)isomer, $K_i = 0.052$ nM), and with an affinity comparable to that of epibatidine (**11**). Thus, the aminoethoxypyridines, or pyridyl ethers, represent a novel class of high-affinity nAChR ligands. At this time it is not known how the AMPs, AEPs, and AXPs bind relative to one another or relative to (-)nicotine. The internitrogen distances in the AEPs and AXPs are longer than that found in nicotine suggesting that these agents may not be binding in the same manner. However, the alkyl chains of the longer-chain compounds are quite flexible and may bend back to mimic a nicotine-like distance. Alternative modes of binding also have been suggested by Abreo et al. (1996). These analogs are certainly distinct from the structurally and conformationally simpler nicotine, and additional study is obviously required.

2.6 QSAR Studies

The availability of a relating equation developed on the basis of quantitative structure-affinity relationship (QSAR) studies would be useful for the design of novel nAChR ligands. This prompted us to undertake such an investigation. On the basis that the chloro-bearing pyridine ring position of epibatidine (**11**) may correspond to the 6-position of nicotine, we prepared and evaluated 6-chloronicotine (**1c**) (Dukat et al. 1994). We subsequently prepared additional analogs including the 6-bromo, 6-fluoro, 6-methyl, and 6-methoxy analogs of nicotine (Dukat et al. 1996). (±)6-Chloronicotine (**1c**; K_i = 0.63 nM) was found to bind with several times the affinity of (-)nicotine itself, and was 15 times more potent than nicotine as an antinociceptive agent in rodents. That the antinociceptive effects were potently antagonized in a dose-related fashion by nicotine antagonists supports involvement of a nAChR-based mechanism. The 6-bromo derivative (**1b**; K_i = 0.45 nM) was found to bind with an affinity similar to its 6-chloro counterpart **1c**, the 6-fluoro and 6-methyl derivatives (K_i = 1.03 and 1.8 nM, respectively) possessed affinities similar to nicotine, whereas the 6-methoxy analog of nicotine (**1h**; K_i = 22 nM) was found to bind with lower affinity. 6-Aminonicotine (**1j**), a precursor for the synthesis of the chloro and bromo derivatives, was also evaluated and found to bind with even lower affinity (K_i = 67.5 nM).

A Hansch analysis was conducted with these seven analogs to determine whether binding (pK_i) was related to either the lipophilic (π) or electronic (σ) nature of the 6-position substituent. Unexpectedly, relating equations were identified suggesting a relationship between both π and σ (r = 0.875 and 0.952, respectively). The number of compounds was insufficient to allow inclusion of both independent variables in the same relating equation. Furthermore, a significant internal correlation (r = 0.896) was found between the π and σ values employed in the Hansch analysis. Thus, only one of the two parameters might explain binding and the second might be simply serving as a surrogate for the first. Consequently, we prepared additional 6-substituted analogs with the following goals in mind: (a) prepare a total of at least

12 analogs so that both substituent parameters can be included in a relating equation, (b) select substituents such that there is no internal correlation between π and σ, and (c) repeat the Hansch analysis to determine if one or a combination of both parameters will explain binding. A total of 17 analogs was eventually prepared (Dowd et al. 1997; Dukat et al. 1998). Two of the analogs, the 6-acetamido and 6-sulfonamido analogs of nicotine (i.e., **1p** and **1q**, respectively), could not be included in the Hansch analysis because they lacked measureable affinity for nAChRs (i.e., K_i > 10,000 nM). The remaining 15 analogs were investigated in detail; some binding data are shown in Table 3. With a sample size of 15 compounds, no internal correlation was found to exist between the substituent constants π and σ. However, it was also found that pK_i was not significantly correlated to either π (r = 0.590) or σ (r = 0.089). Use of both independent variables in the same equation did not improve the correlation.

Table 3. Binding data (observed and predicted) for 6-substituted nicotine analogs[a].

	6-Substituent	K_i, nM	Actual pK_i	Predicted pK_i
1a[b]	-H	1.26	8.90	9.05
1b	-Br	0.45	9.35	8.96
1c	-Cl	0.63	9.20	9.13
1d	-F	1.03	8.99	9.07
1e	$-CH_3$	1.8	8.74	8.66
1f	$-CH_2CH_3$	5.6	8.25	8.13
1g	$-CH_2CH_2CH_3$	17.3	7.76	7.72
1h	$-OCH_3$	22.0	7.67	7.48
1i	-CN	29.0	7.54	7.55
1j	$-NH_2$	67.5	7.09	6.70
1k	$-NO_2$	142	6.95	7.56
1l	-CHO	181	6.93	7.17
1m	$-NHCH_2CH_3$	730	6.14	6.20
1n	$-COOCH_3$	866	6.06	6.36
1o	$-CONH_2$	1,344	5.87	5.53
1p	$-NHCOCH_3$	>10,000	< 5.0	4.90
1q	$-NHSO_2CH_3$	>10,000	< 5.0	4.11

[a] Data from Dukat et al. (1998).

[b] Racemic nicotine.

Because of the small variation in affinity for the original series of compounds, early on we believed that nAChRs might possess a region of bulk tolerance associated with the pyridine 6-position of nicotine. However, the low affinity of 6-methoxynicotine suggested that binding might be influenced by the electronic nature of the substituent, or that the region of bulk tolerance was limited in size and might not accommodate larger substituents (Dukat et al. 1996). The subsequent lack of a correlation between affinity and σ for the series of 15 compounds essentially rules out a significant role for the electronic nature of the 6-position substituent. Consequently, we explored the role of steric bulk and its potential effect on binding. We calculated the volume of each of the 15 nicotine analogs and then subtracted from each the volume of nicotine. The residual amount, termed Δ MOL VOL, was used as an approximation of the steric bulk of the various 6-position substituents. It was found that only a modest correlation exists between pK_i and Δ MOL VOL (i.e., $r = 0.578$). However, a combination of π and Δ MOL VOL afforded the following relating equation:

$$pK_i = 9.05 + 1.19\,\pi - 0.067\,\Delta\,\text{MOL VOL}$$
$$r = 0.970; \quad n = 15$$

That is, affinity appears to be related both to the lipophilicity and steric volume of the 6-position substituent (Dowd et al. 1997; Dukat et al. 1998). In other words, increased lipophilicity results in enhanced affinity, but this effect is modulated by the size of the 6-position substituent. Table 3 provides the actual pK_i values of the 15 6-substituted nicotine analogs and the pK_i values predicted on the basis of the above equation. Compounds **1p** and **1q**, which could not be included in the Hansch analysis due to their indeterminate affinities, are also correctly predicted to bind with low affinity (predicted pK_i = 4.90 and 4.11, respectively). Evidently, this represents the first time that the binding of nicotine analogs at central nAChRs has been related to physicochemical properties of the molecule itself. Additional analogs are currently being synthesized to challenge this relating equation.

3 Conclusion

Although nicotine itself has been available for investigation for many years, it is only recently that radioligands and techniques for the measurement of binding affinities of agents for central nAChRs have become available. Thus, for the first time, questions concerning nAChRs need not be limited solely to examination of peripheral receptors. We and others have availed ourselves of this opportunity by examining the structure-affinity relationships of nicotine. Although the present review focuses primarily on work from our laboratory, the investigations of the Abbott group (e.g., Abreo et al. 1996), the R. J. Reynolds group (e.g., Caldwell et al. 1997), and others should not be overlooked. Some of this work has been already reviewed (Glennon and Dukat 1996; Holladay et al. 1997).

Our studies were initially aimed at identifying the minimal structural features required for the binding of nicotine (**1a**) at central nAChRs. Studies on the pyridine ring portion of nicotine are still in progress. With respect to the pyrrolidine portion, we identified the aminomethylpyridine or AMP moiety as contributing to the binding of nicotine, and further identified the *N*-ethyl-*N*-methyl AMP **3e** as representing the highest affinity member of the series. Investigation of conformationally-restricted analogs led to the naphthyridine **8**. As if to underscore differences between central and peripheral nAChR preparations, compound **8** had been earlier reported to exhibit activity in a rabbit jejunum preparation, but not in a frog rectus muscle assay (Haglid 1967). Compound **8** was ring-expanded to **10**, which was found to bind with reduced affinity. On the basis that certain high-affinity nAChR ligands such as epibatidine (**11**) possess a longer internitrogen distance than that found in **10**, we prepared **12**. Compound **12** was found to bind with enhanced affinity. This led to examination of a series of ring-opened analogs of **12** such as the AEPs. Subsequent study resulted in the synthesis and evaluation of the AXPs. The AEP and AXP analogs represent novel classes of nAChR ligands. Abbott Laboratories independently discovered AXP analogs and are currently exploiting them as potential therapeutic agents. Recently, our attention has returned to the pyridine portion of nicotine. In particular, we are examining the role of 6-position substituents and have found that affinity might be related to their lipophilic nature and steric bulk. A relating equation was identified that accurately predicts the affinities of various 6-substituted nicotine analogs.

Over a decade ago, Sheridan et al. (1986) formulated a "nicotinic pharmacophore." The pharmacophore was developed using a limited series of agents and did not specifically address the issue of central nAChRs. One of the important features of this pharmacophore is that an internitrogen distance of 4.8 Å is seemingly necessary for nicotinic activity; see Glennon and Dukat (1998) for a review of investigations of the nicotinic pharmacophore. However, the discovery of epibatidine (**11**; internitrogen distance = 5.5 Å), which binds with higher affinity than nicotine, questions the importance or at least the exclusivity of the 4.8-Å distance. Although both Dukat et al. (1994) and Abreo et al. (1996) have proposed modes of binding that will accommodate both nicotine and epibatidine, it is not yet known how these agents bind relative to one another. Attempts have been made to address this question by, for example, comparing the steric volumes of epibatidine and nicotine optical isomers, and by conducting comparative molecular field analysis (CoMFA) studies (Dukat et al. 1995; Glennon et al. 1994) The discovery of additional agents, possessing internitrogen distances greater than 4.8 Å, with high affinity for central nAChRs, suggests that not all agents bind in the same manner. This idea is reinforced by the lack of parallel effects on affinity when parallel substituent changes are made in several series of molecules (e.g., see Table 2). The findings presented or cited here represent the most comprehensive structure-affinity and quantitative structure-activity (QSAR) studies on nicotine to date, and open new structural approaches for the development of novel nAChR agents.

Acknowledgments

Work from our laboratory was supported by funding from the Virginia Center for Innovative Technology through the Technology Development Center (R.A.G.), the A. D. Williams Fund (M.D.), and from NIH grant DA 05274 (R.A.G.).

References

Abood LG, Lu X, Banerjee S (1987) Receptor binding characteristics of a ^{3}H-labelled azetidine analogue of nicotine. Biochem Pharmacol 36:2337-2341

Abreo MA, Lin NH, Garvey DS, Gunn DE, Hettinger AM, Wasicak JT, Pavlik PA, Martin YC, Donnelly-Roberts DL, Anderson DJ, Sullivan JP, Williams M, Arneric SP, Holladay MW (1996) Novel 3-pyridyl ethers with subnanomolar affinity for central neuronal nicotinic acetylcholine receptors. J Med Chem 39:817-825

Arneric SP Brioni JD (1998) Neuronal nicotinic receptors: pharmacology and therapeutic opportunities. John Wiley and Sons, New York

Barlow RB, Johnson O (1989) Relations between structure and nicotine-like activity: X-ray crystal structure analysis of (-)-cytisine and (-)-lobeline hydrochloride and a comparison with (-)-nicotine and other nicotine-like compounds. Br J Pharmacol 98:799-808

Beers WH, Reich E (1970) Structure and activity of acetylcholine. Nature (Lond) 225:917-922

Caldwell WS, Benchenif M, Bhatti BS, Deo NM, Dobson GP, Dull GM, Lipiello PM, Lovette ME, Miller CH, Ravard A, Schmitt JD, Crooks PA (1997) Synthesis and structure-activity relationships of analogs of RJR-2403, a CNS-selective nicotinic agonist. Abstracts of International Business Communications Symposium on Nicotinic Acetylcholine Receptors as Pharmaceutical Targets, Washington, DC, July 24-25

Cheng YX, Fiedler W, Dukat M, Damaj I, Martin B, Glennon RA (1995) Conformationally-restricted aminomethylpyridine derivatives as novel nicotine receptor ligands. VA J Sci 46:135

Dowd M, Dukat M, El-Zahaby M, Glennon RA (1997) Role of 6-position substituents on the binding of nicotine at nicotinic cholinergic receptors. VA J Sci 48:137

Dukat M (1994) 208/210 a.k.a. epibatidine. Med Chem Res 4:433-439

Dukat M, Herndon JL, Glennon RA (1995) Epibatidine: reconsideration of the nicotine receptor pharmacophore. NIDA Res Mono 162:286.

Dukat M, Damaj MI, Glassco W, Dumas D, May EL, Martin BR, Glennon RA (1994) Epibatidine: a very high affinity nicotine receptor ligand. Med Chem Res 4:131-139

Dukat M, Fiedler W, Dumas D, Damaj I, Martin BR, Rosecrans JA, James JR, Glennon RA (1996) Pyrrolidine-modified and 6-substituted analogs of nicotine: a structure-affinity investigation. Eur J Med Chem 31:875-888

Dukat M, Dowd M, Damaj MI, Martin BR, El-Zahabi M, Glennon RA (1998) Synthesis, receptor binding, and QSAR studies on 6-substituted nicotine derivatives as cholinergic ligands. Eur J Med Chem, in press

Garvey DS, Wasicak JT, Elliott RL, Lebold SA, Hettinger AM, Carrera GM, Lin NH, He Y, Holladay MW, Anderson DJ, Cadman ED, Raszkiewicz JL, Sullivan JP, Arneric SP (1994) Ligands for brain cholinergic channel receptors: synthesis and in vitro characterization of novel isoxazoles and isothiazoles as biosteric replacements for the pyridine ring of nicotine. J Med Chem 37: 4455-4463

Glassco W, May EL, Damaj MI, Martin BR (1993a) In vivo and in vitro activity of some N-substituted (±)-nornicotine analogs. Med Chem Res 4: 273-282

Glassco W, Suchocki J, George C, Martin BR, May EL (1993b) Synthesis, optical resolution, absolute configuration, and preliminary pharmacology of (+)- and (-)-*cis*-2,3,3a,4,5,9b-hexahydro-1-methyl-1H-pyrrolo[3,2-h]isoquinoline,a structural analog of nicotine. J Med Chem 36: 3381-3385

Glennon RA, Dukat M (1996) Nicotine receptor ligands. Med Chem Res 6:465-486

Glennon RA, Dukat M (1998) Nicotinic cholinergic receptor pharmacophores. In: Arneric SP, Brioni JD (Eds) Neuronal nicotinic receptors: pharmacology and therapeutic opportunities. John Wiley and Sons, New York, pp 271-284

Glennon RA, Herndon JL, Dukat M (1994) Epibatidine-aided studies toward definition of a nicotinic receptor pharmacophore. Med Chem Res 4:461-473

Glennon RA, Maarouf A, Fahmy S, Martin B, Fan F, Yousif Y, Shafik RM, Dukat M (1993) Structure-activity relationships of simple nicotine analogs. Med Chem Res 2:546-551

Haglid F (1967) Studies on pyridine alkaloidsand their analogues. Acta Pharm Suec 4:117-138.

Holladay MW, Dart MJ, Lynch JK (1997) Neuronal nicotinic acetylcholine receptors as targets for drug discovery. J Med Chem 40:4169-4194

Lin N-H, Carrera GM, Anderson DJ (1994) Synthesis and evaluation of nicotine analogs as neuronal nicotinic acetylcholine receptor ligands. J Med Chem 37:3542-3553

Shacka JJ, Robinson SE (1996) Central and peripheral anatomy of nicotine sites. Med Chem Res 6:444-464

Sheridan RP, Nilakantan R, Dixon JS, Venkataraghavan R (1986) The ensemble approach to distance geometry: Application to the nicotinic pharmacophore. J Med Chem 29:899-906

Spande TF, Garraffo HM, Edwards MW, Yeh HJC, Pannell L, Daly JW (1992) Epibatidine: a novel (chloropyridyl)azabicycloheptane with potent analgesic activity from an Ecuadoran poison frog. J Am Chem Soc 114:3475-3478

Tomizawa M, Latli B, Casida JE (1996) Novel neonicotinoid-agarose affinity column for *Drosophila* and *Musca* nicotinic acetylcholine receptors. J Neurochem 67:1667-1676

12 Managing Resistance to the Chloronicotinyl Insecticides - Rhetoric or Reality?

Matthew Cahill and Ian Denholm
Department of Biological & Ecological Chemistry, IACR-Rothamsted, Harpenden, Hertfordshire, AL5 2JQ, England

1 Introduction

When the launch of a novel, safe, and highly efficacious insecticide such as imidacloprid (Elbert et al. 1990) provides agricultural producers with a much needed new tool for crop protection, and also heralds a whole new class with the potential to rival the pyrethroids, organophosphates, carbamates and cyclodienes, the optimism and opportunities are vast. For each of these classes, however, the reality has followed a familiar sequence of discovery, development, launch, marketing, deployment, gradual loss of field performance as a result of resistance brought on in most cases by indiscriminate overuse, and finally the urgent need for a new, safe, and efficacious insecticide.

The potential for insecticide discovery and development has apparently not diminished, and the diversity of chemistry for controlling key pests is now probably greater than ever (Ishaaya and Horowitz 1998). It is also widely accepted that available insecticides should be conserved, if only for the simple economics of making a return on the vast investment in their development. As discussed later, this diversity of insecticides must be exploited to reduce the reliance on each particular product although it appears that multinational company marketing strategies, and in some cases the attitude of regulators, seems designed to introduce potentially useful rotation partners sequentially rather than simultaneously.

The paradox, therefore, is that those agro-ecosystems that have the greatest need for new insecticides are also those with the recidivist, multiresistant pests that constitute the greatest resistance risk. It is these systems in which sequential registration of pesticides has led to the intense cycle of introduction, overuse, and loss to resistance within brief periods of time. Many examples abound but the chronology of insecticide resistance in the Colorado potato beetle (*Leptinotarsa decemlineata*, Coleoptera: Chrysomelidae) illustrates the point well (Forgash

1985; Roush and Tingey 1991).

Principles of insecticide resistance management have been extensively reviewed (Denholm and Rowland 1992; Georghiou 1994; Roush 1989) as have general considerations for novel insecticides (Denholm et al. 1998), and initial guidelines for imidacloprid (Elbert et al. 1996). In this chapter we aim to present ideas more specific to the chloronicotinyls, keeping in mind the attributes of the class that make it unique. These characteristics include the novel mode of action, the physicochemical properties that contribute to the deployment options, as well as the current understanding of resistance risks and the tactics for management, which collectively leave few excuses for not conserving susceptibility to this class of insecticide.

2 Attributes of the Chloronicotinyls

2.1 A Novel Biochemical Target

In spite of the use of nicotine and its derivatives for centuries, the nicotinic acetylcholine receptor (nAChR) has been an underexploited biochemical target of modern insecticides with an estimated market share of 2% or less (Leicht 1996). Chemicals active on this receptor include nereistoxin analogues such as cartap, bensultap, and thiocyclam, as well as the nicotinoids. However, the nAChR is a complex of subunits, and although these insecticides act on the same overall receptor, the specific site of binding may differ. For example, it is proposed that nicotine and nitenpyram act at two positions of the nAChR while imidacloprid and acetamiprid act on only one (Tomizawa et al. 1995) and that imidacloprid, cartap, and bensultap all have distinct and different types of action on the nAChR (D Sattelle, 1998 personal communication; see also the chapters by G.G. Lunt and R.A. Glennon, this volume).

The nereistoxin analogues have been most commonly used for controlling Colorado potato beetle, stem borers such as *Chilo* spp. (Lepidoptera: Pyralidae), diamondback moth (*Plutella xylostella*, Lepidoptera: Yponomeutidae), and some cicadellid rice pests. Of these, the Colorado potato beetle and rice planthoppers are major targets of the first commercial chloronicotinyls. Nicotine is regularly used in greenhouses as a fumigant against aphids and whiteflies, which are also major targets of the chloronicotinyls. However, the overlap between the traditional nAChR insecticide market share and the pests targetted by the chloronicotinyls accounts for the relative novelty of this biochemical target.

2.2 Options for Deployment

For most insecticides, the major operational choices that have an impact on resistance management are the timing, frequency, placement, and rate of application; i.e., when, where, and how often the insecticide should be applied and at what concentration. These factors also need consideration with the chloronicotinyls, and many of the same conclusions apply. However, the physicochemical properties of various members of the class also allow the insecticides to be applied in unusually diverse ways, e.g., as a foliar spray, or systemically as a drench or in irrigation water, as a granular soil application, as a seed dressing, or as a paint-on formulation. Such versatility is one of the single most important features of this new class of insecticides, especially with regard to resistance management.

The dual attributes of biochemical novelty and operational versatility improve prospects for proactive resistance management but also raise fundamental questions to be addressed before developing recommendations appropriate to particular pest and cropping systems. Initially it is necessary to consider if the novel mode of action does render them invulnerable from existing resistance mechanisms, and if the deployment options differ intrinsically in their resistance risk.

3 Genetic Considerations

Estimates of unselected equilibrium frequencies of alleles responsible for resistance to xenobiotics cover a range too large (10^{-3} to 10^{-13}) (McKenzie 1996) to be of direct predictive value. Empirical measurements require the considerable effort of screening large numbers of field-collected individuals or their progeny (ffrench-Constant et al. 1990; Gould et al. 1997). However, because insecticide resistance management strategies have a greater chance of success when gene frequencies are low, the factors that influence initial gene frequency need careful consideration. The most important factor affecting resistance gene frequencies is, of course, insecticide selection, exerted by the specific product under consideration or by other insecticides that select for a common mechanism of resistance. Primary candidates for cross-resistance are those insecticides acting on the same target site as the chloronicotinyls, although unexpected cross-resistance patterns between apparently unrelated insecticides are not uncommon.

3.1 Cross-Resistance to nAChR Insecticides

Reports of resistance to the nereistoxin analogues are mostly confined to cartap resistance in *P. xylostella* (Chen et al. 1993), with some reports of cross-resistance to either bensultap or thiocyclam (Hama 1986). On the whole, however, resistance to these insecticides in *P. xylostella* was far lower than for other insecticides even though they were sometimes relied upon quite heavily (Sakai 1985). Of the major pest targeted by chloronicotinyls, bensultap resistance has been reported in *L. decemlineata* (Pap et al. 1997). Despite considerable use over the years, there are only a few reports of low levels of nicotine resistance in aphids (Devine et al. 1996; Dhingra 1994) and one anecdotal report of resistance in the whitefly *Trialeurodes vaporariorum* (Cary 1903). Interestingly, the report by Cary stating "During the past two or three years tobacco [fumigation] seems to have been less effective than. . .eight years ago" predates the often quoted first report of insecticide resistance (Melander 1914) by more than a decade. Recent work has shown considerable nicotine resistance in the tobacco whitefly *Bemisia tabaci* (M. Cahill, unpublished data).

Strains of the aphid *Myzus nicotianae* with tolerance to nicotine, probably as a result of feeding on tobacco, also showed a low level of tolerance to imidacloprid (Devine, et al. 1996; Nauen et al. 1996). This tolerance was more pronounced to the antifeeding than to the lethal effects of imidacloprid, and in the second study, slight tolerance to cartap was also reported. In a French strain of *M. nicotianae* exhibiting resistance to both imidacloprid and nicotine, resistance factors varied between bioassay methods but reached 192 and >22 to imidacloprid and nicotine respectively (Nauen and Elbert 1997). Collectively, these aphid studies imply a degree of cross-tolerance between the chloronicotinyl and nicotine but do not exclude the possibility of other, more specific resistance mechanisms as well. Similarly, a positive relationship between nicotine and imidacloprid mortality has been demonstrated for populations of *Bemisia tabaci* tested with both compounds (Figure 1). As for aphids, the underlying mechanism(s) have not yet been identified.

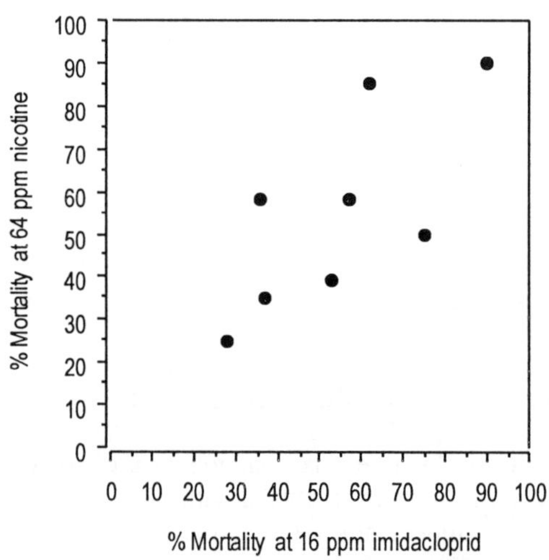

Fig. 1. Relationship between percent adult motality at 16 ppm imidacloprid tested in the systemic-leaf bioassay and 64 ppm nicotine tested in the glass-vial to leaf bioassay for eight *Bemisia tabaci* strains

3.2 Cross-Resistance to Other Insecticides

Cross-resistance patterns between chemically unrelated classes of insecticides are difficult to predict and can only be established by rigorous bioassays against strains of insect with well-identified and contrasting resistance profiles. To this end, substantial imidacloprid bioassay data have been generated for populations of target pests such as Colorado potato beetle, whiteflies, aphids, lygus bugs, and leafhoppers multiresistant to older, established insecticides (Dennehy and Russell 1996; Grafius and Bishop 1996; Olson et al. 1996; Williams et al. 1996).

Populations of aphids, whiteflies, leafhoppers and cat fleas with resistance to organophosphates (OPs), carbamates, endosulfan, and pyrethroids did not show any apparent resistance to imidacloprid (Bardt and Schein 1996; Cahill et al. 1996; Elbert, et al. 1990). Imidacloprid LC_{50}s of 35 collections of *L. decemlineata* from USA and Canada ranged from 0.28 ppm to 4.4 ppm however, there was no clear evidence for a correlation with resistance to either esfenvalerate or azinphosmethyl (Olson, et al. 1996).

In contrast, a population of small brown planthopper (*Laodelphax striatellus*, Hemiptera: Delphacidae) reared under malathion and propoxur pressure showed about 20-fold tolerance to imidacloprid, and some field strains also showed 5- to 6- fold tolerance (Sone et al. 1995). In the same study a positive correlation was established between LC_{50}s for imidacloprid and those for propoxur or disulfuton. Populations of lygus bug (*Lygus hesperus*, Hemiptera: Miridae), a key cotton pest in Arizona and a target for chloronicotinyls have shown extraordinary between-population variation in response to imidacloprid. The LC_{50} of a lygus bug population collected from Maricopa in 1995 was more than 100 fold that of a population from Safford when tested in adult vial bioassays (Dennehy and Russell 1996). The most tolerant population was simultaneously highly resistant to aldicarb, dimethoate, malathion, and oxamyl but not resistant to oxydemeton-methyl, endosulfan, or methamidophos.

A population of Western flower thrips, *Frankliniella occidentalis* (Thysanoptera: Thripidae), selected repeatedly with diazinon showed a five fold increase in imidacloprid LC_{50} compared to a laboratory strain (Zhao et al. 1995), as well as resistance to other OPs and pyrethroids. Although not highly toxic to either houseflies or cockroaches, imidacloprid was tested on multi-resistant strains of both species and low-level tolerance demonstrated (Wen and Scott 1997).

Overall, data on cross-resistance threats to chloronicotinyls are inconsistent and difficult to interpret. The clearest evidence is for possible cross-resistance between older insecticides that target the nAChR and imidacloprid. At worst this appears to afford low-level tolerance, which may have little or no impact on field efficacy of the product. Cross-resistance between insecticides that target acetylcholine esterase (AChE) and imidacloprid is also implied from some of the data above but this is neither consistent nor conclusive. Clearly further research is required, which should be supported where possible by mechanistic studies rather than based solely on correlations in responses.

3.3. Chloronicotinyl Resistance

Absence of substantive cross-resistance data must not, however, be equated with a low risk of resistance arising *de novo* from intensive field selection with chloronicotinyls. Continuous laboratory selection with imidacloprid of a population of *Bemisia argentifolii* (B-type *B. tabaci*) originally collected from melons near Brawley, California (U.S.A.) in May 1993, resulted in >80-fold resistance after 24 generations (Prabhaker et al. 1997), implying that imidacloprid resistance genes were present at a surprisingly high frequency in the founding population. Similarly, in adult systemic-leaf bioassays more than 80% of female *B. tabaci* from a population collected near Almeria, Spain, in May 1995 survived a diagnostic concentration of imidacloprid expected to kill more than 90% of fully susceptible individuals (Cahill, et al. 1996). Over the last 3 years, whiteflies collected from Arizona have also exhibited a tendency to survive high concentrations of imidacloprid and, more importantly, the frequency of survivors has increased significantly each season (Williams, et al. 1998). Adult *L. decemlineata* survivors from an imidacloprid-treated potato field in Michigan were collected and reared in the laboratory. When tested in both topical and feeding assays, the proportion of F_1 progeny survival from this field strain was significantly higher than the susceptible strain (Grafius and Bishop 1996).

3.4. Cross-Resistance Between Chloronicotinyls

As well as considering cross-resistance between the chloronicotinyls and other classes of insecticides, it is important to recognise the risk of cross-resistance within the class. There are, to date, insufficient data to either support or reject the null hypothesis that resistance to one chloronicotinyl will also confer resistance to others. Strains of *M. nicotianae* from Greece with tolerance to imidacloprid were also more tolerant to acetamiprid and nitenpyram (Nauen et al. 1997) , and similarly *B. tabaci* from Almeria with resistance to imidacloprid also showed cross-resistance to other chloronicotinyls (M. Cahill, unpublished data).

Even though the major pest targets of the novel chloronicotinyls do not appear severely compromised by the use of other insecticides with the same putative mode of action or by unexpected cross-resistance to insecticides with other modes of action, resistance can and has been selected by at least one of the chloronicotinyls in either the laboratory and the field. Although we might expect to introduce the first chloronicotinyls into systems where the initial resistance gene frequency has not been substantially increased by previous insecticide use, the overuse of any one of the chloronicotinyls may not only select for resistance to that product but also probably have serious implications for other members of the class.

4. Operational Considerations

Some of the most vigorous debates regarding the deployment of imidacloprid have centered on its use as either a foliar or soil-applied insecticide. In some cases the practical constraints may be over riding, however, numerous researchers have cautioned against prophylactic overuse of the biologically persistent soil applications.

4.1. Foliar vs Soil

Imidacloprid applied as a sugar beet seed treatment provided effective *M. persicae* control for 50 days after sowing (Dewar and Read 1990), while potato seed tuber treatment with imidacloprid caused 99% mortality of *M. persicae* for 80 days post planting (Meredith and Heatherington 1992). Similarly, subfurrow treatment of lettuce with granular imidacloprid prevented aphid colonisation for 100 days (Palumbo and Kerns 1994). The residual activity of foliar imidacloprid is, however, considerably shorter with most reports indicating good control of various pests for ~1 week (e.g., Boiteau et al. 1997). Such outstanding persistence of systemic treatments provides a very useful tool for the grower but could increase the risk of resistance selection for many target pests.

Similar circumstances existed for deployment of the synthetic pyrethroids for housefly (*Musca domestica*) control in pig farms in the early 1980s when comparisons were made between persistent surface applications and short-lived spacesprays. Comparative trials showed unequivocally that residual applications selected far more rapidly for resistance (Denholm et al. 1983; Keiding and Jesperson 1986; Sawicki 1986). Data such as these coupled with theoretical evidence (Mani and Wood 1984; Roush 1989) indicate that persistent treatments pose a higher resistance risk than nonpersistent ones applied intermittently, especially when the target pest is fecund and invasive and goes through a number of generations during the period of insecticide persistence.

Large cage experiments with imidacloprid-resistant *B. tabaci* have explored interactions between application method, population control, and expression of resistance. Four hundred females from either a susceptible or a resistant *B. tabaci* strain were used to establish replicate populations on cotton plants within each of four cages. Ten days later, each cage was treated with imidacloprid by either a foliar application at the equivalent of 450 l/ha of 1 ml/l Confidor 200SL or 20 ml of the same solution to the soil at the base of the plants. Adult numbers were monitored regularly. Initially, soil-applied imidacloprid gave better control of adults than foliar applications but the discrimination between phenotypes was greater under soil treatments. Foliar-applied imidacloprid also appeared to have a greater impact on very young nymphs than the soil treatment, and the accumulation of these effects demonstrated that expression and possibly selection of resistance was greater for soil-applied imidacloprid than for foliar (Figure 2).

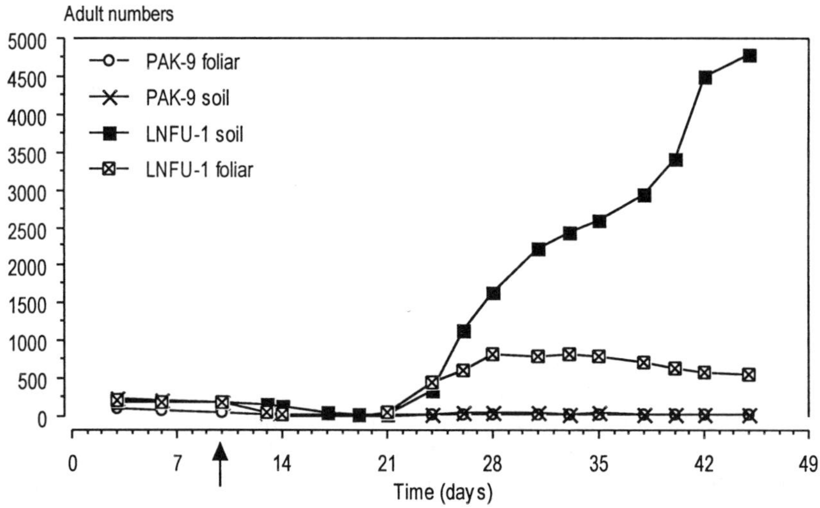

Fig. 2. Adult numbers for imidacloprid-susceptible (PAK-9) or -resistant (LNFU-1) *Bemisia tabaci* treated once (arrow) with either foliar or soil-applied imidacloprid during a 6-week large cage experiment

These data indicated that although resistance in these populations with this experimental method appeared to be expressed more strongly under systemic application, even resistant populations may be controlled by repeated foliar applications of imidacloprid. Within the cropping system of Almeria, in southern Spain where imidacloprid-resistant populations were collected (Cahill, et al. 1996), the economics of production allow growers to use insecticides at a frequency that would be uneconomic in many other cropping systems. This may in fact be masking the resistance episode.

Conversely, there may be practical integrated pest management (IPM) advantages to systemic applications, e.g., when bees are required for pollination or when biological control would be severely disrupted by a foliar application but not affected by a systemic. The example above, however, shows the potential for systemic applications to select for resistance, despite the high insecticide concentrations that can be achieved within the plant. The aim of insecticide applications is to achieve acceptable control, and therefore they should be used only when and for as long as necessary. When addressing the persistence debate, Roush (1989) emphasised the dangers of unnecessary persistence, i.e., continuing to select when control is no longer needed. The operational judgement should be to only use systemic applications when they are justified by a practical pest control requirement.

The concerns and challenges to resistance management posed by very persistent systemic insecticides are analogous to the debate surrounding the appropriate deployment of transgenic crops where it is not possible to impose simple operational tactics such as within-season rotations between modes of action. The practicalities of season-long, high expression of the *B. thuringiensis* toxin in transformed crops require an ecological approach to resistance management with emphasis on ensuring the presence on untreated refugia (Roush 1997). Similar principles may be appropriate for the chloronicotinyl insecticides, especially for relatively short season crops in which systemic applications of the insecticide have effectively season-long expression within the plant.

No one familiar with even the most basic aspects of resistance management would contemplate expressing the same *B. thuringiensis* toxin in a range of crops attacked by the same pest within a regional ecosystem for obvious and well-accepted reasons. Similarly, for the equally potent and valuable resource as the chloronicotinyls, we should consider whether blanket registration and use for all the hosts of a target pest within a region is appropriate. If more than one insecticide is available for the same pest then registration authorities, local growers, and scientists as well as the agrochemical representatives should consider restricting registration by matching the best characteristics of the insecticide with the pest control requirements of the cropping system.

5. Practical Examples

It is clearly difficult to make blanket statements and recommendations for resistance management tactics appropriate for the diverse systems in which the chloronicotinyls will be utilised. To illustrate some principles, four examples of chloronicotinyl use are discussed and some alternatives and ideas presented.

5.1. Damson-Hop Aphid in the U.K.

Phorodon humuli is a major hop pest throughout the Paleoarctic region, both as a pest in its own right and as a vector of hop mosaic virus. During the summer *P. humuli* is virtually restricted to wild and cultivated hops with by far the largest proportion of the population on the latter, which is heavily treated with insecticides. The combination of high pesticide use on almost the entire gene pool has selected for resistance to almost all available aphicides (Hrdy et al. 1986; Lewis and Madge 1984). During the winter, aphids retreat to their primary host (*Prunus* spp.) and then recolonise hops the following spring. This annual redistribution has the potential to enhance the spread of insecticide-resistant aphids.

Hop production in the U.K. is confined to two major growing areas, Worcestershire and Kent. Currently 100% of mature hop-bines are treated at least once with imidacloprid to achieve season-long control (J. Blackman, personal communication). In view of the biology of *P. humuli*, the intensity of imidacloprid use, and the history of resistance development, this cannot be considered sustainable. Monitoring methods and baseline responses of European *P. humuli* response to imidacloprid have been established at IACR-Rothamsted and show that existing resistance mechanisms do not affect this insecticide.

Without reforms to the current practises, however, these data will merely provide the means to document another resistance episode and the loss of a valuable pest management option.

Unfortunately there are very few effective alternatives to imidacloprid for incorporation into a resistance management strategy in U.K. hops. As and when other insecticides are introduced into the hop production system, there will be opportunities to alternate between at least two active ingredients. It will be essential to deploy these in a manner (e.g., annual or regional rotation) that will avoid continuous reliance on just one class.

5.2. Colorado Potato Beetle in Northeastern U.S.A.

In parts of northeastern U.S.A., *L. decemlineata* it is the most destructive pest of potatoes and has developed resistance to virtually all the pesticides so far used for control (Roush and Tingey 1991). Its ecology is well understood; adults emerging from overwintering pupae recolonise newly planted fields in the spring and undergo two to three generations on potatoes, after which adults retreat for the winter. The most important nonchemical control practice is crop rotation, which delays colonisation, reduces immigrant density, and may decrease the number of summer generations (Roush et al. 1990). There are few alternative hosts and, as for damson hop aphid, by far the largest proportion of the population within an area is confined to potatoes.

Imidacloprid has been registered on potatoes since 1995 as a granular formulation for use at planting and as a foliar spray. The lack of cost-effective alternatives meant that within two years, 90% of the potatoes in Michigan were treated with one or other formulation, and already there are indications of a low frequency of tolerant individuals (Grafius and Bishop 1996). Transgenic NewLeaf potatoes expressing the Cry3A protein of *Bacillus thuringiensis* are also available and offer an excellent alternative control option for Colorado potato beetle. The concurrent availability of these two novel tools for Colorado potato beetle control should decrease the risk of resistance to both, especially when combined with other practices.

Areawide management tactics are currently under evaluation in the Northeast (G. Dively, 1997, personal communication). These include good general agronomic and pest control practises, monitoring for resistance to important insecticides, ranking fields for risk of Colorado potato beetle infestation, and tailoring control options to suit. The recommendations for 'high-risk' fields include rotating transgenic and nontransgenic imidacloprid-treated crops, use of entomopathogens for Colorado potato beetle control, and manipulating crop cutout to force Colorado potato beetle survivors into refuge areas prior to mating. In 'low-risk' fields, it is proposed to treat only crop borders with granular imidacloprid rather than whole fields and in all cases to avoid using foliar application of imidacloprid on a systemically treated crop. The important principles are to reduce the need to treat and the selection pressure on both the chloronicotinyl and the *B. thuringiensis* toxin.

These tactics exploit knowledge of the pest ecology with the appropriate use of more than one control option. Importantly the strategy is based on areawide management and is underpinned by extension and ongoing monitoring for any changes in resistance levels to either the insecticide or the NewLeaf potato so that the impact of the tactics can be evaluated and modified accordingly.

5.3. *Bemisia tabaci* in Arizona

Since the late 1980s, B-type *B. tabaci* has been a serious threat to the production of cotton, melons, vegetables, and ornamental crops in the desert southwest of the U.S.A. Insecticides are still the dominant whitefly control option, but as in many parts of the world, *Bemisia* populations in have rapidly developed resistance to practically all the insecticides registered for use (Dennehy and Williams 1997). Two contrasting agro-ecosystems within Arizona serve to illustrate some points.

In Maricopa County, central Arizona, there are approximately 225,000 acres of whitefly hosts, 65% of which is cotton, 25% alfalfa, and 10% vegetables and melons. In Yuma County, western Arizona, the total area of whitefly hosts is 125,000 acres, 25% of which is cotton, 25% alfalfa, and the remaining 50% vegetables and melons. Pesticide use is negligible in alfalfa, while the primary insecticide for whitefly control in melons and vegetables is imidacloprid as a granular formulation applied in the soil. For the 1996 cotton season, a whitefly resistance management strategy was implemented that relied on strictly controlled early-season use of the insect growth regulators (IGRs) buprofezin and pyriproxyfen and, when necessary, late-season use of other chemical classes including OP/pyrethroid combinations. The strategy has been designed to make full use of the available diversity of chemistry, while restricting the use of all insecticides. Since imidacloprid has proved more effective against whiteflies on vegetables than on cotton, there has been a *de facto* separation of insecticide use between the chloronicotinyls on vegetables and the IGRs on cotton.

The combination of differences in cropping systems and preferred pesticide use means that the amount of imidacloprid used in Yuma County is more than 10 times that in Maricopa, while conversely IGR use in Yuma is much less (Williams, et al. 1996). Importantly, there is obligatory whitefly migration from cotton to melons and vegetables in the autumn and from melons and vegetables to cotton in the late spring. The between-crop pesticide use has created a spatial and temporal insecticide rotation while the significant area of untreated alfalfa provides a source of unselected individuals. This strategy therefore satisfies a number of requirements for successful resistance management, namely, high population dispersal and mating, a significant proportion of the population untreated with either one or other of the important insecticides, alternation of unrelated insecticide classes based on the most appropriate insecticides for each crop, and ongoing monitoring to evaluate the strategy and provide early warning of any increase in the frequency of resistance.

The greatest dangers to the sustainability of this strategy may come from the introduction of another chloronicotinyl into cotton for either whitefly control or for management of another pest if that coincided with significant whitefly populations. *L. hesperus* is the primary early season pest of cotton in Arizona and management is presently also difficult. Foliar-applied chloronicotinyls provide excellent control of lygus bugs and may therefore precipitate a major increase in chloronicotinyl use in cotton. Alternatively, any increase in IGR use in cotton by recommending more than one application of each per season, or any significant increase in IGR use in melons and vegetables, may have an impact on the longevity of these insecticides. This would not only be of concern specifically, but as a consequence would have an impact on chloronicotinyl use. The key to the strategy is maintaining a diversity of chemistry and to exploit it to best effect.

5.4. *Bemisia tabaci* in Southern Spain

The agricultural production system in the region near Almeria in southern Spain comprises some 20,000 ha of polythene-covered greenhouses primarily growing tomatoes, melons, peppers, and some other vegetables (Moreno et al. 1994). It is an intensive industry with year-round, high-value, export-driven production. Growers rely extensively on insecticides to control the pest complex, which includes both *Trialeurodes vaporariorum* (Westwood) (Homoptera: Aleyrodidae) and *B. tabaci*. Whitefly pressures are high throughout the year and the continuous cropping provides an ideal environment for generating large populations. Resistance to OPs, pyrethroids, carbamates, and endosulfan is widespread and buprofezin resistance has also been documented from the area (Cahill et al. 1996). There have so far been no significant attempts at insecticide resistance management and the pressure for IPM is not yet intense enough to alter the practices of the majority of growers.

Nurseries that supply propagation material are required to provide plants entirely free of pests and diseases. As a result, insecticide use in the nursery greenhouses is excessively high with reports of twice-weekly foliar applications of imidacloprid and other insecticides as well as fungicides for disease control. Even though the nurseries themselves may not have any insect survivors of such a regime, the consequences are that the synchronised planting of both the summer and winter crops, particularly of tomatoes and peppers, will have almost without exception been exposed as small plants to imidacloprid. Therefore, a large proportion of the 20,000 ha are planted twice per year with plants whose rootball has been soaked with imidacloprid from the runoff of nursery treatment and that will therefore be selecting an equally large proportion of the resident whitefly population within a brief window at the beginning of each season. The implications for resistance selection under these conditions are substantial.

Additionally, tomatoes are vulnerable to the endemic tomato yellow leaf curl virus (TYLCV), which dramatically lowers the action thresholds for *B. tabaci* and consequently increases the pesticide use. Many growers routinely apply one or more pesticides per week for whitefly control, and a significant number of these are imidacloprid, either in the drip irrigation system or more frequently as a foliar spray. The combination of greenhouse agriculture, with the attendant closed populations plus synchronised early-season imidacloprid use and repeated foliar applications of the same insecticide, creates a very high resistance risk. It was from this region that field-selected resistance to imidacloprid was first reported (Cahill, et al. 1996).

Imidacloprid is registered for use on all the whitefly hosts within the region and the continuity of hosts, large whitefly populations, and lack of alternatives are the reasons for over reliance on this insecticide. However, tomatoes have the lowest action thresholds for *B. tabaci* and arguably the greatest need for regular early-season protection to reduce TYLCV transmission. If whitefly insecticides other than the chloronicotinyls were to be registered or incorporated into a regional resistance management strategy, e.g., pyriproxyfen, buprofezin, and pymetrozine, then these could be confined to use on hosts other than tomatoes and imidacloprid and other chloronicotinyls retained for that crop. This would provide an ecological separation of pesticide use, similar to that in Arizona, which would significantly reduce the selection pressure on each insecticide class.

6. Stewardship of the Chloronicotinyls

Stewardship of the chloronicotinyls, as for all insecticides, is a collective responsibility. While public sector researchers and advisors are typically responsible for advice and extension to growers, managing a class of insecticide will prove impossible without the full and concerted support from the manufacturers and distributors. In our experience, technical experts in the major multinational agrochemical companies have a clear understanding of the need for and principles of resistance management, and indicate a genuine willingness to participate (e.g., in the industry based Insecticide Resistance Action Committee, IRAC) to establish and promote guidelines. Implementing these recommendations "on the ground", however, is often entirely different.

Where, for example, the responsibility for insecticide distribution and sales rests with a company that has not been part of the long gestation of the product, then the motivation for product sales may not include the long-term view of retaining susceptibility. Alternatively, when distribution centres are independent franchises supplied by the manufacturer, then the sole business motive at the distribution level is product sales. These marketing strategies make the implementation of resistance management guidelines extremely difficult. Recommendations for restricting use that have been determined by the manufacturer at company headquarters may be sound and worthwhile but irrelevant if there is either no mechanism for implementation or a clear conflict of interest by the most influential link in the chain – the pesticide salesman.

An alternative to agrochemical industry self-regulation is public sector regulation, probably via the registration process. Forthcoming European Union legislation to determine and combat resistance risks for pesticides undergoing registration and to legislate accordingly (Rotteveel et al. 1997) is not entirely favoured by the agrochemical producers who propose that the crop protection industry through IRAC has the motivation, means, and commitment to address the problem (Leonard 1997). However, neither industry nor legislators are likely to succeed in isolation, and IRAC readily acknowledges that collaboration is the key to successful resistance management (Leonard 1997).

A partnership between those with the resources (the chemical industry), those with the powers to enforce (legislators), and those with the knowledge, skills, and more importantly the independence (researchers and growers) may provide a workable solution. The best examples of successful resistance management strategies to date are those in which the lead has been taken by an individual or small group of public sector scientists over a cohesive agricultural industry with either full co-operation or legislation of the chemical industry at the local or national level. This paradigm provides clear focus, local knowledge, obvious independence and lack of vested interests together with a long-term view consistent with insecticide resistance management.

7. Conclusions

The importance of conserving the chloronicotinyls by preventing loss of efficacy through resistance cannot be overstated. The nicotinic acetylcholine receptor has so far been under-exploited and the chloronicotinyls to date have favourable environmental profiles. These attributes, coupled with extensive theoretical and practical knowledge of the principles of resistance management provide the reasons and the knowledge for sustainable management of these pesticides. Such sentiments have been expressed many times before (Herve 1985) but with apparently limited impact. Already populations of chloronicotinyl target pests with significant proportions of resistant individuals have been detected. The threat to field performance and probable cross-resistance among the class implies an urgency of response, particularly as momentum gathers for the release of new chloronicotinyls.

For each region and pest complex it is critical to develop clear, consistent and rational guidelines for practical, cost-effective Resistance Management Strategies as a component of Integrated Pest Management. These guidelines should be based on a solid understanding of the pest ecology, exploiting the operational attributes of as wide a diversity of insecticides as possible and underpinned by ongoing research including systematic resistance monitoring. Such goals can be achieved through a partnership between the agrochemical industry, growers, and scientists. The greatest challenge is not to design strategies but to implement them, and failure to do so will expose the boldest statements as mere rhetoric.

Acknowledgment

The authors thank especially Drs Tim Dennehy, Galen Dively, Alfred Elbert, Rami Horowitz, Isaac Ishaaya, Ralf Nauen and Rick Roush for stimulating and productive discussions on this subject as well as other colleagues too numerous to mention. IACR-Rothamsted receives grant aided support from the Biotechnology and Biological Sciences Research Council of the United Kingdom.

References

Bardt D, Schein E (1996) Drug resistant flea populations using the example of the "Cottontail" strain. Kleintierpraxis 41:561-566

Boiteau G, Osborn WPL, Drew ME (1997) Residual activity of imidacloprid controlling Colorado potato beetle (Coleoptera: Chrysomelidae) and three species of potato colonizing aphids (Homoptera: Aphidae). J Econ Entomol 90:309-319

Cahill M, Gorman K, Day S, Denholm I, Elbert A, Nauen R (1996) Baseline determination and detection of resistance to imidacloprid in *Bemisia tabaci* (Homoptera: Aleyrodidae). Bull Entomol Res 86:343-349

Cahill M, Jarvis W, Gorman K, Denholm I (1996) Resolution of baseline responses and documentation of resistance to buprofezin in *Bemisia tabaci* (Homoptera: Aleyrodidae). Bull Entomol Res 86:117-122

Cary LR (1903) Plant-house Aleyrodes. Nineteenth Annual Report of the Maine Agricultural Experiment Station, Augusta

Chen ZH, Liu CX, Li FL, Han ZJ (1993) Development of diamond-back moth strains resistant to dimehypo and cartap with reference to the mechanism of resistance. Acta Entomol Sin 36:409-418

Denholm I, Rowland MW (1992) Tactics for managing pesticide resistance in arthropods: theory and practice. Ann Rev Entomol 37:91-112

Denholm I, Farnham AW, O' Dell K, Sawicki RM (1983) Factors affecting resistance to insecticides in house-flies, *Musca domestica* L. (Diptera: Muscidae). I. Long-term control with bioresmethrin of flies with strong pyrethroid-resistance potential. Bull Entomol Res 73:481-489

Denholm I, Horowitz AR, Cahill M, Ishaaya I (1998) Management of resistance to novel insecticides. In: Ishaaya I, Degheele D (Eds) Insecticides with novel modes of action. Springer-Verlag, Heidelberg,

Dennehy TJ, Russell JS (1996) Susceptibility of lygus bug populations in Arizona to acephate (Orthene) and bifenthrin (Capture) with related contrasts of other insecticides. In: Proceedings Beltwide Cotton Conferences, Nashville, TN, USA, Vol 2, pp771-776

Dennehy TJ, Williams III L (1997) Management of resistance in *Bemisia* in Arizona cotton. Pestic Sci 51:398-406

Devine GJ, Harling ZK, Scarr AW, Devonshire AL (1996) Lethal and sublethal effects of imidacloprid on nicotine-tolerant *Myzus nicotianae* and *Myzus persicae*. Pestic Sci 48:57-62

Dewar AM, Read LA (1990) Evaluation of an insecticidal seed treatment, imidacloprid, for controlling aphids on sugar beet. In: Proceedings Brighton Crop Protection Conference, Pests and Diseases, Brighton, England, Vol 2, pp721-726

Dhingra S (1994) Development of resistance in the bean aphid, *Aphis craccivora* Koch. to various insecticides used for nearly a quarter century. J Entomol Res 18:105-108

Elbert A, Nauen R, Cahill M, Devonshire AL, Scarr AW, Sone S, Steffens R (1996) Resistance management with chloronicotinyl insecticides using imidacloprid as an example. Pflanz Nachr Bayer (Engl Ed) 49:5-54

Elbert A, Overbeck H, Iwaya K, Tsuboi S (1990) Imidacloprid, a novel systemic nitromethylene analogue insecticide for crop protection. In: Proceedings Brighton Crop Protection Conference, Pests and Diseases, Brighton, England, Vol 1, pp21-28

ffrench-Constant RH, Roush RT, Mortlock D, Dively GP (1990) Isolation of dieldrin resistance from field populations of *Drosophila melanogaster* (Diptera: Drosophilidae). J Econ Entomol 83:1733-1737

Forgash AJ (1985) Insecticide resistance in the Colorado potato beetle. Research Bulletin, Massachusetts Agricultural Experiment Station, pp33-52

Georghiou GP (1994) Principles of insecticide resistance management. Phytoprotection 75:51-59

Gould F, Anderson A, Jones A, Sumerford D, Heckel DG, Lopez J, Micinski S, Leonard R, Laster M (1997) Initial frequency of alleles for resistance to *Bacillus thuringiensis* toxins in field populations of *Heliothis virescens*. Proc Natl Acad Sci USA 94:3519-3523

Grafius EJ, Bishop BA (1996) Resistance to imidacloprid in Colorado potato beetles from Michigan. Resist Pest Manage Newsl 8:21-25

Hama H (1986) Resistance spectrum to various insecticides in the diamondback moth, *Plutella xylostella* Linne (Lepidoptera: Yponomeutidae). Jpn J Appl Entomol Zool 30:277-284

Herve JJ (1985) Agricultural, public health and animal health usage. In: Leahey JP (Ed) The pyrethroid insecticides. Taylor and Francis, London, pp343-425

Hrdy I, Kremheller HT, Kuldova J, Luders W, Sula J (1986) Insecticide resistance of the hop aphid *Phorodon humuli* in the Bohemian, Bavarian and Baden Wurttemberg hop-growing areas. Acta Entomol Bohemoslov 83:1-9

Ishaaya I, Horowitz AR (1998) Insecticides with novel modes of action, an overview. In: Ishaaya I, Degheele D (Eds) Insecticides with novel modes of action, Mechanism and Application. Springer-Verlag, Heidelberg, pp11-24

Keiding J, Jesperson JB (1986) Effect of different control strategies on the development of insecticide resistance by houseflies: experience from Danish farms. Proceedings Brighton Crop Protection Conference, Pests and Diseases, Brighton, England, Vol 2, pp607-612

Leicht W (1996) Imidacloprid - a chloronicotinyl insecticide, biological activity and agricultural significance. Pflanz Nachr Bayer (Engl Ed) 49:71-84

Leonard PK (1997) There's never been a better time or a greater need for resistance management. Pestic Sci 51:387-390

Lewis GA, Madge DS (1984) Esterase activity and associated insecticide resistance in the damson-hop aphid, *Phorodon humuli* (Schrank) (Hemiptera: Aphididae). Bull Entomol Res 74:227-238

Mani GS, Wood RJ (1984) Persistence and frequency of application of an insecticide in relation to the rate of evolution of resistance. Pestic Sci 15:325-336

McKenzie JA (1996) Ecological and evolutionary aspects of insecticide resistance. Landes, Austin

Melander AL (1914) Can insects become resistant to sprays? J Econ Entomol 7:167-173

Meredith RH, Heatherington PJ (1992) Aphid control in potatoes from imidacloprid, a new systemic insecticide for application to seed tubers or in-furrow at planting. In: Proceedings Brighton Crop Protection Conference, Pests and Diseases, Brighton, England, Vol 2, pp551-556

Moreno R, Garijo C, Rodriguez MD, Garcia S, Tellez MM, Garcia E, Rodriguez MP (1994) IPM development on tomato under plastic in the south of Spain. Bull IOBC/WPRS 17:21-26

Nauen R, Elbert A (1997) Apparent tolerance of a field-collected strain of *Myzus nicotianae* to imidacloprid due to strong antifeeding responses. Pestic Sci 49:252-258

Nauen R, Lawson B, Elbert A (1997) Resistance status of some field strains of the tobacco aphid *Myzus nicotianae* from tobacco in Greece. In: Proceedings, Resistance '97, IACR-Rothamsted, Harpenden

Nauen R, Strobel J, Tietjen K, Otsu Y, Erdelen C, Elbert A (1996) Aphicidal activity of imidacloprid against a tobacco feeding strain of *Myzus persicae* (Homoptera: Aphididae) from Japan closely related to *Myzus nicotianae* and highly resistant to carbamates and organophosphates. Bull Entomol Res 86:165-171

Olson ER, Dively GP, Nelson JO (1996) Survey of susceptibility to imidacloprid (Admire) in Colorado potato beetle (Coleoptera: Chrysomelidae). Resist Pest Manage Newsl 8:39-41

Palumbo JC, Kerns DL (1994) Effects of imidacloprid as a soil treatment on colonization of green peach aphid and marketability of lettuce. Southwest Entomol 19:339-346

Pap L, Toth A, Karikas S (1997) A survey of the insecticide resistance status of the Colorado potato beetle, *Leptinotarsa decemlineata*, in Hungary between 1987 and 1991. Pestic Sci 49:389-392

Prabhaker N, Toscano NC, Castle SJ, Henneberry TJ (1997) Selection for imidacloprid resistance in silverleaf whiteflies from the Imperial Valley and development of a hydroponic bioassay for resistance monitoring. Pestic Sci 51:419-428

Rotteveel TJW, de Goeij JWFW, van Gemerden AF (1997) Towards the construction of a resistance risk evaluation scheme. Pestic Sci 51:407-411

Roush RT (1989) Designing resistance management programs: how can you choose? Pestic Sci 26:423-441

Roush RT (1997) Bt-transgenic crops: just another pretty insecticide or a chance for a new start in resistance management? Pestic Sci 51:328-334

Roush RT, Hoy CW, Ferro DN, Tingey WM (1990) Insecticide resistance in the Colorado potato beetle (Coleoptera: Chrysomelidae): influence of crop rotation and insecticide use. J Econ Entomol 83:315-319

Roush RT, Tingey WM (1991) Evolution and management of resistance in Colorado potato beetle. In: Denholm I, Devonshire AL, Holloman DW (Eds) Resistance '91: achievements and developments in combating pesticide resistance. Elsevier, pp61-74

Sakai M (1985) Chemical control of diamondback moth in Japan with special reference to cartap. In: Diamondback Moth Management. Proceedings of the First International Workshop, Tainan, Taiwan

Sawicki RM (1986) Resistance to synthetic pyrethroids can be countered successfully. Agribusiness Worldwide 8:22-25

Sone S, Hattori Y, Tsuboi S, Otsu Y (1995) Difference in susceptibility to imidacloprid of the populations of the small brown planthopper, *Laodelphax striatellus* Fallen, from various localities in Japan. J Pestic Sci 20:541-543

Tomizawa M, Otsuka H, Miyamoto T, Eldefrawi ME, Yamamoto I (1995) Pharmacological characteristics of insect nicotinic acetylcholine-receptor with its ion-channel and the comparison of the effect of nicotinoids and neonicotinoids. J Pestic Sci 20:57-64

Wen Z, Scott JG (1997) Cross-resistance to imidacloprid in strains of German cockroach (*Blattella germanica*) and house fly (*Musca domestica*). Pestic Sci 49:367-371

Williams III L, Dennehy TJ, Palumbo JC (1996) Whitefly control in Arizona: development of a resistance management program for imidacloprid. In: Proceedings Beltwide Cotton Conferences, Nashville, TN, USA, Vol 2, pp752-758

Williams III L.; Dennehy TJ, Palumbo, JC (1998) Can resistance to chloronicotinyl insecticides be averted in Arizona field crops? In: Proceedings Beltwide Cotton Conferences, Memphis, TN, USA, Vol 2, pp1250-1253

Zhao G, Liu W, Brown JM, Knowles CO (1995) Insecticide resistance in field and laboratory strains of Western flower thrips (Thysanoptera: Thripidae). J Econ Entomol 88:1164-1170

13 Structure and Function of Insect Nicotinic Acetylcholine Receptors Studied with Nicotinoid Insecticide Affinity Probes

Motohiro Tomizawa, Bachir Latli, and John E. Casida
Environmental Chemistry and Toxicology Laboratory, Department of Environmental Science, Policy and Management, University of California, Berkeley, CA 94720-3112, USA

1 Introduction

The insect nicotinic acetylcholine (ACh) receptor (nAChR) is the target not only for the botanical insecticide nicotine but also for the synthetic nicotinoid insecticides such as imidacloprid (IMI) (Bai et al. 1991; Tomizawa and Yamamoto 1992, 1993; Liu and Casida 1993; Liu et al. 1993, 1995; Tomizawa et al. 1995a; Matsuo et al. 1998). Knowledge of the structure-activity relationship (SAR) of nicotinoids contributes to an understanding of the functional architecture of the nicotinic acetylcholine receptor (nAChR), a generic interrelationship of SAR and target site research applicable to any agrochemical or pharmaceutical. This approach has helped us to understand the action and selectivity of the new and increasingly important synthetic nicotinoid insecticides useful in controlling important pests and in resistance management programs (Casida and Quistad 1998).

Insect nAChR research was facilitated greatly by the discovery of the IMI-type insecticides (Kagabu et al. 1992; Moriya et al. 1992; Takahashi et al. 1992; Minamida et al. 1993), thereby creating both the need and the means to explore their mode of action. α-Bungarotoxin (α-BGT), a classical competitive antagonist from the Elapidae snake *Bungarus multicinctus* (Mebs et al. 1971), is a major radiolabeled probe for the nAChRs from insects and mammals. Although α-BGT shares the same insect nAChR target with nicotine and IMI (Tomizawa and Yamamoto 1992; Liu and Casida 1993), it is not suitable, as an 8-kDa polypeptide, to investigate the target site for an 0.2-kDa nonionized insecticide. The only way to fully understand the insecticide binding site is to use the nicotinoid itself as the radioligand. This review therefore begins with the synthesis of radiolabeled ligands such as [^{3}H]IMI for use as direct probes of the insecticide binding site. The preparation and use of affinity probes is then considered for isolation and identification of the target molecule. Finally, the knowledge gained from their use is combined with that from other approaches in considering the structure and function of the insect nAChRs.

2 Chemistry of Nicotinoid Insecticide Affinity Probes

2.1 Radiolabeled Ligands

The nicotinoid insecticides were radiolabeled with tritium at high specific activity (Fig. 1) to study their specific binding site, mode of action, and selective toxicity. Four 6-chloro-3-pyridinyl derivatives were selected for labeling, i.e., [^{3}H]IMI, its desnitro metabolite [^{3}H]DN-IMI, and the highly insecticidal nitromethylene and cyanoacetamidine analogs, [^{3}H]CH-IMI and acetamiprid [^{3}H]AAP, respectively (Latli and Casida 1992; Latli et al. 1996). Also labeled was a highly potent chlorothiazolyl analog of CH-IMI, [^{3}H]Cl-TMNI (Liu et al. 1994). Particularly important in these syntheses was the availability of tritium-enriched reducing agents at the National Tritium Labelling Facility of the Lawrence Berkeley National Laboratory in Berkeley, California. Sodium borotritide has been usually used to reduce aldehydes to tritium-labeled alcohols with specific activities never exceeding 25% of that of the $NaBH_4$. Syntheses were therefore developed from 6-chloronicotinoyl chloride (**1**) to incorporate a maximum of two tritons (50% of the

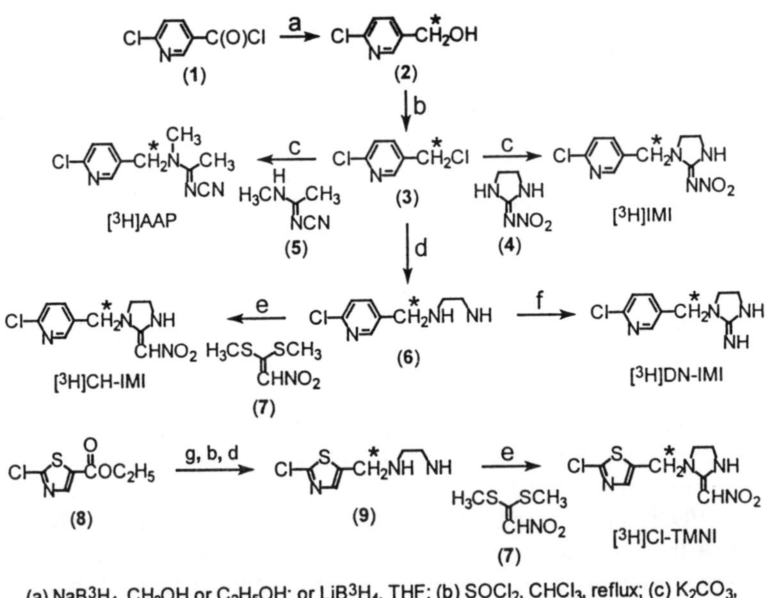

(a) NaB^3H_4, CH_3OH or C_2H_5OH; or LiB^3H_4, THF; (b) $SOCl_2$, $CHCl_3$, reflux; (c) K_2CO_3, CH_3CN, reflux; (d) $H_2N(CH_2)_2NH_2$, CH_3CN, NaOH; (e) C_2H_5OH, reflux; (f) CNBr, $C_6H_5CH_3$; (g) $LiAl^3H_4$, THF.

Fig. 1. Radiosynthesis of [^{3}H]imidacloprid ([^{3}H]IMI, 25 Ci/mmol), its desnitro metabolite ([^{3}H]DN-IMI, 55 Ci/mmol), and three insecticidal analogues ([^{3}H]CH-IMI (55 Ci/mmol), [^{3}H]Cl-TMNI (55 Ci/mmol), and [^{3}H]acetamiprid ([^{3}H]AAP, 45 Ci/mmol).

specific activity of the reducing agent) on reduction with either NaB^3H_4 or LiB^3H_4 (prepared freshly and with up to 97% isotopic enrichment) in methanol, ethanol, or tetrahydrofuran (THF) at room temperature. The labeled alcohol (**2**) thus produced was then refluxed with thionyl chloride in chloroform to give labeled 6-chloro-3-chloromethylpyridine ([^{3}H]**3**) as the key intermediate. To prepare [^{3}H]IMI, [^{3}H]**3** was coupled with 4,5-dihydro-*N*-nitro-1*H*-imidazol-2-amine (**4**) in acetonitrile in the presence of potassium carbonate. [^{3}H]AAP was obtained by refluxing an acetonitrile solution of [^{3}H]**3** with *N*-cyano-*N*-methylacetamidine (**5**) as above. Finally, [^{3}H]**3** was reacted with ethylenediamine in acetonitrile and an aqueous solution of sodium hydroxide to furnish *N*-(6-chloro-3-pyridinylmethyl)-ethylenediamine (**6**). This compound was then refluxed with 1,1-bis(methylthio)-2-nitroethylene (**7**) in ethanol to provide [^{3}H]CH-IMI or reacted in toluene with a solution of cyanogen bromide to give [^{3}H]DN-IMI. [^{3}H]Cl-TMNI was made by reducing the arylcarboxylate (**8**) with freshly prepared $LiAl^3H_4$ (>95% enrichment) to the alcohol, which was converted, as indicated above, to tritium-labeled 2-chloro-5-chloromethylthiazole for coupling to ethylenediamine to give [^{3}H]**9**, which was finally coupled with **7** to obtain the desired product.

2.2 Affinity Chromatography

The SAR and high affinity of synthetic nicotinoids made it possible to develop a novel nicotinoid-agarose (nic-agar) matrix for affinity chromatography isolation of insect nAChRs (Fig. 2). High affinity for the [^{3}H]IMI binding site of fruit fly

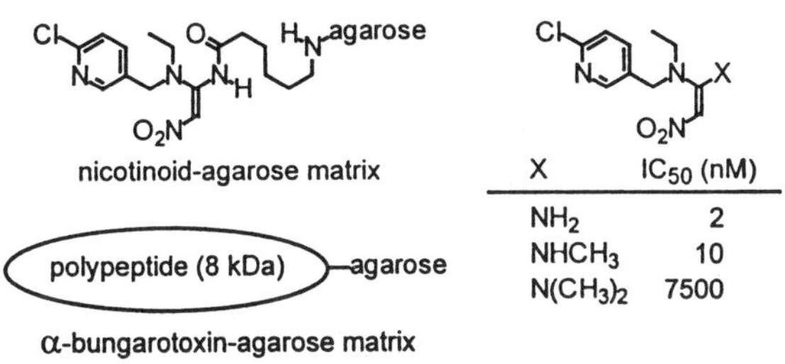

Fig. 2. Affinity chromatography matrices of nicotinoid and α-BGT with agarose for isolation of insect nAChRs. Development of the nic-agar system was based on the indicated SAR for the nitenpyram series as inhibitors of [^{3}H]IMI binding averaged for *Drosophila* and *Musca* head membranes (Tomizawa et al. 1996).

(*Drosophila melanogaster*) and housefly (*Musca domestica*) nAChRs in the nitenpyram (NTP) series (Minamida et al. 1993) is conferred by $-NH_2$ and $-NHCH_3$ moieties, whereas the potency diminishes remarkably with the $-N(CH_3)_2$ analogue (Tomizawa et al. 1996). A similar relationship is seen for IMI and *N*-methyl-IMI as inhibitors for binding of [^{3}H]IMI and [^{3}H]α-BGT (Liu et al. 1993; Tomizawa and Yamamoto 1993). These observations led to coupling the $-NH_2$

NTP analogue with the aminohexanoic acid derivative of agarose using 1-ethyl-3-(3-dimethylaminopropyl)carbodiimide as the coupling agent (Tomizawa et al. 1996). This chromatography matrix with the 7-atom spacer arm distance from an active NTP moiety proved to be very successful in isolation of insect nAChRs, with advantages over the α-BGT-agarose (α-BGT-agar) matrix discussed later.

2.3 Photoaffinity Ligand

Photoaffinity labeling is a biochemical approach to access and characterize the ligand-binding site in the receptor, i.e., the structural and functional features of the target protein. In principle, the photoaffinity radioligand is specifically bound to the receptor site, then a covalent bond is formed by photochemical reaction, i.e., it must have high affinity and a suitable photoactivatable substituent such as an aryl azide. The free radical generated by irradiation (light exposure) is then coupled with an amino acid residue in the specific target site (reviewed in Kotzyba-Hibert et al. 1995).

The hexahydronitroimidazopyrimidine analogues (Fig. 3) prepared by addition of formaldehyde and a primary amine to CH-IMI have potent insecticidal activity (Kishida et al. 1992). These compounds with a wide range of substituents inhibit [^{3}H]IMI binding to the *Drosophila* or *Musca* nAChRs by 50% at 0.7-24 nM, thereby providing a method to optimize a candidate photoaffinity probe retaining high affinity to the insect receptor (Latli et al. 1997). Ultimately, the 2-azido-5-[^{125}I]iodobenzoyl derivative designated as [^{125}I]azidonicotinoid or [^{125}I]AN was prepared via the iododestannylation of its tin precursor (IC_{50} of 38 nM in *Drosophila* nAChR) using Na^{125}I and chloramine-T. The optimized photoaffinity probe assayed as [^{127}I]AN (without radiolabeling) displays an IC_{50} of 8 or 25 nM for [^{3}H]IMI binding in the *Drosophila* or *Musca* nAChR, respectively.

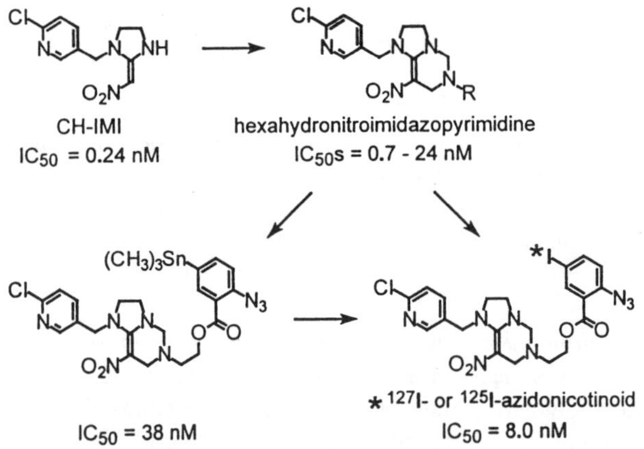

Fig. 3. Photoaffinity ligand [^{125}I]azidonicotinoid development and synthesis. IC_{50} values are for [^{3}H]IMI binding to *Drosophila* or *Musca* nAChR (Latli et al. 1997).

3 Nicotinoid Insecticide Binding Site

3.1 Insects

Characterization of the target site for the nicotinoid insecticides was greatly facilitated by using [^{3}H]IMI as a probe. This radioligand displays saturable and reversible binding to *Musca* head membranes with >95% specific binding (Fig. 4). There is a single high-affinity binding component and fast kinetic constants determined by time-course experiments of association and dissociation (Liu and Casida 1993). [^{3}H]Cl-TMNI is similar to [^{3}H]IMI in binding properties with high affinity at the *Musca* site (Liu et al. 1994). [^{3}H]IMI also binds to membrane preparations of *Drosophila* head and whitefly (*Bemisia argentifolii*) whole body with a single high-affinity site in each case and K_D values of 2.4 and 2.0 nM, respectively (Tomizawa et al. 1996; Chao et al. 1997).

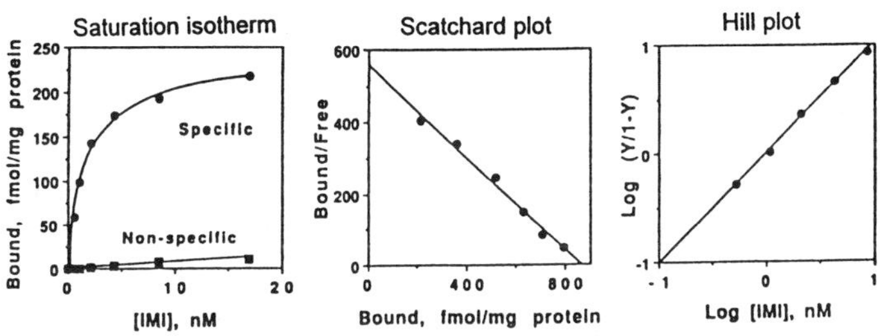

Fig. 4. Saturation isotherm and Scatchard and Hill plots for specific binding of [^{3}H]IMI to *Musca* head membranes. A single high-affinity binding site is observed: K_D, 1.2 nM; B_{max}, 853 fmol/mg protein; and n_H, 1.0 (Liu and Casida 1993).

Pharmacological profiles of the [^{3}H]IMI and [^{3}H]Cl-TMNI binding sites of *Musca* and *Bemisia* are compared with the [^{3}H]α-BGT binding site of honeybee (*Apis mellifera*) in Table 1. These binding characteristics and the ease of displacement by nicotinic ligands strongly indicate that the target site for IMI is the ACh-binding site of the nAChR in the central nervous system (CNS) of insects. Muscarinic ligands such as atropine and quinuclidinyl benzilate at high levels also displace [^{3}H]IMI binding, probably by acting directly on the nAChR rather than binding of [^{3}H]IMI to the muscarinic receptor, because very similar profiles are reported for American cockroach (*Periplaneta americana*) nAChR from electrophysiological studies (David and Sattelle 1984; Buckingham et al. 1997). IMI and α-BGT share the same binding site or region in the insect nAChR based on competitive studies of IMI with [^{3}H]α-BGT and α-BGT with [^{3}H]IMI (Tomizawa and Yamamoto 1992; Liu and Casida 1993).

Table 1. Pharmacological characterization of [^{3}H]IMI, [^{3}H]Cl-TMNI, and [^{3}H]α-BGT binding sites in several insect membrane preparations

	IC$_{50}$ (μM)			
	[^{3}H]IMI		[^{3}H]Cl-TMNI,	[^{3}H]α-BGT,
Ligand	Housefly[a]	Whitefly[b]	Housefly[c]	Honeybee[d]
Acetylcholine[e]	0.2	400	1.8	NT[f]
Carbachol	1.9	2.9	80	72
(-)-Nicotine	0.6	37	3.4	0.9
Imidacloprid	0.002	0.002	0.02	1.5
d-Tubocurarine	30	NT[f]	34	2.6
α-Bungarotoxin	2.2	1.3	4.8	0.007
Atropine	90	789	80	26
Quinuclidinyl benzilate	275	NT[f]	NT[f]	>300[g]

Data from [a]Liu and Casida 1993; [b]Chao et al. 1997; [c]Liu et al. 1994; [d]Tomizawa 1994.
[e]Assayed in the presence of 1 mM paraoxon to inhibit acetylcholinesterase.
[f]Not tested.
[g]Tomizawa M, 1998, unpublished.

The toxicological relevance of the target site requires validation by comparing the binding affinities of IMI analogues at the nAChR with their intrinsic insecticidal activity (Table 2; structures of the chemicals considered are given in Fig. 5). The potencies of the insecticidal analogues as inhibitors of [^{3}H]IMI binding to *Musca* head membranes are a good predictor of their intrinsic neurotoxicities measured as injected 50% knockdown activity (KD$_{50}$) against synergist-pretreated *Musca*; the calculated correlation coefficient for 22 analogues is $r = 0.90$ (Fig. 6). This relationship generally coincides with that measured in terms of the minimum lethal doses and the frequency of spontaneous discharges (excitation effects) in nerve cords conferred by the nicotinoid insecticides in *Periplaneta* (Nishimura et al. 1994). Clearly, [^{3}H]IMI is an excellent probe for characterizing the toxicologically relevant binding site in the insect nAChR for the synthetic nicotinoid insecticides.

No.	Ar	n	X-Y
1.	6-Cl-3-pyridinyl	1	N-NO$_2$
2.	6-Cl-3-pyridinyl	1	NH
3.	6-Cl-3-pyridinyl	1	CH-NO$_2$
4.	6-Cl-3-pyridinyl	2	CH-NO$_2$
5.	5-(2-Cl-thiazolyl)	1	CH-NO$_2$

No.	R	X-Y	Z
6.	CH$_3$	N-CN	CH$_3$
7.	C$_2$H$_5$	CH-NO$_2$	NHCH$_3$

Fig. 5. Structures of nicotinoid insecticide analogues including one metabolite and number designations used in Table 2.

Table 2. Toxicological characterization of nicotinoid insecticides in radioligand binding and injected toxicity

		Housefly[b]		Rodent		
No.	Nicotinoid[a]	[^{3}H]IMI binding (IC_{50}, nM)	Toxicity with PPP[c] (KD_{50}, μg/g)	[^{3}H]NIC binding (IC_{50}, μM)[d]	[^{125}I]α-BGT binding (IC_{50}, μM)	Toxicity rating[d,f]
1.	IMI	2.4	0.016	0.81	42	+
2.	DN-IMI	720	>5	0.015	2.6	+ + +
3.	CH-IMI	0.42	0.006	0.033	0.63	+ + + +
4.	THPCH-IMI	0.49	0.058	0.012	0.31	+ + + +
5.	Cl-TMNI	0.37	0.004	0.26	4.1	+ + + +
6.	AAP	3.2	0.095	0.70[e]	19	+ +
7.	NTP	5.1	NT[g]	27[e]	130	-
8.	(-)-Nicotine	600	NT[g]	0.009	1.9	+ + + +

[a]Chemical structures are given in Fig. 5, except for (-)-nicotine.
[b]Data from Liu and Casida 1993; Liu et al. 1993, 1995; Tomizawa et al. 1996.
[c]*O*-Propyl *O*-(2-propynyl) phenylphosphonate (PPP), a synergist (Casida 1970).
[d]Mouse (Chao and Casida 1997) and [e]rat brain membranes (Tomizawa M, Casida JE, 1998, unpublished).
[f]LD_{50} (ip, mg/kg) ranges: -, ≥50; +, 35-49; + +, 25-34; + + +, 16-24; + + + +, 7-15 (Chao and Casida 1997; Tomizawa M, Casida JE, 1998, unpublished).
[g]Not tested.

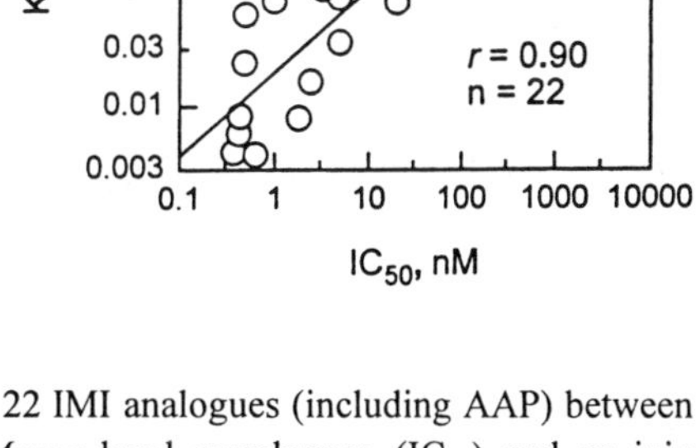

Fig. 6. Correlation for 22 IMI analogues (including AAP) between potency as inhibitors of [^{3}H]IMI binding to *Musca* head membranes (IC_{50}) and as injected knockdown agents (KD_{50}) against PPP-pretreated *Musca* (plotted with data from Liu et al. 1993, 1995).

3.2 Mammals

The very favorable selective toxicity of IMI, AAP, and NTP is apparently attributable in part to their lower affinity for the mammalian nAChR than that of sensitive insects (Liu and Casida 1993; Zwart et al. 1994; Tomizawa et al. 1995b; Yamamoto et al. 1995, 1998). The first direct evidence for the low sensitivity was the failure to recognize the [^{3}H]IMI binding site(s) in the brain from several mammalian and avian species and the electric organ of the electric eel (Liu and Casida 1993). These observations are verified for IMI by (1) the low potency as an inhibitor of [^{3}H]nicotine binding in rat brain and [^{3}H]α-BGT binding to the muscle-type nAChR from *Torpedo* (Tomizawa 1994; Tomizawa et al. 1995b; Yamamoto et al. 1995); (2) the poor activation (agonistic) potency compared to ACh with the rat α4β2 and α7 subtypes of neuronal nAChRs expressed in *Xenopus* oocytes (Yamamoto et al. 1998); and (3) the weak agonistic action in mouse N1E-115 neuroblastoma and BC3H1 muscle cells (Zwart et al. 1994).

Toxicological characterization of nicotinoid insecticides revealed that several IMI analogues and at least one of its metabolites are similar to nicotine in toxicity to mammals with high affinity to the neuronal type of mammalian nAChR (Table 2). In the [^{3}H]nicotine ([^{3}H]NIC)-binding portion of mouse or rat brain [which reflects the predominant subtype of α-BGT-insensitive nAChR (Sargent 1993; Lindstrom 1997)], DN-IMI, CH-IMI, and THPCH-IMI display high affinity comparable to that of nicotine, and there is a moderate affinity with Cl-TMNI but not with IMI, AAP, and NTP (Chao and Casida 1997; Tomizawa, M, Casida, JE, 1998, unpublished). The affinity of [^{3}H]NIC binding to mouse brain membranes is significantly altered by DN-IMI at 10 and 20 nM or CH-IMI at 15 and 30 nM assayed *in vitro*, or DN-IMI administered intraperitoneally (ip) at 10 and 100 mg/kg to the mouse before [^{3}H]NIC binding assay with the dissected brain *ex vivo* (Fig. 7). These observations suggest that the nicotinoid insecticides share

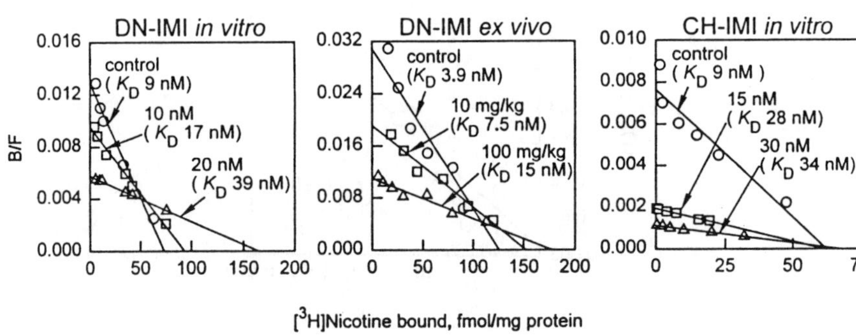

Fig. 7. Scatchard plots of specific [^{3}H]NIC binding to membranes from mouse brain (without cerebellum) and its inhibition by DN-IMI assayed *in vitro* and *ex vivo* (30 min after ip administration) and CH-IMI assayed *in vitro* (Chao and Casida 1997).

essentially the same site or mode of action as that for nicotine (Chao and Casida 1997). Furthermore, [^{3}H]DN-IMI and [^{3}H]CH-IMI, as novel radioligands, bind to mouse brain membranes each with a high affinity (K_D values, 13 and 16 nM, respectively) (Fig. 8) comparable to that of [^{3}H]NIC (Chao and Casida 1997).

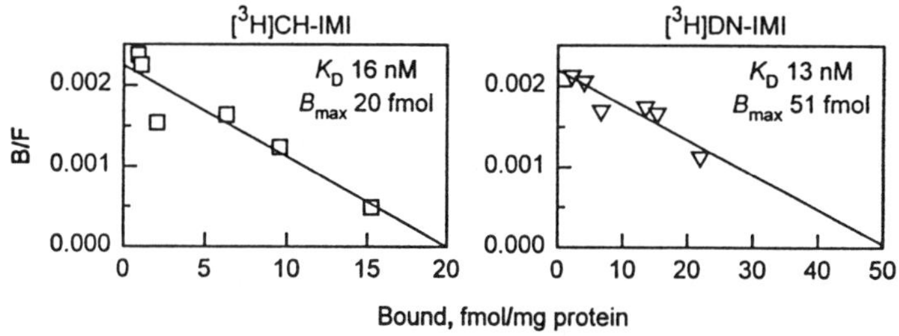

Fig. 8. Scatchard plots for specific radioligand binding of [^{3}H]CH-IMI and [^{3}H]DN-IMI to mouse brain (without cerebellum) membranes (Chao and Casida 1997).

Another prominent form of nAChR in mammalian brain is the α-BGT-sensitive subtype assayed with [^{125}I]α-BGT binding (Lindstrom 1997). The affinities of synthetic nicotinoids for this site are classified into three types: (1) CH-IMI and THPCH-IMI show greater potency than that of nicotine; (2) DN-IMI and Cl-TMNI are comparable in potency to nicotine; and (3) IMI, AAP, and NTP are less potent than nicotine (Tomizawa M, Casida JE, 1998, unpublished).

Interestingly, these potencies of nicotinoid insecticides derived from both types of radioligand-binding experiments generally coincide with those for toxicity to mice with ip administration. These studies clearly show that binding sites for the nicotinoid insecticides are present in the mammalian CNS. It is proposed that IMI is bioactivated on conversion to its metabolite DN-IMI with high affinity to mammalian nAChRs (Chao and Casida 1997). Further, it should be emphasized that differential nAChR subtype selectivity is conferred by only minor structural modification of the nicotinoid insecticides.

4 Affinity Chromatography of Insect Receptor

Advances in nicotinoid insecticide chemistry and SAR led to a new method to isolate nAChRs from the CNS of *Drosophila* and *Musca* in a single purification step from solubilized head membrane to relatively pure native receptor. Affinity chromatography of Triton X-100 detergent extracts of *Drosophila* or *Musca* head membranes on a nicotinoid-agarose matrix (see Fig. 2) and elution with IMI gives only three putative subunit proteins of the native nAChR with molecular masses corresponding to 61, 66, and 69 kDa. Moreover, the identical three putative subunits are isolated from the *Musca* preparation through the α-BGT-agarose matrix (Fig. 9). This is the first successful isolation of native nAChRs in the CNS of dipterous insects accomplished by two steps conferring high specificity; i.e., first, affinity chromatography on a nicotinoid agarose matrix to retain the nAChR, and second, specific displacement with an IMI-type elutrient with suitable physical properties (i.e., nonionized nature and moderate water solubility). Nicotinoid affinity chromatography therefore has distinct advantages over the α-BGT affinity matrix for isolation of native insect nAChRs, at least from dipterous insects (Tomizawa et al. 1996).

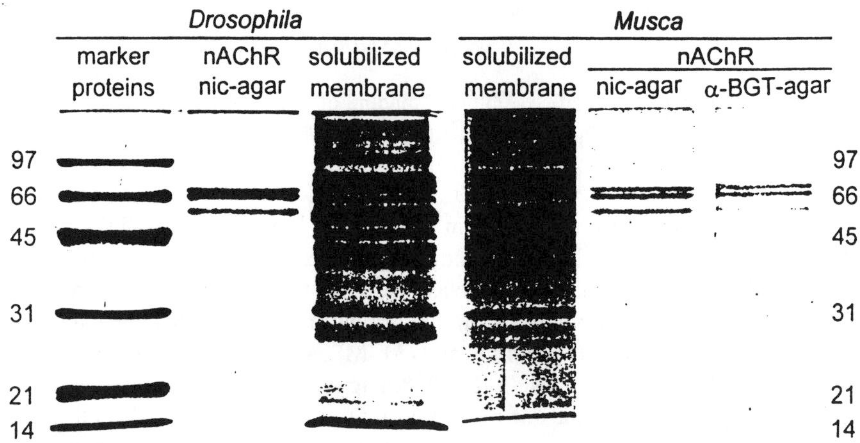

Fig. 9. Polyacrylamide gel electrophoresis of the native nAChR from *Drosophila* and *Musca* head membranes (61-, 66-, and 69-kDa subunits) isolated from detergent extracts of membrane preparations with the nicotinoid-agarose (nic-agar) and α-BGT-agarose (α-BGT-agar) affinity columns. Specific elutrients for the two columns are 50 μM IMI and 50 μM CH-IMI, respectively. Protein patterns are from three lithium dodecyl sulfate polyacrylamide gels: one for *Drosophila* solubilized membrane and nAChR subunits from the nicotinoid-agarose column with marker proteins; a second for *Musca* solubilized membrane and nAChR subunits from the nicotinoid column; and the third for *Musca* nAChR subunits from the α-BGT column, indicating by number the position of marker proteins. [This figure is reproduced from Tomizawa et al. (1996) with permission from Lippincott-Raven Publishers, Philadelphia, PA, USA.]

5 Photoaffinity Labeling of Insect Receptor

Photoaffinity labeling of the insect receptor requires high affinity and sensitive detection because of the small receptor abundance, prompting the use of α-BGT-based and nicotinoid-insecticide-type photoaffinity probes. The ligand-binding subunit of the isolated native *Drosophila* and *Musca* nAChRs is recognized with an azidosalicylate derivative of [^{125}I]α-BGT ([^{125}I]α-BGT-ASA) prepared by coupling [^{125}I]α-BGT with *N*-hydroxysuccinimidyl-4-azidosalicylic acid. Photoaffinity labeling of the *Drosophila* and *Musca* receptors (isolated from the nicotinoid-agarose column) with [^{125}I]α-BGT-ASA gives a labeled polypeptide derivative at the apparent molecular mass of 66-69 kDa, which is protectable in the presence of unlabeled α-BGT (Fig. 10). This labeled polypeptide is considered to be either the 61-kDa subunit migrating at 66-69 kDa when coupled with the α-BGT (molecular mass, 8 kDa) photoaffinity label or the 66-kDa subunit undergoing little change in apparent mass on adduct formation (Tomizawa et al. 1996). This observation points out the importance of adapting the nicotinoid insecticide itself for use as a photoaffinity probe.

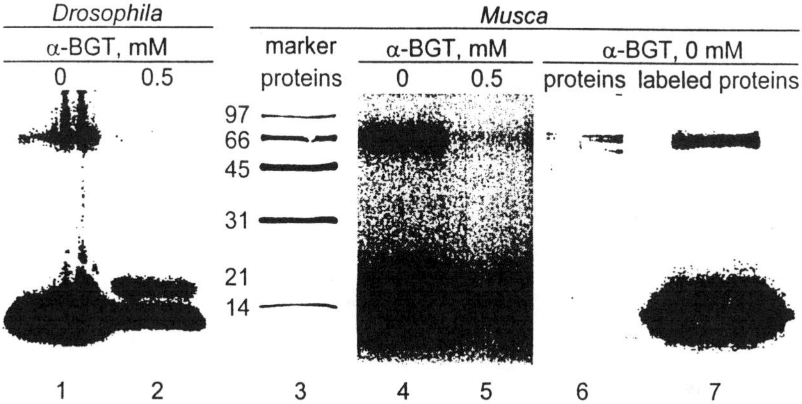

Fig. 10. Polyacrylamide gel electrophoresis of the α-BGT-binding subunit of the native nAChR from *Drosophila* and *Musca* head membranes isolated by nicotinoid-agarose affinity chromatography and then photoaffinity labeled with [^{125}I]α-BGT-ASA. Lanes 1, 2, 4, 5, and 7 of the lithium dodecyl sulfate polyacrylamide gel are autoradiograms; lane 3 shows the marker proteins, and lane 6 shows the putative subunits stained by Coomassie blue. Unlabeled α-BGT completely inhibits specific labeling (lanes 2 and 5). The high-mobility label (*bottom*) is unreacted [^{125}I]α-BGT-ASA or a dimer thereof. [This figure from Tomizawa et al. (1996) is reproduced with permission from Lippincott-Raven Publishers, Philadelphia, PA,USA.]

[^{125}I]Azidonicotinoid ([^{125}I]AN with 0.6-kDa molecular mass) photoaffinity labels only a single polypeptide in *Drosophila* head membranes, corresponding to 66 kDa at a specific site strongly inhibited by various cholinergic drugs including nicotine, cytisine, carbachol, *d*-tubocurarine, and α-BGT as well as the insecticides IMI and AAP (Fig. 11). This protection experiment for [^{125}I]AN photoaffinity labeling reveals that the 66-kDa polypeptide is pharmacologically consistent with the ligand- and insecticide-binding subunit of the native *Drosophila* nAChR. The isolated native *Drosophila* receptor with the three putative subunits (61, 66, and 69 kDa) is labeled by [^{125}I]AN primarily with the 66-kDa subunit and secondarily with the 61-kDa subunit (Fig. 11), inferring that the binding site for the nicotinoid insecticide is located at the interface between the 66- and 61-kDa subunits (Tomizawa and Casida 1997).

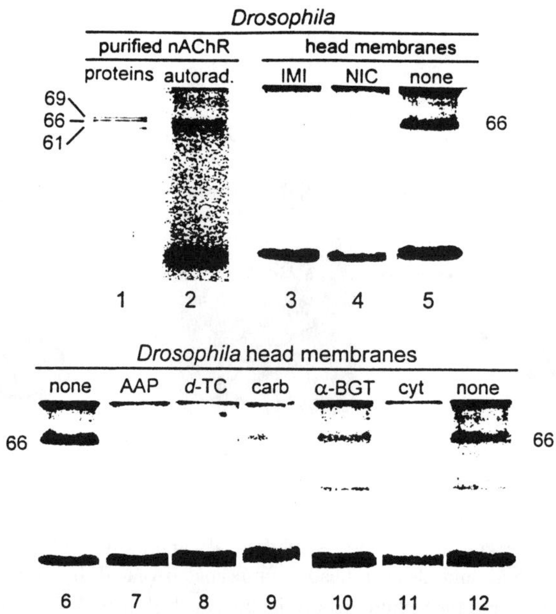

Fig. 11. Photoaffinity labeling with [^{125}I]azidonicotinoid ([^{125}I]AN) of the *Drosophila* nAChR. Lanes 1 and 2 show labeling of isolated native *Drosophila* nAChR as subunit proteins and autoradiogram, respectively. The subunit bands appear at 61, 66, and 69 kDa relative to marker proteins on the same gel. Lanes 5, 6, and 12 are autoradiograms of photoaffinity-labeled nAChR at 66 kDa from the head membranes of *Drosophila* reacted with [^{125}I]AN in the absence of cholinergic ligands (none). Lanes 3, 4, and 7-11 establish specific labeling protected by IMI, nicotine (NIC), AAP, *d*-tubocurarine (*d*-TC), carbachol (carb), α-BGT, and cytisine (cyt), respectively (Tomizawa and Casida 1997).

6 Structure and Function Studies of Insect Nicotinic Acetylcholine Receptors

The functional architecture of nAChRs is much better understood for mammals and other vertebrates than for insects. An outline of the mammalian nAChR subtypes and structural features will therefore serve as a comparison for the insect receptors (Fig. 12).

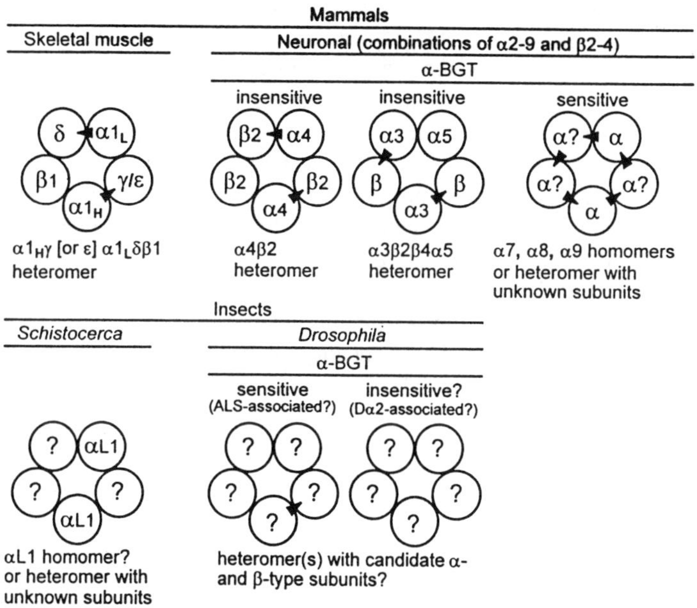

Fig. 12. Knowledge of structure and function for mammalian nAChR subtypes far exceeds that for insects. *Triangle* designates putative ligand-binding site.

6.1 Mammalian Receptors

Mammalian nAChRs involve diverse subtypes formed from five homologous subunits in combinations of nine α, four β, γ, δ, and ε subunits. The skeletal muscle-type nAChR is a heteromer consisting of $\alpha 1_H \gamma \alpha 1_L \delta \beta 1$ subunits (or an ε subunit in adult replacing the γ subunit in the fetal form), and the ACh-binding sites (high and low affinities; designated as H and L in subscript style) are located at the interface of the $\alpha 1_H$-γ and $\alpha 1_L$-δ subunits. Detailed information on the binding site environment for agonist or competitive and noncompetitive antagonists is available only in the case of receptors from muscle and *Torpedo* electric organ (Unwin 1995; Karlin and Akabas 1995; Arias 1997).

Neuronal nAChR subtypes in brain and ganglia are assembled in combinations of α2-9 and β2-4 and are pharmacologically classified into two main groups. The first is the α-BGT-insensitive subtypes formed from combinations of α2, α3, α4, and α6 subunits with β2 or β4 subunits, sometimes including α5 or β3 (Fig. 12). Two putative ACh-binding sites are suggested to be at the interface between α2, α3, α4, or α6 and β2 or β4 subunits. The most prominent subtype of this group is α4β2 (consisting of two α4 and three β2), which represents >90% of the nicotine-binding portion in the brain. The expressed amounts of α3β2β4α5 type nAChRs are thought to be smaller in more limited regions of brain than in peripheral ganglia. The second group of neuronal nAChR subtypes is associated with α7, α8, and α9 subunits, which are antagonized by α-BGT. The abundance of α7-associated nAChRs in brain is about as many as α4β2 nAChRs, and α7 receptors far predominate over the α3 receptors in ciliary ganglia. The α8 has been found only in chickens and the α9 in limited regions in the rat nervous system. These α-BGT-sensitive receptors are considered to be formed either as homomers of the α7, α8, or α9 subunit or as heteromers consisting of the α7, α8, or α9 subunit with unidentified subunit(s). The α7α8 heteromer is found with unknown subunit(s) in chicken brain and retina. Five putative ACh-binding sites are estimated when the receptor forms an α7 functional homomeric complex. Homomeric and native α7 nAChRs exhibit remarkably similar but not identical pharmacological profiles. Comprehensive reviews are available for these neuronal nAChR subtypes (Papke 1993; Sargent 1993; Lindstrom 1997).

6.2 Insect Receptors

6.2.1 Orthoptera and Blattodea

nAChRs from the CNS of the migratory locust (*Locusta migratoria*) (Breer et al. 1985) and *Periplaneta* (Sattelle and Breer 1985) purified by α-BGT affinity chromatography give only a single subunit protein with a molecular mass of 65 kDa in both cases (Table 3). The isolated *Locusta* receptor protein is partially confirmed as the nAChR when reconstituted in planar lipid bilayers by the functional ion channel properties with a picoampere level of electrophysiological responses (Hanke and Breer 1986). Alternatively, in the desert locust (*Schistocerca gregaria*), a cloned cDNA encoding the αL1 subunit of the nAChR can be expressed in *Xenopus* oocytes as a functional homomeric complex (Marshall et al. 1990) that differs in some properties from those of the native insect nAChRs, suggesting that heterogeneity of the locust nAChR with further unknown subunits(s) contributes to assembling the native receptor-ion channel complex (Amar et al. 1995).

Table 3. Current understanding of structural and functional features of insect nAChRs revealed by protein biochemistry and molecular biology approaches

Approach, species	Parameter[a]	Molecular mass (kDa)[b]
Protein biochemistry		
Drosophila	nic-agar	61[c], 66[c], 69
Musca	nic-agar, α-BGT-agar	61, 66[c], 69
Locusta	α-BGT-agar	65[d]
Periplaneta	α-BGT-agar	65
Molecular biology		
Drosophila	ALS	62[e]
	Dα2/SAD	61[e]
	ARD	57.3[f]
	SBD	57.3[f]
Schistocerca	αL1	60.6[g]

[a]Affinity chromatography for protein isolation; subunit for molecular biology. Designations for affinity matrices are nic-agar and α-BGT-agar for nicotinoid-agarose and α-BGT-agarose, respectively.
[b]Measured for protein isolation (Breer et al. 1985; Sattelle and Breer 1985; Tomizawa et al. 1996); deduced for molecular biology (Bossy et al. 1988; Hermans-Borgmeyer et al. 1986; Jonas et al. 1990; Lansdell et al. 1997; Sawruk et al. 1990a,b).
[c]Ligand- and insecticide-binding subunits revealed by azidosalicylate derivative of [^{125}I]α-BGT and [^{125}I]AN (Tomizawa et al. 1996; Tomizawa and Casida 1997).
[d]Electrophysiological assay with isolated receptor protein reconstituted in planar lipid bilayers (Hanke and Breer 1986).
[e]Putative ligand-binding subunit (α-type) based on conserved sequence and functional heterologous expression in *Xenopus* oocytes, HEK-293, or S2 cell lines with chick and rat β2, or with rat β4 genes confirmed with electrophysiology and radioligand binding (Bertrand et al. 1994; Bossy et al. 1988; Jonas et al. 1990; Lansdell et al. 1997; Sawruk et al. 1990a).
[f]Putative structural β-type subunit based on conserved sequence (Hermans-Borgmeyer et al. 1986; Lansdell et al. 1997; Sawruk et al. 1990b).
[g]Putative ligand-binding subunit based on conserved sequence and functional expression of αL1 homomer in *Xenopus* oocytes confirmed with electrophysiology (Marshall et al. 1990; Amar et al. 1995).

6.2.2 Diptera

Molecular biology (Table 3). In *Drosophila melanogaster*, two genes putatively encoding the ligand-binding α-type subunits referred to as ALS and Dα2 or SAD (Bossy et al. 1988; Jonas et al. 1990; Sawruk et al. 1990a) and encoding the structural β-type subunits (ARD and SBD) (Hermans-Borgmeyer et al. 1986;

Lansdell et al. 1997; Sawruk et al. 1990b) are identified as components of the neuronal nAChR. The ALS- and Dα2 (SAD)-like subunit genes are widely distributed in many insect species based on the polymerase chain reaction technique (Sgard et al. 1993). The recombinant ligand-binding site [expressed as a specific region 184-204 in the ALS gene (from N-terminal sequence)] in the transformed bacterial clone is recognized by [^{125}I]α-BGT with a K_D value of 1700 nM (Ohana and Gershoni 1990). The cRNA of the SAD (Dα2) gene can be expressed in *Xenopus* oocytes, but this SAD homomer only responds to a high concentration of nicotine (Sawruk et al. 1990a). On heterologous expression in *Xenopus* oocytes, human embryonic kidney (HEK-293) and *Drosophila* (S2) cell lines of the *Drosophila* α- and β-type subunits in various combinations do not produce any electrophysiological response or radioligand binding. However, the functional ion channel property of [^{3}H]epibatidine (an nAChR ligand) binding is observed when either of the two α-type subunits (ALS or Dα2) is coexpressed in *Xenopus* oocytes, HEK-293, or S2 cell lines with chick and rat β2 or with rat β4 (weaker response than those with β2) subunits. These observations strongly indicate the importance of the β-type subunit and heteromeric status of the *Drosophila* nAChR with involvement of unidentified subunit(s) (Bertrand et al. 1994; Lansdell et al. 1997). Interestingly, the ALS/chick β2 receptor is sensitive to α-BGT while the Dα2/chick β2 receptor is insensitive (Bertrand et al. 1994), suggesting the presence of multiple *Drosophila* nAChR subtypes with diversity in pharmacological and functional properties. It is proposed for *Periplaneta* that the α-BGT-sensitive and -insensitive nAChRs are expressed in the dorsal unpaired median neurons and that both subtypes are affected by IMI, based on electrophysiology studies (Lapied et al. 1990; Buckingham et al. 1997).

Immunohistochemistry. Further evidence for the heterogeneity of the *Drosophila* nAChR-ion channel complex is revealed by immunohistochemical studies. Two antibodies raised against fusion constructs encompassing specific regions of the ALS (α-type) and ARD (β-type) subunits can in each case immunoprecipitate the [^{125}I]α-BGT binding component in detergent extracts of *Drosophila* head membranes (Schloss et al. 1988, 1991). Moreover, each of the three antibodies against ALS, Dα2 (α-type), and ARD (β-type) subunits can visualize specific regions in the *Drosophila* CNS, and their distribution patterns are similar (Schuster et al. 1993; Jonas et al. 1994).

Protein isolation (Table 3). The nicotinoid-agarose matrix for affinity isolation of *Drosophila* and *Musca* nAChRs gives three putative subunits of the native receptors, each with molecular masses of 61, 66, and 69 kDa, which are comparable to those deduced from amino acid sequences of the subunit genes. The identical subunit polypeptides are also isolated with the α-BGT-agarose matrix (Tomizawa et al. 1996). The *Musca* nAChR was purified earlier by gel filtration followed by an affinity column coupled with α-BGT or concanavalin A to give major polypeptides with molecular masses of 26 and 42 kDa, which are considered to be proteolysis products thereof (March et al. 1982).

Identifying ligand-binding subunit (Table 3). $[^{125}I]\alpha$-BGT-ASA (molecular mass, 8 kDa) apparently recognizes a single polypeptide among three putative subunits of the isolated native *Drosophila* and *Musca* nAChRs; in each case either the 61- or 66- kDa subunit is considered to contain the α-BGT-binding site without clear definition of which one is labeled (Tomizawa et al. 1996). In contrast, Schloss et al. (1992) reported a 50-kDa polypeptide from cross-linking of *Drosophila* head membranes with $[^{125}I]\alpha$-BGT (using 1-ethyl-3-[3-(dimethylamino)propyl]-carbodiimide), and they calculated by difference (50 minus 8 kDa) that a 42-kDa polypeptide is a major ligand-binding component of the *Drosophila* nAChR. They also recognized that there is considerable difference between molecular masses of the α-type subunits (ALS and SAD/Dα2) as calculated from deduced amino acid sequences of the genes (61 and 62 kDa) (Bossy et al. 1988; Jonas et al. 1990; Sawruk et al. 1990a) and the 42-kDa polypeptide (Schloss et al. 1992). Although α-BGT shares the same binding region with IMI in the insect nAChR, based on competition studies with radiolabeled ligand-binding assays (Tomizawa and Yamamoto 1992; Liu and Casida 1993), the site for insecticide binding can only be partially defined using α-BGT due to its large molecular mass (8 kDa). $[^{125}I]$AN (molecular mass, 0.6 kDa) photoaffinity labels a 66-kDa polypeptide in *Drosophila* head membranes, and the labeled polypeptide is pharmacologically consistent with the ligand- and insecticide-binding subunit. Interestingly, $[^{125}I]$AN photoaffinity labels both the 61- (minor) and 66- (major) kDa subunits of the native *Drosophila* nAChR isolated by nicotinoid-agarose affinity chromatography. This result suggests that possibly the insecticide binding site is at the interface between the 61- and 66-kDa subunits (Tomizawa and Casida 1997), as described for the ACh or *d*-tubocurarine binding site located in the α-γ or α-δ interface of muscle-type and α-β interface of neuronal-type nAChRs in mammals (Pedersen and Cohen 1990; Papke 1993; Karlin and Akabas 1995; Arias 1997; Lindstrom 1997). The differences in molecular masses of the labeled polypeptide (66 kDa) with $[^{125}I]$AN or $[^{125}I]\alpha$-BGT-ASA in the isolated native *Drosophila* nAChR and two α-type subunits (ALS and Dα2/SAD) deduced from DNA sequences (61 and 62 kDa) (Bossy et al. 1988; Jonas et al. 1990; Sawruk et al. 1990a) may be due to glycosylated regions. A similar relationship is observed with locust nAChR in the molecular mass of the isolated native subunit (65 kDa) (Breer et al. 1985), compared with that calculated from the αL1 subunit gene (60.6 kDa) (Marshall et al. 1990). This also may explain differences in the molecular masses of structural β-type subunits deduced from molecular biology studies (57.3 kDa) (Hermans-Borgmeyer et al. 1986; Lansdell et al. 1997; Sawruk et al. 1990b) and determined as the non-ligand-binding subunits following isolation (Tomizawa et al. 1996). These protein biochemical approaches indicate that the three isolated subunit polypeptides (61, 66, and 69 kDa) are perhaps the minimal number of putative nAChR subunits in the native heteromeric *Drosophila* and *Musca* receptors.

7 Summary

The insect nAChR is the target for novel and increasingly important synthetic nicotinoids such as IMI and related high-potency insecticides, thereby providing an incentive to advance the knowledge on insect receptors. These nicotinoids in turn are candidate radiolabeled ligands and affinity probes to explore the structure and function of the insect nAChR. [^{3}H]IMI undergoes high-affinity specific binding to head membranes from *Musca* and a few other insect species with pharmacological and toxicological profiles appropriate for the nAChR as the target of insecticide action. The *Drosophila* and *Musca* nAChRs are isolated by nicotinoid-agarose affinity chromatography, in each case with three putative subunits (comparable to molecular biology studies for *Drosophila*), and the ligand- and insecticide-binding subunit(s) of *Drosophila* nAChR is recognized by photoaffinity labeling with [^{125}I]AN and [^{125}I]α-BGT-ASA. Clearly, nicotinoid insecticides and their analogues serve as excellent radiolabeled ligands and affinity probes for defining SAR, determining mode of insecticidal action, and exploring the structure and function of the insect nAChRs. Much remains to be learned on the assembly mechanism for the native receptor-ion channel complex and the pharmacological actions of diverse ligands and insecticides. More generally, an understanding of the comparative features of the insect and mammalian nAChRs will contribute to the design of safe insecticides for the future.

Acknowledgment

The authors thank their coworkers Ming-Yie Liu and Shirley Lee Chao for advancing the knowledge in this field, and Gary B. Quistad and other laboratory associates for advice, assistance, or encouragement. Research considered in this review was supported by grants P01 ES00049 and R01 ES08424 from the National Institute of Environmental Health Sciences (NIEHS), NIH, and its contents are solely the responsibility of the authors and do not necessarily represent the official views of the NIEHS, NIH.

References

Amar M, Thomas P, Wonnacott S, Lunt GG (1995) A nicotinic acetylcholine receptor subunit from insect brain forms a non-desensitising homo-oligomeric nicotinic acetylcholine receptor when expressed in *Xenopus* oocytes. Neurosci Lett 199:107-110

Arias HR (1997) Topology of ligand binding sites on the nicotinic acetylcholine receptor. Brain Res Rev 25:133-191

Bai D, Lummis SCR, Leicht W, Breer H, Sattelle DB (1991) Actions of imidacloprid and a related nitromethylene on cholinergic receptors of an identified insect motor neurone. Pestic Sci 33:197-204

Bertrand D, Ballivet M, Gomez M, Bertrand S, Phannavong B, Gundelfinger ED (1994) Physiological properties of neuronal nicotinic receptors reconstituted from the vertebrate β2 subunit and *Drosophila* α subunits. Eur J Neurosci 6:869-875

Bossy B, Ballivet M, Spierer P (1988) Conservation of neuronal nicotinic acetylcholine receptors from *Drosophila* to vertebrate central nervous systems. EMBO J 7:611-618

Breer H, Kleene R, Hinz G (1985) Molecular forms and subunit structure of the acetylcholine receptor in the central nervous system of insects. J Neurosci 5:3386-3392

Buckingham SD, Lapied B, Corronc HL, Grolleau F, Sattelle DB (1997) Imidacloprid actions on insect neuronal acetylcholine receptors. J Exp Biol 200:2685-2692

Casida JE (1970) Mixed-function oxidase involvement in the biochemistry of insecticide synergists. J Agric Food Chem 18:753-772

Casida JE, Quistad GB (1998) Golden age of insecticide research: past, present, or future? Annu Rev Entomol 43:1-16

Chao SL, Casida JE (1997) Interaction of imidacloprid metabolites and analogs with the nicotinic acetylcholine receptor of mouse brain in relation to toxicity. Pestic Biochem Physiol 58:77-88

Chao SL, Dennehy TJ, Casida JE (1997) Whitefly (Hemiptera: Aleyrodidae) binding site for imidacloprid and related insecticides: a putative nicotinic acetylcholine receptor. J Econ Entomol 90:879-882

David JA, Sattelle DB (1984) Actions of cholinergic pharmacological agents on the cell body membrane of the fast coxal depressor motorneuron of the cockroach (*Periplaneta* americana). J Exp Biol 108:119-136

Hanke W, Breer H (1986) Channel properties of an insect neuronal acetylcholine receptor protein reconstituted in planar lipid bilayers. Nature 321:171-174

Hermans-Borgmeyer I, Zopf D, Ryseck R-P, Hovemann B, Betz H, Gundelfinger ED (1986) Primary structure of a developmentally regulated nicotinic acetylcholine receptor protein from *Drosophila*. EMBO J 5:1503-1508

Jonas P, Baumann A, Merz B, Gundelfinger ED (1990) Structure and developmental expression of the Dα2 gene encoding a novel nicotinic acetylcholine receptor protein of *Drosophila melanogaster*. FEBS Lett 269:264-268

Jonas PE, Phannavong B, Schuster R, Schröder C, Gundelfinger ED (1994) Expression of the ligand-binding nicotinic acetylcholine receptor subunit Dα2 in the *Drosophila* central nervous system. J Neurobiol 25:1494-1508

Kagabu S, Moriya K, Shibuya K, Hattori Y, Tsuboi S, Shiokawa K (1992) 1-(6-Halonicotinyl)-2-nitromethylene imidazolidines as potential new insecticides. Biosci Biotechnol Biochem 56:362-363

Karlin A, Akabas MH (1995) Toward a structural basis for the function of nicotinic acetylcholine receptors and their cousins. Neuron 15:1231-1244

Kishida H, Sakamoto N, Umeda K, Fujimoto H (1992) Preparation of nitropyrimidine derivatives as insecticides. Jpn Kokai Tokkyo Koho JP04,173,788; Chem Abstr 118, 22251q

Kotzyba-Hibert F, Kapfer I, Goeldner M (1995) Recent trends in photoaffinity labeling. Angew Chem Int Ed Engl 34:1296-1312

Lansdell SJ, Schmitt B, Betz H, Sattelle DB, Millar NS (1997) Temperature-sensitive expression of *Drosophila* neuronal nicotinic acetylcholine receptors. J Neurochem 68:1812-1819

Lapied B, Corronc HL, Hue B (1990) Sensitive nicotinic and mixed nicotinic-muscarinic receptors in insect neurosecretory cells. Brain Res 533:132-136

Latli B, Casida JE (1992) [^{3}H]Imidacloprid: synthesis of a candidate radioligand for the nicotinic acetylcholine receptor. J Labelled Compd Radiopharm 31:609-613

Latli B, Than C, Morimoto H, Williams PG, Casida JE (1996) [6-Chloro-3-pyridylmethyl-[^{3}H]neonicotinoids as high-affinity radioligands for the nicotinic acetylcholine receptor: preparation using NaB^3H_4 and LiB^3H_4. J Labelled Comp Radiopharm 38:971-978

Latli B, Tomizawa M, Casida JE (1997) Synthesis of a novel [^{125}I]neonicotinoid photoaffinity probe for the *Drosophila* nicotinic acetylcholine receptor. Bioconjugate Chem 8:7-14

Lindstrom J (1997) Nicotinic acetylcholine receptors in health and disease. Mol Neurobiol 15:193-222

Liu M-Y, Casida JE (1993) High affinity binding of [^{3}H]imidacloprid in the insect acetylcholine receptor. Pestic Biochem Physiol 46:40-46

Liu M-Y, Lanford J, Casida JE (1993) Relevance of [^{3}H]imidacloprid binding site in house fly head acetylcholine receptor to insecticidal activity of 2-nitromethylene- and 2-nitroimino-imidazolidines. Pestic Biochem Physiol 46:200-206

Liu M-Y, Latli B, Casida JE (1994) Nitomethyleneimidazolidine radioligand ([^{3}H]NMI): high affinity and cooperative binding for house fly acetylcholine receptor. Pestic Biochem Physiol 50:171-182

Liu M-Y, Latli B, Casida JE (1995) Imidacloprid binding site in *Musca* nicotinic acetylcholine receptor: interactions with physostigmine and a variety of nicotinic agonists with chloropyridyl and chlorothiazolyl substituents. Pestic Biochem Physiol 52:170-181

March CS, Cattell KJ, Donnellan JF (1982) Pharmacologicaal characteristics of a putative nicotinic acetylcholine receptor from *Musca domestica*. In: Evered D, O'Connor M, Whelan J (eds) Ciba Foundation Symposium 88: Neuropharmacology of Insects. Pitman, London, pp 118-136

Marshall J, Buckingham SD, Shingai R, Lunt GG, Goosey MW, Darlison MG, Sattelle DB, Barnard EA (1990) Sequence and functional expression of a single α subunit of an insect nicotinic acetylcholine receptor. EMBO J 9:4391-4398

Matsuo H, Tomizawa M, Yamamoto I (1998) Structure-activity relationships of acyclic nicotinoids and neonicotinoids for insect nicotinic acetylcholine receptor/ion channel complex. Arch Insect Biochem Physiol 37:17-23

Mebs D, Narita K, Iwanaga S, Samejima Y, Lee CY (1971) Amino acid sequence of α-bungarotoxin from the venom of *Bungarus multicinctus*. Biochem Biophys Res Commun 44:711-716

Minamida I, Iwanaga K, Tabuchi T, Uneme H, Dantsuji H, Okauchi T (1993) Synthesis and insecticidal activity of acyclic nitroethene compounds containing a 3-pyridyl-methylamino group. J Pestic Sci 18:31-40

Moriya K, Shibuya K, Hattori Y, Tsuboi S, Shiokawa K, Kagabu S (1992) 1-(6-Chloronicotinyl)-2-nitroimino-imidazolidines and related compounds as potential new insecticides. Biosci Biotechnol Biochem 56:364-365

Nishimura K, Kanda Y, Okazawa A, Ueno T (1994) Relationship between insecticidal and neurophysiological activities of imidacloprid and related compounds. Pestic Biochem Physiol 50:51-59

Ohana B, Gershoni JM (1990) Comparison of the toxin binding sites of the nicotinic acetylcholine receptor from *Drosophila* to human. Biochemistry 29:6409-6415

Papke RL (1993) The kinetic properties of neuronal nicotinic receptors: genetic basis of functional diversity. Prog Neurobiol 41:509-531

Pedersen SE, Cohen JB (1990) *d*-Tubocurarine binding sites are located at α-γ and α-δ subunit interfaces of the nicotinic acetylcholine receptor. Proc Natl Acad Sci USA 87:2785-2789

Sargent PB (1993) The diversity of neuronal nicotinic acetylcholine receptors. Annu Rev Neurosci 16:403-443

Sattelle DB, Breer H (1985) Purification by affinity-chromatography of a nicotinic acetylcholine receptor from the CNS of the cockroach *Periplaneta americana*. Comp Biochem Physiol [C] 82:349-352

Sawruk E, Schloss P, Betz H, Schmitt B (1990a) Heterogeneity of *Drosophila* nicotinic acetylcholine receptors: SAD, a novel developmentally regulated α-subunit. EMBO J 9:2671-2677

Sawruk E, Udri C, Betz H, Schmitt B (1990b) SBD, a novel structural subunit of the *Drosophila* nicotinic acetylcholine receptor, shares its genomic localization with two α-subunits. FEBS Lett 273:177-181

Schloss P, Hermans-Borgmeyer I, Betz H, Gundelfinger ED (1988) Neuronal acetylcholine receptors in *Drosophila*: the ARD protein is a component of a high-affinity α-bungarotoxin binding complex. EMBO J 7:2889-2894

Schloss P, Betz H, Schröder C, Gundelfinger ED (1991) Neuronal nicotinic acetylcholine receptors in *Drosophila*: antibodies against an α-like and a non-α-subunit recognize the same high-affinity α-bungarotoxin binding complex. J Neurochem 57:1556-1562

Schloss P, Mayser W, Gundelfinger ED, Betz H (1992) Cross-linking of ^{125}I-α-bungarotoxin to *Drosophila* head membranes identifies a 42 kDa toxin binding polypeptide. Neurosci Lett 145:63-66

Schuster R, Phannavong B, Schröder C, Gundelfinger ED (1993) Immunohistochemical localization of a ligand-binding and a structural subunit of nicotinic acetylcholine receptors in the central nervous system of *Drosophila melanogaster*. J Comp Neurol 335:149-162

Sgard F, Obosi LA, King LA, Windass JD (1993) ALS and SAD-like nicotinic acetylcholine receptor subunit genes are widely distributed in insects. Insect Mol Biol 2:215-223

Takahashi H, Mitsui J, Takakusa N, Matsuda M, Yoneda H, Suzuki J, Ishimitsu K, Kishimoto T (1992) NI-25, a new type of systemic and broad spectrum insecticide. Brighton Crop Prot Conf Pests Dis 1:89-96

Tomizawa M (1994) Structure-activity relationships of nicotinoids and the related compounds. J Pestic Sci 19:S229-S240

Tomizawa M, Casida JE (1997) [^{125}I]Azidonicotinoid photoaffinity labeling of insecticide-binding subunit of *Drosophila* nicotinic acetylcholine receptor. Neurosci Lett 237:61-64

Tomizawa M, Yamamoto I (1992) Binding of nicotinoids and the related compounds to the insect nicotinic acetylcholine receptor. J Pestic Sci 17:231-236

Tomizawa M, Yamamoto I (1993) Structure-activity relationships of nicotinoids and imidacloprid analogs. J Pestic Sci 18:91-98

Tomizawa M, Otsuka H, Miyamoto T, Eldefrawi ME, Yamamoto I (1995a) Pharmacological characteristics of insect nicotinic acetylcholine receptor with its ion channel and the comparison of the effect of nicotinoids and neonicotinoids. J Pestic Sci 20:57-64

Tomizawa M, Otsuka H, Miyamoto T, Yamamoto I (1995b) Pharmacological effects of imidacloprid and its related compounds on the nicotinic acetylcholine receptor with its ion channel from the *Torpedo* electric organ. J Pestic Sci 20:49-56

Tomizawa M, Latli B, Casida JE (1996) Novel neonicotinoid-agarose affinity column for *Drosophila* and *Musca* nicotinic acetylcholine receptors. J Neurochem 67:1669-1676

Unwin N (1995) Acetylcholine receptor channel imaged in the open state. Nature 373:37-43

Yamamoto I, Yabuta G, Tomizawa M, Saito T, Miyamoto T, Kagabu S (1995) Molecular mechanism for selective toxicity of nicotinoids and neonicotinoids. J Pestic Sci 20:33-40

Yamamoto I, Tomizawa M, Saito T, Miyamoto T, Walcott EC, Sumikawa K (1998) Structural factors contributing to insecticidal and selective actions of neonicotinoids. Arch Insect Biochem Physiol 37:24-32

Zwart R, Oortgiesen M, Vijverberg HPM (1994) Nitromethylene heterocycles: selective agonists of nicotinic receptors in locust neurons compared to mouse N1E-115 and BC3H1 cells. Pestic Biochem Physiol 48:202-213

Footnotes

Correspondence should be addressed to Dr. John E. Casida at Environmental Chemistry and Toxicology Laboratory, Department of Environmental Science, Policy and Management, University of California, Berkeley, California 94720-3112, USA. E-mail: ectl@nature.berkeley.edu

*Abbreviations used: ACh, acetylcholine; α-BGT or [^{125}I]α-BGT, α-bungarotoxin or its 125-iodine labeled derivative; α-BGT-agar, α-BGT-agarose affinity matrix; [^{125}I]α-BGT-ASA, azidosalicylate derivative of [^{125}I]α-BGT; B_{max}, maximal binding capacity; carb, carbachol; CNS, central nervous system; cyt, cytisine; *d*-TC, *d*-tubocurarine; HEK-293, human embryonic kidney cell line; IC_{50}, the concentration of test ligand for 50% inhibition of specific radioligand binding; ip, intraperitoneal; K_D, dissociation constant; KD_{50} , the dose for 50% knockdown; LD_{50}, the dose for 50% mortality; nAChR, nicotinic ACh receptor; n_H, Hill coefficient; NIC or [^{3}H]NIC, nicotine or its tritiated ligand; nic-agar, nicotinoid-agarose affinity matrix; PPP, *O*-propyl *O*-(2-propynyl) phenylphosphonate; S2, *Drosophila* cell line; SAR, structure-activity relationship; THF, tetrahydrofuran.

Abbreviations used for nicotinoid insecticides: AAP or [^{3}H]AAP, acetamiprid or its tritiated ligand; AN, [^{127}I]AN or [^{125}I]AN, azidonicotinoid, its 127- or 125-iodine labeled photoaffinity ligand; CH-IMI or [^{3}H]CH-IMI, nitromethylene analogue of IMI or its tritiated ligand; Cl-TMNI or [^{3}H]Cl-TMNI, chlorothiazolyl analogue of CH-IMI or its tritiated ligand; DN-IMI or [^{3}H]DN-IMI, desnitro metabolite of IMI or its tritiated ligand; IMI or [^{3}H]IMI, imidacloprid or its tritiated ligand; NTP, nitenpyram; THPCH-IMI, tetrahydropyrimidine analogue of CH-IMI.

Abbreviations used for insect nAChR subunits: *Drosophila*: ALS, α-like subunit; ARD, first structural β-type subunit; Dα2, α-like subunit 2; SAD, alternate name for Dα2; SBD, second structural β-type subunit. *Schistocerca*: αL1, α-subunit.

INDEX

B

C

K

L

M

N

O

P

Q

R

S

T

U

V

W

X